S*pringer* **M***onographs in* **M***athematics*

For further volumes:
http://www.springer.com/series/3733

Goro Shimura

Modular Forms: Basics and Beyond

Springer

Goro Shimura
Department of Mathematics
Princeton University
Princeton, NJ 08544-1000
USA
goro@math.princeton.edu

ISSN 1439-7382
ISBN 978-1-4614-2124-5 e-ISBN 978-1-4614-2125-2
DOI 10.1007/978-1-4614-2125-2
Springer New York Dordrecht Heidelberg London

Library of Congress Control Number: 2011941793

Mathematics Subject Classification (2010): 11F11, 11F27, 11F37, 11F67, 11M36, 11M41, 14K25

Printed on acid-free paper

Springer is part of Springer Science+Business Media (www.springer.com)

PREFACE

It was forty years ago that my "Introduction to the arithmetic theory of automorphic functions" appeared. At present the terminology "modular form" can be counted among those most frequently heard in the conversations of mathematicians, and indeed, there are many textbooks on this topic. However, almost all of them are at the elementary level, and not so interesting from the viewpoint of the reader who already knows the basics. So, my intention in the present book is to offer something new that may satisfy the desire of such a reader. Therefore we naturally assume that the reader has at least rudimentary knowledge of modular forms of integral weight with respect to congruence subgroups of $SL_2(\mathbf{Z})$, though we state every definition and some basic theorems on such forms.

One of the principal new features of this book is the theory of modular forms of half-integral weight, another the discussion of theta functions and Eisenstein series of holomorphic and nonholomorphic types. Thus we have written the book so that the reader can learn such theories systematically. However, we present them with the following two themes as the ultimate aims:

(I) The correspondence between the forms of half-integral weight and those of integral weight.

(II) The arithmeticity of various Dirichlet series associated with modular forms of integral or half-integral weight.

The correspondence of (I) associates a cusp form of weight k with a modular form of weight $2k-1$, where k is half an odd positive integer. I gave such a correspondence in my papers in 1973. In the present book I prove a stronger, perhaps the best possible, result with different methods.

As for (II), a typical example is a Dirichlet series

$$\mathscr{D}(s;\, f,\, g) = L(2s+2,\, \omega) \sum_{n=1}^{\infty} a_n b_n n^{-s-(k+\ell)/2}$$

obtained from a cusp form $f(z)=\sum_{n=1}^{\infty} a_n \exp(2\pi inz)$ of weight k and another form $g(z) = \sum_{n=0}^{\infty} b_n \exp(2\pi inz)$ of weight ℓ, where $L(s, \omega)$ is the L-function of a Dirichlet character ω determined by f and g. In the crudest form, our main results show that there exists a constant $A(f)$ that depends on f, k, ℓ, ω, and an integer κ such that $D(\kappa;\, f,\, g)/A(f)$ is algebraic if a_n and b_n are algebraic, for infinitely many different g. We can of course consider $D(s;\, f,\, \chi) = \sum_{n=1}^{\infty} \chi(n) a_n n^{-s}$ with a Dirichlet character χ and ask about the nature of $D(m;\, f,\, \chi)$ for certain integers m.

Though we eventually restrict our modular forms to functions of one complex variable, some of our earlier sections provide an easy introduction to the theory of Siegel modular forms, since that gives a good perspective and makes our proofs of various facts more transparent. Also, since our second theme concerns the arithmeticity, we naturally discuss the rationality of the Fourier coefficients of a modular form, and how the form behaves under the action of an automorphism of the field to which the coefficients belong. This is a delicate problem, particularly when it is combined with the group action. Therefore, a considerable number of pages are spent on this problem. Another essential aspect of our theory is the involvement of the class of functions which we call nearly holomorphic modular forms, especially nonholomorphic Eisenstein series.

As for $D(m;\, f,\, \chi)$, we only state the results without proof, and cite two of my papers published in 1976 and 1977. My original plan was to make the book self-contained even in this respect by including the proof, but an unexpected accident made me abandon the idea. Possibly I may be excused by saying that once the reader acquires some elementary results in earlier sections of the present book, those two papers will be easy to read, and so the exclusion of the proof is not a great loss. Also, I allowed myself to quote some standard facts discussed in my books of 1971 and 2007 without proof, since I thought it awkward to reproduce the proof of every quoted fact.

It is my great pleasure to express my heartfelt thanks to my friends Koji Doi, Tomokazu Kashio, Kamal Khuri-Makdisi, Kaoru Okada, and Hiroyuki Yoshida, who read my manuscript and helped me eliminate many misprints and improve the exposition.

Princeton
September 2011

Goro Shimura

TABLE OF CONTENTS

NOTATION AND TERMINOLOGY

0.1. For a set X we denote by $\#X$ or $\#(X)$ the number of elements of X when it is finite, and put $\#X = \#(X) = \infty$ otherwise.

We denote by $\mathbf{Z}$, $\mathbf{Q}$, $\mathbf{R}$, and $\mathbf{C}$ the ring of rational integers and the fields of rational numbers, real numbers, and complex numbers, respectively. Also, $\overline{\mathbf{Q}}$ and $\mathbf{Q}_{\mathrm{ab}}$ mean the algebraic closure of $\mathbf{Q}$ in $\mathbf{C}$ and the maximal abelian extension of $\mathbf{Q}$ in $\overline{\mathbf{Q}}$, respectively. We put

$$\mathbf{T} = \{z \in \mathbf{C} \mid |z| = 1\}. \tag{0.1}$$

We denote by $\mathrm{Aut}(\mathbf{C})$ the group of all ring-automorphisms of $\mathbf{C}$.

Given an associative ring A with identity element and an A-module X, we denote by $A^\times$ the group of all invertible elements of A, and by X_n^m the A-module of all $m \times n$-matrices with entries in X; we put $X^m = X_1^m$ for simplicity. For an element y of X_1^m or X_m^1 we denote by y_i the ith entry of y. The zero element of A_n^m is denoted by 0_n^m or simply by 0. When we view A_n^n as a ring, we usually denote it by $M_n(A)$. We denote the identity element of $M_n(A)$ by 1_n or simply by 1. The transpose, determinant, and trace of a matrix x are denoted by ${}^t x$, $\det(x)$, and $\mathrm{tr}(x)$. For $X \in A_n^m$ and an ideal B of A we write $X \prec B$ if all the entries of X belong to B. For square matrices $x_1, \ldots, x_r$ we denote by $\mathrm{diag}[x_1, \ldots, x_r]$ the square matrix with $x_1, \ldots, x_r$ in the diagonal blocks and 0 in all other blocks.

We put $GL_n(A) = M_n(A)^\times$, $SL_n(A) = \{\alpha \in GL_n(A) \mid \det(\alpha) = 1\}$ if A is commutative, and

$$S_n(A) = \{T \in M_n(A) \mid {}^t T = T\}. \tag{0.2}$$

For $T \in S_n(A)$ and $X \in A_m^n$ we put $T[X] = {}^t XTX$; we also put $T(x, y) = {}^t xTy$ for $x, y \in A^n$. For $h = {}^t\overline{h} \in M_n(\mathbf{C})$ we write $h > 0$ if h is positive definite, and we write $h > k$ if $h - k > 0$. Throughout the book we put

$$\mathbf{e}(c) = \exp(2\pi i c) \qquad (c \in \mathbf{C}). \tag{0.3}$$

0.2. We define the Legendre-Jacobi symbol $\left(\dfrac{m}{n}\right)$ for $0 \le n - 1 \in 2\mathbf{Z}$ and $m \in \mathbf{Z}$ as follows. Let $n = q_1 \cdots q_s$ with odd prime numbers q_j. Then we put

$$\left(\frac{m}{n}\right) = \prod_{j=1}^{s} \left(\frac{m}{q_j}\right), \tag{0.4}$$

where $\left(\dfrac{m}{q_j}\right)$ is the quadratic residue symbol, and we understand that the product means 1 if $n=1$ (even when $m=0$). Clearly $\left(\frac{m}{n}\right)=0$ if $m\mathbf{Z}+n\mathbf{Z}\neq\mathbf{Z}$ and $\left(\frac{mm'}{n}\right)=\left(\frac{m}{n}\right)\left(\frac{m'}{n}\right)$. If $n<0$ and n is prime to $2m$, we put

$$\left(\frac{m}{n}\right)=\frac{|m|}{m}\left(\frac{m}{|n|}\right), \tag{0.5}$$

where we understand that $|0|/0=1$. We also put, for every odd integer d,

$$\varepsilon_d=\begin{cases}1 & \text{if } d-1\in 4\mathbf{Z},\\ i & \text{if } d+1\in 4\mathbf{Z}.\end{cases} \tag{0.6}$$

Thus $\varepsilon_d^2=\left(\frac{-1}{d}\right)$ for odd d.

0.3. For a finite-dimensional vector space V over $\mathbf{Q}$, by a **Z-lattice** in V we mean a finitely generated $\mathbf{Z}$-submodule of V that spans V over $\mathbf{Q}$. We also denote by $\mathscr{L}(V)$ the set of all $\mathbf{C}$-valued functions λ for which there exist two $\mathbf{Z}$-lattices L and M in V such that $\lambda(x)=0$ for $x\notin L$ and $\lambda(x)$ for $x\in L$ depends only on the coset $x+M$.

For example, take $V=\mathbf{Q}^n$ and put $L=\mathbf{Z}^n$. Given $r,\ s\in\mathbf{Q}^n$, define a function $\lambda_{r,s}$ on V by $\lambda_{r,s}(x)=0$ if $x-r\notin L$ and $\lambda_{r,s}(x)=\mathbf{e}({}^t xs)$ for $x-r\in L$. Then clearly $\lambda_{r,s}\in\mathscr{L}(V)$. Let us now show that *$\mathscr{L}(V)$ is spanned over $\mathbf{C}$ by $\lambda_{r,s}$ for all $(r,\ s)$.*

Given $\lambda\in\mathscr{L}(V)$, we can find positive integers g and h such that $\lambda(x)=0$ for $x\notin g^{-1}L$ and $\lambda(x)$ for $x\in g^{-1}L$ depends only on $x+hL$. Let R and S be complete sets of representatives for $g^{-1}L/L$ and $h^{-1}L/L$, respectively. For $r\in R$ let $\mu_r=\varepsilon_r\lambda$, where ε_r is the characteristic function of $r+L$. Then $\lambda=\sum_{r\in R}\mu_r$, and $\mu_r(r+y)$ for $y\in L$ depends only on $y+hL$. Now for each $s\in S$ the map $y\mapsto\mathbf{e}({}^t ys)$ defines a function on L/hL, and the space of functions on L/hL is spanned by such functions $\mathbf{e}({}^t ys)$ for all $s\in S$, and so we can put $\mu_r(r+y)=\sum_{s\in S}c_{r,s}\mathbf{e}({}^t ys)$ for $y\in L$ with $c_{r,s}\in\mathbf{C}$. Consequently $\lambda=\sum_{r\in R}\sum_{s\in S}c_{r,s}\mathbf{e}(-{}^t rs)\lambda_{r,s}$, which proves the italicized statement above.

0.4. If σ is an isomorphism of a field F onto K, then for $x\in F$ we denote by x^σ the image of x under σ. If τ is an isomorphism of K onto another field, then $\sigma\tau$ denotes the composite of σ and τ defined by $x^{\sigma\tau}=(x^\sigma)^\tau$. We will define the action of σ on various objects X such as Dirichlet characters and modular forms, but the action will always be written X^σ, and the rule $X^{\sigma\tau}=(X^\sigma)^\tau$ is universal.

CHAPTER I

PRELIMINARIES

1. Symplectic groups and symmetric domains

1.1. Though the principal aim of this book is to discuss various topics on modular forms of one complex variable, we first introduce the so-called Siegel modular forms defined on a certain space $\mathfrak{H}_n$, called the **Siegel upper half space of degree** n and defined by (1.12) below, since these will make our exposition easier. Besides, what we need about them are some formal identities, which are not complicated, and we find no reason for avoiding them.

For $0 < n \in \mathbf{Z}$ and a commutative ring A with identity element we put

$$\text{(1.1a)}\qquad Sp(n,\,A) = \{\alpha \in GL_{2n}(A) \mid {}^t\alpha\iota\alpha = \iota\},\quad \iota = \iota_n = \begin{bmatrix} 0 & -1_n \\ 1_n & 0 \end{bmatrix},$$

$$\text{(1.1b)}\qquad Gp(n,\,A) = \{\alpha \in GL_{2n}(A) \mid {}^t\alpha\iota\alpha = \nu(\alpha)\iota \ \textit{with}\ \nu(\alpha) \in A^\times\}.$$

Clearly $Sp(n,\,A)$ and $Gp(n,\,A)$ are subgroups of $GL_{2n}(A)$. In particular, the group $Sp(n,\,A)$ is called the **symplectic group of degree** n **over** A. Notice that ${}^t\alpha\iota\alpha = \nu(\alpha)\iota$ if and only if $\alpha\iota\cdot{}^t\alpha = \nu(\alpha)\iota$, since $\iota^{-1} = -\iota$. We easily see that

$$\text{(1.2a)}\quad \gamma = \begin{bmatrix} a & b \\ c & d \end{bmatrix} \in Gp(n,\,A) \Longrightarrow \begin{cases} a\cdot{}^td - b\cdot{}^tc = {}^tda - {}^tbc = \nu(\gamma), \\ a\cdot{}^tb,\ c\cdot{}^td,\ {}^tac,\ {}^tbd \in S_n(A), \end{cases}$$

$$\text{(1.2b)}\quad \gamma = \begin{bmatrix} a & b \\ c & d \end{bmatrix} \in Gp(n,\,A) \Longrightarrow \gamma^{-1} = \nu(\gamma)^{-1}\begin{bmatrix} {}^td & -{}^tb \\ -{}^tc & {}^ta \end{bmatrix},$$

$$\text{(1.2c)}\quad \begin{bmatrix} a & b \\ c & d \end{bmatrix} \in Sp(n,\,A) \Longleftrightarrow {}^tac = {}^tca,\ {}^tbd = {}^tdb,\ {}^tda - {}^tbc = 1,$$
$$\Longleftrightarrow a\cdot{}^tb = b\cdot{}^ta,\ c\cdot{}^td = d\cdot{}^tc,\ a\cdot{}^td - b\cdot{}^tc = 1,$$

$$\text{(1.2d)}\qquad \gamma \in Gp(n,\,A) \Longrightarrow {}^t\gamma \in Gp(n,\,A),\ \nu({}^t\gamma) = \nu(\gamma),$$

$$\text{(1.2e)}\qquad {}^t\iota = \iota^{-1} = -\iota,\ \iota^2 = -1_{2n},\ \text{and}\ \ \iota \in Sp(n,\,A).$$

Given $\alpha \in Gp(n,\,A)$, put $\gamma = \mathrm{diag}[\nu(\alpha)1_n,\,1_n]$. Then clearly $\gamma \in Gp(n,\,A)$ and $\nu(\gamma) = \nu(\alpha)$, and so $\gamma^{-1}\alpha \in Sp(n,\,A)$. In this way we see that

(1.3) *Every element of* $Gp(n,\,A)$ *is the product of an element of* $Sp(n,\,A)$ *and an element of the form* $\mathrm{diag}[e1_n,\,1_n]$ *with* $e \in A^\times$.

G. Shimura, *Modular Forms: Basics and Beyond*, Springer Monographs in Mathematics, DOI 10.1007/978-1-4614-2125-2_1, © Springer Science+Business Media, LLC 2012

If $n = 1$, noting that $\iota^{-1} \cdot {}^t\alpha\iota\alpha = \det(\alpha)1_2$ for every $\alpha \in M_2(A)$, we see that

$$Sp(1, A) = SL_2(A), \quad Gp(1, A) = GL_2(A), \tag{1.4a}$$

$$\nu(\alpha) = \det(\alpha) \text{ for } \alpha \in Gp(1, A). \tag{1.4b}$$

1.2. We will eventually define the action of $Sp(n, \mathbf{R})$ on $\mathfrak{H}_n$, but we first define more generally the action (in a weak sense) of $Gp(n, \mathbf{C})$ on $S_n(\mathbf{C})$. For $\alpha = \begin{bmatrix} a & b \\ c & d \end{bmatrix} \in M_{2n}(\mathbf{C})$ with a of size n we hereafter put $a = a_\alpha = a(\alpha)$, $b = b_\alpha = b(\alpha)$, $c = c_\alpha = c(\alpha)$, and $d = d_\alpha = d(\alpha)$, whenever there is no fear of confusion. Then for $z \in S_n(\mathbf{C})$ and $\alpha \in Gp(n, \mathbf{C})$ we put

$$\mu_\alpha(z) = \mu(\alpha, z) = c_\alpha z + d_\alpha, \quad j_\alpha(z) = j(\alpha, z) = \det\left[\mu_\alpha(z)\right], \tag{1.5a}$$

$$\alpha(z) = \alpha z = (a_\alpha z + b_\alpha)(c_\alpha z + d_\alpha)^{-1}, \tag{1.5b}$$

where $\alpha(z)$ is defined only when $\mu_\alpha(z)$ is invertible. We will often write αz for $\alpha(z)$. To see the nature of $\alpha(z)$, put $p = a_\alpha z + b_\alpha$ and $q = \mu_\alpha(z)$. Using the relations in (1.2a), we easily see that ${}^tpq = {}^tqp$, and so $pq^{-1} \in S_n(\mathbf{C})$ if q is invertible. Therefore, if $\alpha(z)$ is defined, then $\alpha(z) \in S_n(\mathbf{C})$, and

$$\alpha \begin{bmatrix} z \\ 1 \end{bmatrix} = \begin{bmatrix} \alpha z \\ 1 \end{bmatrix} \mu_\alpha(z). \tag{1.6}$$

Given $z \in S_n(\mathbf{C})$ and $\alpha \in Gp(n, \mathbf{C})$, suppose $\alpha \begin{bmatrix} z \\ 1 \end{bmatrix} = \begin{bmatrix} w \\ 1 \end{bmatrix} \lambda$ with $w \in \mathbf{C}_n^n$ and $\lambda \in GL_n(\mathbf{C})$; then we easily see that αz is defined, $\alpha z = w$, $\lambda = \mu_\alpha(z)$, and (1.5b) holds.

Next suppose $\alpha, \beta \in Gp(n, \mathbf{C})$ and both αz and $\beta(\alpha z)$ are meaningful; then applying β to (1.6), we obtain

$$\beta\alpha \begin{bmatrix} z \\ 1 \end{bmatrix} = \beta \begin{bmatrix} \alpha z \\ 1 \end{bmatrix} \mu_\alpha(z) = \begin{bmatrix} \beta(\alpha z) \\ 1 \end{bmatrix} \mu_\beta(\alpha z)\mu_\alpha(z),$$

and so $(\beta\alpha)(z)$ is meaningful and

$$\mu_{\beta\alpha}(z) = \mu_\beta(\alpha z)\mu_\alpha(z), \quad j_{\beta\alpha}(z) = j_\beta(\alpha z)j_\alpha(z), \quad \text{and} \quad (\beta\alpha)(z) = \beta(\alpha z). \tag{1.7}$$

Moreover, if $z' \in S_n(\mathbf{C})$ and $\alpha z'$ is defined, then

$$\alpha \begin{bmatrix} z' & z \\ 1 & 1 \end{bmatrix} = \begin{bmatrix} \alpha z' & \alpha z \\ 1 & 1 \end{bmatrix} \begin{bmatrix} \mu_\alpha(z') & 0 \\ 0 & \mu_\alpha(z) \end{bmatrix}. \tag{1.8}$$

Calling this product W and forming ${}^tW\iota W$, we obtain

$$\nu(\alpha)(z' - z) = {}^t\mu_\alpha(z')(\alpha z' - \alpha z)\mu_\alpha(z). \tag{1.9}$$

If $dz = (dz_{ij})$ denotes the differential of the variable matrix $z = (z_{ij})$ on $S_n(\mathbf{C})$, then from (1.9) we obtain

$$\nu(\alpha)dz = {}^t\mu_\alpha(z)d(\alpha z)\mu_\alpha(z). \tag{1.10}$$

We now put

(1.11) $$Gp^+(n, \mathbf{R}) = \{\alpha \in Gp(n, \mathbf{R}) \mid \nu(\alpha) > 0\},$$

(1.11a) $$Gp^+(n, \mathbf{Q}) = GL_{2n}(\mathbf{Q}) \cap Gp^+(n, \mathbf{R}),$$

(1.12) $$\mathfrak{H}_n = \{z \in S_n(\mathbf{C}) \mid \mathrm{Im}(z) > 0\}.$$

Every element z of $\mathfrak{H}_n$ can be written $z = x + iy$ with $x \in S_n(\mathbf{R})$ and $0 < y \in S_n(\mathbf{R})$, and vice versa. Hereafter, whenever we write $z = x + iy$ for $z \in \mathfrak{H}_n$, we always take x and y in that sense.

Lemma 1.3. *Let $\mathfrak{X}$ be the set of all $X \in \mathbf{C}_n^{2n}$ such that*

(1.13) $${}^tX\iota X = 0 \quad \text{and} \quad i \cdot {}^tX\iota\bar{X} > 0.$$

Then the following assertions hold:

(i) *The map $(z, \mu) \mapsto \begin{bmatrix} z \\ 1_n \end{bmatrix}\mu$ gives a bijection of $\mathfrak{H}_n \times GL_n(\mathbf{C})$ onto $\mathfrak{X}$.*

(ii) *$\alpha\mathfrak{X}\beta \subset \mathfrak{X}$ for every $\alpha \in Gp^+(n, \mathbf{R})$ and $\beta \in GL_n(\mathbf{C})$.*

PROOF. That the image of the map of (i) is indeed in $\mathfrak{X}$ can easily be seen. Given $X = \begin{bmatrix} g \\ h \end{bmatrix}$ with $g, h \in \mathbf{C}_n^n$, we have (1.13) if and only if

(*) $${}^thg = {}^tgh \quad \text{and} \quad i({}^th\bar{g} - {}^tg\bar{h}) > 0.$$

Therefore, for $0 \neq x \in \mathbf{C}_1^n$ we have

$$0 < i \cdot {}^tx({}^th\bar{g} - {}^tg\bar{h})\bar{x} = i({}^t(hx)\bar{g}\bar{x} - {}^t(gx)\bar{h}\bar{x}),$$

and so $gx \neq 0$ and $hx \neq 0$. Thus both g and h are invertible. Put $z = gh^{-1}$. Then (*) shows that ${}^tz = z$ and $i(\bar{z} - z) = i \cdot {}^th^{-1}({}^th\bar{g} - {}^tg\bar{h})\bar{h}^{-1} > 0$, and so $z \in \mathfrak{H}_n$. Since $X = \begin{bmatrix} z \\ 1 \end{bmatrix} h$, we see that our map is surjective. The injectivity and (ii) are obvious.

1.4. Let $z \in \mathfrak{H}_n$ and $\alpha \in Gp^+(n, \mathbf{R})$. Then by Lemma 1.3(i), $\begin{bmatrix} z \\ 1 \end{bmatrix} \in \mathfrak{X}$. By (ii) of the same lemma, $\alpha\begin{bmatrix} z \\ 1 \end{bmatrix} \in \mathfrak{X}$, and so by (i) we can put $\alpha\begin{bmatrix} z \\ 1 \end{bmatrix} = \begin{bmatrix} w \\ 1 \end{bmatrix}\mu$ with unique $w \in \mathfrak{H}_n$ and $\mu \in GL_n(\mathbf{C})$. Thus αz is meaningful and $w = \alpha z$; consequently (1.5b) and (1.6) hold. Since we have $1 = \mu(\alpha\alpha^{-1}, z) = \mu_\alpha(\alpha^{-1}z)\mu(\alpha^{-1}, z)$, we obtain

(1.14) $$\mu(\alpha^{-1}, z) = \mu_\alpha(\alpha^{-1}z)^{-1} \quad \text{and} \quad j(\alpha^{-1}, z) = j_\alpha(\alpha^{-1}z)^{-1}.$$

Taking the complex conjugate of (1.6) in this case, we obtain

(1.15) $$\alpha\begin{bmatrix} \bar{z} \\ 1 \end{bmatrix} = \begin{bmatrix} \overline{\alpha z} \\ 1 \end{bmatrix}\overline{\mu_\alpha(z)},$$

since α is a real matrix. This means that we can let an element α of $Gp^+(n, \mathbf{R})$ act on the set $\{\bar{z} \mid z \in \mathfrak{H}_n\}$, and view $\overline{\alpha z}$ as the image of $\bar{z}$ under α. Thus $\alpha(\bar{z}) = \alpha\bar{z} = \overline{\alpha z}$, $\mu_\alpha(\bar{z}) = c_\alpha \bar{z} + d_\alpha = \overline{\mu_\alpha(z)}$, and $j_\alpha(\bar{z}) = \det\left[\mu_\alpha(\bar{z})\right] = \overline{j_\alpha(z)}$. Taking $\bar{z}$ as z' in (1.9), we obtain

$$\text{(1.16)} \qquad \operatorname{Im}(\alpha z) = {}^t\overline{\mu_\alpha(z)}^{-1} \operatorname{Im}(z)\mu_\alpha(z)^{-1} \quad \text{if} \quad \alpha \in Sp(n, \mathbf{R}).$$

1.5. We note here how an element of $Gp^+(n, \mathbf{R})$ belonging to several special types acts on $z \in \mathfrak{H}_n$:

$$\text{(1.17a)} \qquad \begin{bmatrix} r1_n & 0 \\ 0 & 1_n \end{bmatrix} : z \mapsto rz \qquad (0 < r \in \mathbf{R}),$$

$$\text{(1.17b)} \qquad \begin{bmatrix} a & 0 \\ 0 & d \end{bmatrix} : z \mapsto az \cdot {}^t a \qquad (d = {}^t a^{-1} \in GL_n(\mathbf{R})),$$

$$\text{(1.17c)} \qquad \begin{bmatrix} 1_n & b \\ 0 & 1_n \end{bmatrix} : z \mapsto z + b \qquad (b \in S_n(\mathbf{R})),$$

$$\text{(1.17d)} \qquad \begin{bmatrix} a & b \\ 0 & d \end{bmatrix} : z \mapsto az \cdot {}^t a + bd^{-1} \qquad (d = {}^t a^{-1} \in GL_n(\mathbf{R}),\ {}^t bd \in S_n(\mathbf{R})),$$

$$\text{(1.17e)} \qquad \begin{bmatrix} 0 & b \\ c & 0 \end{bmatrix} : z \mapsto -bz^{-1} \cdot {}^t b \qquad (c = -{}^t b^{-1} \in GL_n(\mathbf{R})).$$

We also note that $Sp(n, \mathbf{R})$ *acts transitively on* $\mathfrak{H}_n$, that is, *given* z *and* w *in* $\mathfrak{H}_n$, *there exists an element* α *of* $Sp(n, \mathbf{R})$ *such that* $\alpha z = w$. Indeed, given $z = x + iy \in \mathfrak{H}_n$, take $a \in GL_n(\mathbf{R})$ so that $a \cdot {}^t a = y$ and put $\beta = \begin{bmatrix} 1 & x \\ 0 & 1 \end{bmatrix} \begin{bmatrix} a & 0 \\ 0 & {}^t a^{-1} \end{bmatrix}$. Then $\beta \in Sp(n, \mathbf{R})$ and $\beta(i1_n) = z$. Similarly we can find $\gamma \in Sp(n, \mathbf{R})$ such that $\gamma(i1_n) = w$. Then $\alpha z = w$ with $\alpha = \gamma\beta^{-1}$.

2. Some algebraic and arithmetic preliminaries

2.1. We begin with some easy facts on $\mathbf{Z}_n^1$ and $SL_n(\mathbf{Z})$. We call an element $v = (v_1, \ldots, v_n)$ of $\mathbf{Z}_n^1$ **primitive** if the v_i have no common divisors other than ± 1. Given such a v, put $M = \mathbf{Z}_n^1/\mathbf{Z}v$. Then M has no torsion elements and is finitely generated over $\mathbf{Z}$, and so it must be a free $\mathbf{Z}$-module. Thus M has a $\mathbf{Z}$-basis $\{y_i\}_{i=1}^m$. Let u_i be an element of $\mathbf{Z}_n^1$ that represents y_i. Then we can easily verify that $\{u_1, \ldots, u_m, v\}$ is a $\mathbf{Z}$-basis of $\mathbf{Z}_n^1$. Clearly $m = n - 1$. Let α be the square matrix whose rows are $u_1, \ldots, u_m, v$. Then $\alpha \in GL_n(\mathbf{Z})$, which proves the "only-if"-part of the first of the following statements:

(2.1) *An element* v *of* $\mathbf{Z}_n^1$ *is primitive if and only if* v *is the last row of an element of* $GL_n(\mathbf{Z})$. *Moreover, if* $n > 1$, *this is true with* $SL_n(\mathbf{Z})$ *in place of* $GL_n(\mathbf{Z})$.

The "if"-part is obvious. If $n > 1$ and $\det(\alpha) = -1$, then by replacing u_1 by $-u_1$, we obtain an element of $SL_n(\mathbf{Z})$.

Generalizing the idea of (2.1), we call an element x of $\mathbf{Z}_n^m$ with $m < n$ **primitive** if there is an element y of $\mathbf{Z}_n^{n-m}$ such that $\begin{bmatrix} y \\ x \end{bmatrix} \in GL_n(\mathbf{Z})$. Notice that if an element x of $\mathbf{Z}_n^m$ is primitive, then $\alpha x \beta$ is primitive for every $\alpha \in GL_m(\mathbf{Z})$ and $\beta \in GL_n(\mathbf{Z})$.

Lemma 2.2. (i) *Let Q_n denote the subgroup of $GL_n(\mathbf{Q})$ consisting of all the upper triangular matrices. Then $GL_n(\mathbf{Q}) = Q_n SL_n(\mathbf{Z})$.*

(ii) *Let W_n be the set of all primitive elements of $\mathbf{Z}_{2n}^n$ such that $w\iota \cdot {}^t w = 0$. Then $W_n = [0 \quad 1_n] Sp(n, \mathbf{Z})$.*

(iii) *Let P_n, or simply P, denote the subgroup of $Sp(n, \mathbf{Q})$ consisting of all the elements of the form $\begin{bmatrix} a & b \\ 0 & d \end{bmatrix}$, where a, b, d are of size n. Then $Sp(n, \mathbf{Q}) = P \cdot Sp(n, \mathbf{Z})$.*

(iv) *$Sp(n, \mathbf{Q})$ is generated by P and ι.*

PROOF. We first prove (i) by induction on n. It is trivial if $n = 1$. Given $\xi \in GL_n(\mathbf{Q})$, $n > 1$, let x be the last row of ξ. Then $x = qy$ with $0 \neq q \in \mathbf{Q}$ and a primitive element y of $\mathbf{Z}_n^1$. By (2.1) we can find an element α of $SL_n(\mathbf{Z})$ whose last row is y. Then $y = [0_{n-1}^1 \quad 1]\alpha$, so that $x\alpha^{-1} = [0_{n-1}^1 \quad q]$. Thus we can put $\xi\alpha^{-1} = \begin{bmatrix} r & s \\ 0 & q \end{bmatrix}$ with $r \in \mathbf{Q}_{n-1}^{n-1}$ and $s \in \mathbf{Q}_1^{n-1}$. By induction we find $\tau \in Q_{n-1}$ and $\sigma \in SL_{n-1}(\mathbf{Z})$ such that $r = \tau\sigma$. Then $\xi\alpha^{-1} \cdot \mathrm{diag}[\sigma^{-1}, 1] \in Q_n$, which proves (i).

As for (ii), clearly $[0 \quad 1_n]Sp(n, \mathbf{Z}) \subset W_n$. Let $x \in W_n$. Since x is primitive, there exists an element $\alpha \in SL_{2n}(\mathbf{Z})$ of the form $\alpha = \begin{bmatrix} y \\ x \end{bmatrix}$ with some $y \in \mathbf{Z}_{2n}^n$. Then $\alpha\iota \cdot {}^t\alpha = \begin{bmatrix} u & v \\ -{}^t v & 0 \end{bmatrix}$ with $u, v \in \mathbf{Z}_n^n$. Since $\alpha\iota \cdot {}^t\alpha \in GL_{2n}(\mathbf{Z})$, we see that $v \in GL_n(\mathbf{Z})$. Put $\beta = \mathrm{diag}[-v^{-1}, 1_n]$. Then $\beta\alpha\iota \cdot {}^t\alpha \cdot {}^t\beta = \begin{bmatrix} z & -1_n \\ 1_n & 0 \end{bmatrix}$ with $z \in \mathbf{Z}_n^n$. If $[a \quad b]$ is the upper half of $\beta\alpha$, then $z = -a \cdot {}^t b + b \cdot {}^t a$. Put $\gamma = \begin{bmatrix} 1_n & -b \cdot {}^t a \\ 0 & 1_n \end{bmatrix}$. Then $\gamma\beta\alpha\iota \cdot {}^t\alpha \cdot {}^t\beta \cdot {}^t\gamma = \iota$, and so $\gamma\beta\alpha \in Sp(n, \mathbf{Z})$. Now we see that $[0 \quad 1_n]\gamma\beta = [0 \quad 1_n]$, and so $[0 \quad 1_n]\gamma\beta\alpha = [0 \quad 1_n]\alpha = x$, which proves (ii).

To prove (iii), let $\xi \in Sp(n, \mathbf{Q})$. By (i) we have $\xi = \eta\alpha$ with $\eta \in Q_{2n}$ and $\alpha \in SL_{2n}(\mathbf{Z})$. Put $\eta = \begin{bmatrix} a & b \\ 0 & d \end{bmatrix}$ and $\xi = \begin{bmatrix} p & q \\ r & s \end{bmatrix}$ with $a, b, \ldots, r, s$ of size n. Then $d^{-1}[r \quad s] = [0 \quad 1_n]\alpha$, which is primitive. Since $\xi \in Sp(n, \mathbf{Q})$, we easily see that $d^{-1}[r \quad s] \in W_n$. By (ii) we can put $[0 \quad 1_n]\alpha = [0 \quad 1_n]\beta$ with $\beta \in Sp(n, \mathbf{Z})$. Put $\gamma = \alpha\beta^{-1}$. Then $[0 \quad 1_n]\gamma = [0 \quad 1_n]$, and so $\gamma = \begin{bmatrix} u & v \\ 0 & 1_n \end{bmatrix}$

with u, v of size n. Thus $\eta\gamma = \begin{bmatrix} w & z \\ 0 & d \end{bmatrix}$ with w, z of size n. Now $\xi = \eta\alpha = \eta\gamma\beta$, and so $\eta\gamma = \xi\beta^{-1} \in Sp(n, \mathbf{Q})$. Therefore, $\eta\gamma \in P$, which proves (iii).

Finally, as will be shown in Lemma A1.3(i) of the Appendix, $Sp(n, \mathbf{Z})$ is generated by ι and $P \cap Sp(n, \mathbf{Z})$. This fact combined with (iii) proves (iv) and completes the proof.

Lemma 2.3. *Let f be an odd integer and s an element of $S_n(\mathbf{Z})$. Then there exist an element u of $M_n(\mathbf{Z})$ and a diagonal element d of $M_n(\mathbf{Z})$ such that ${}^t usu - d \prec f\mathbf{Z}$ and $\det(u)$ is a positive integer prime to f.*

PROOF. We can reduce the problem to the following statement, in which p is an odd prime number and $\mathbf{Z}_p$ is the ring of p-adic integers:

(2.2) *Given $s \in S_n(\mathbf{Z}_p)$, there exists an element τ of $GL_n(\mathbf{Z}_p)$ such that ${}^t\tau s\tau$ is diagonal.*

Indeed, given s and f as in our lemma, employing (2.2), for each prime factor p of f we take $\tau_p \in GL_n(\mathbf{Z}_p)$ such that ${}^t\tau_p s\tau_p$ is diagonal. We can find $u \in M_n(\mathbf{Z})$ and a diagonal matrix d such that $u - \tau_p \prec f\mathbf{Z}_p$ and $d - {}^t\tau_p s\tau_p \prec f\mathbf{Z}_p$ for every $p|f$. Replacing u by $u + fv$ with a suitable $v \in M_n(\mathbf{Z})$, we may assume that $\det(u) > 0$. Then u has the required properties of our lemma.

We now prove (2.2) by induction on n. The case $n = 1$ or $s = 0$ is trivial. Assuming that $n > 1$ and $s \neq 0$, put $s = (s_{ij})$ and $\sum_{i,j} s_{ij}\mathbf{Z}_p = \sigma\mathbf{Z}_p$ with $0 \neq \sigma \in \mathbf{Z}_p$. Replacing s by $\sigma^{-1}s$, we may assume that $\sigma = 1$. Suppose $s_{ii} \in \mathbf{Z}_p^\times$ for some i; we may assume that $i = 1$. Then we can put $s = \begin{bmatrix} a & b \\ {}^tb & d \end{bmatrix}$ with $a = s_{11}$. Put $\gamma = \begin{bmatrix} 1 & -a^{-1}b \\ 0 & 1_{n-1} \end{bmatrix}$. Then $\gamma \in GL_n(\mathbf{Z}_p)$ and ${}^t\gamma s\gamma = \mathrm{diag}[a, e]$ with $e = d - {}^tba^{-1}b$. Clearly $e \in S_{n-1}(\mathbf{Z}_p)$. Applying induction to e, we obtain the desired conclusion. Next suppose $s_{ii} \notin \mathbf{Z}_p^\times$ for every i. Then changing the numbering, we may assume that $s_{12} \in \mathbf{Z}_p^\times$. Put $a = \begin{bmatrix} s_{11} & s_{12} \\ s_{21} & s_{22} \end{bmatrix}$ and $s = \begin{bmatrix} a & b \\ {}^tb & d \end{bmatrix}$. Then $a \in GL_2(\mathbf{Z}_p)$. Putting $\gamma = \begin{bmatrix} 1_2 & -a^{-1}b \\ 0 & 1_{n-2} \end{bmatrix}$, we see that $\gamma \in GL_n(\mathbf{Z}_p)$, and ${}^t\gamma s\gamma = \mathrm{diag}[a, d - {}^tba^{-1}b]$. Thus again induction gives the desired fact. This completes the proof.

2.4. We need two types of Gauss sums. The first type is defined with respect to a **Dirichlet character** (or simply, a **character**) **modulo** a positive integer r, by which we mean a function $\chi : \mathbf{Z} \to \mathbf{T}$ belonging to one of the following two types:

(i) $r = 1$ and $\chi(a) = 1$ for every $a \in \mathbf{Z}$;

(ii) $r > 1$ and χ is a homomorphism of $(\mathbf{Z}/r\mathbf{Z})^\times$ into $\mathbf{T}$, which we view as a function on $\mathbf{Z}$ by putting

$$\chi(a) = \begin{cases} \chi(a \pmod{r\mathbf{Z}}) & \text{if } a \text{ is prime to } r, \\ 0 & \text{if } a \text{ is not prime to } r. \end{cases}$$

We call χ of type (i) **the principal character**. We say that χ is **trivial** if it is of type (i) or $\chi(a) = 1$ for every a prime to r. We call χ **primitive** if it is of type (i), or it is a nontrivial character for which there is no character ξ modulo a proper divisor s of r such that $\chi(a) = \xi(a)$ for every a prime to r. We will say more about characters in §2.7, but we first define Gauss sums.

We put $\mathbf{e}(z) = \exp(2\pi i z)$ for $z \in \mathbf{C}$ as we did in (0.3). Let χ be a primitive Dirichlet character modulo an integer $r > 1$. Then we put

$$G(\chi) = \sum_{a=1}^{r} \chi(a)\mathbf{e}(a/r) \tag{2.3}$$

and call it **the Gauss sum of** χ. Since $\chi(r) = 0$, we can use $\sum_{a=1}^{r-1}$ in place of $\sum_{a=1}^{r}$. We note here three well-known facts:

$$\sum_{a=1}^{r} \chi(a)\mathbf{e}(sa/r) = \overline{\chi}(s)G(\chi) \quad \text{for every } s \in \mathbf{Z}, \tag{2.3a}$$

$$rG(\chi)^{-1} = \overline{G(\chi)} = \chi(-1)G(\bar{\chi}), \tag{2.3b}$$

$$|G(\chi)|^2 = r. \tag{2.3c}$$

The proof is easy; see [S71, Lemma 3.63] or [S10, (3.8a, c, d)], for example.

There is a classical result about $G(\chi)$ when $\overline{\chi} = \chi$. Namely,

$$G(\chi) = \begin{cases} \sqrt{r} & \text{if } \chi(-1) = 1, \\ i\sqrt{r} & \text{if } \chi(-1) = -1. \end{cases} \tag{2.4a}$$

In particular, if $\chi(x) = \left(\dfrac{x}{p}\right)$ with an odd prime number p, then

$$G(\chi) = \begin{cases} \sqrt{p} & \text{if } p - 1 \in 4\mathbf{Z}, \\ i\sqrt{p} & \text{if } p - 3 \in 4\mathbf{Z}. \end{cases} \tag{2.4b}$$

The simplest proof of (2.4a) follows from the functional equation of the Dirichlet series $\sum_{n=1}^{\infty} \chi(n)n^{-s}$. This is explained in [S07, p. 40] when $\chi(-1) = -1$, but the same method is applicable to the other case. In fact, by the same technique Hecke determined the Gauss sum of a Hecke character associated with a quadratic extension of an arbitrary algebraic number field; see [S97, (A6.3.4)].

2.5. The second type of Gauss sum, denoted by $G(a, b)$, is defined for relatively prime integers a and b and given by

(2.5) $$G(a,\,b) = \sum_{x=1}^{|b|} \mathbf{e}(ax^2/b),$$

where $\sum_{x=1}^{|b|}$ can be replaced by $\sum_{x\in\mathbf{Z}/b\mathbf{Z}}$.

Define the Legendre-Jacobi symbol $\left(\frac{m}{n}\right)$ as in (0.4). For two positive odd integers m and n we have

(2.6) $$\left(\frac{m}{n}\right)\left(\frac{n}{m}\right) = -1 \iff m \equiv n \equiv 3 \pmod 4,$$

(2.7) $$\left(\frac{-1}{n}\right) = -1 \iff n \equiv 3 \pmod 4.$$

To prove these, we first note that

(2.8) $$\left(\frac{p}{q}\right)\left(\frac{q}{p}\right) = -1 \iff p \equiv q \equiv 3 \pmod 4$$

for two different odd prime numbers p and q. This follows easily from the quadratic reciprocity law. To prove (2.6) in general, put $m = p_1 \cdots p_r$ and $n = q_1 \cdots q_s$ with odd prime numbers p_i and q_j. Then

$$\left(\frac{m}{n}\right)\left(\frac{n}{m}\right) = \prod_{i=1}^{r}\prod_{j=1}^{s}\left(\frac{p_i}{q_j}\right)\left(\frac{q_j}{p_i}\right).$$

By (2.8) we can eliminate from the right-hand side any p_i or q_j that is $\equiv 1 \pmod 4$. Let ρ resp. σ be the number of i's resp. j's such that $p_i - 3 \in 4\mathbf{Z}$ resp. $q_j - 3 \in 4\mathbf{Z}$. By (2.8), $\left(\frac{m}{n}\right)\left(\frac{n}{m}\right) = (-1)^{\rho\sigma}$, which is -1 if and only if both ρ and σ are odd, that is, if and only if $m \equiv n \equiv 3 \pmod 4$. Thus we obtain (2.6). Similarly we have $\left(\frac{-1}{n}\right) = \prod_{j=1}^{s}\left(\frac{-1}{q_j}\right) = (-1)^{\sigma}$, and so we obtain (2.7).

Theorem 2.6. *If b is odd and positive, then*

(2.9) $$G(a,\,b) = \left(\frac{a}{b}\right)\varepsilon_b\sqrt{b},$$

where $\varepsilon_b = 1$ if $b - 1 \in 4\mathbf{Z}$ and $\varepsilon_b = i$ if $b - 3 \in 4\mathbf{Z}$.

PROOF. We first prove

(2.10) $$G(a,\,cd) = G(ad,\,c)G(ac,d) \quad \text{if} \quad c,\,d \in \mathbf{Z} \text{ and } c\mathbf{Z} + d\mathbf{Z} = \mathbf{Z}.$$

Indeed, the map $\bigl(y \pmod c),\, z \pmod d)\bigr) \mapsto dy + cz \pmod{cd}$ gives a bijection of $(\mathbf{Z}/c\mathbf{Z}) \times (\mathbf{Z}/d\mathbf{Z})$ onto $\mathbf{Z}/cd\mathbf{Z}$, and so for $b = cd$ we have

$$\begin{aligned} G(a,\,b) &= \sum_{y\in\mathbf{Z}/c\mathbf{Z}}\ \sum_{z\in\mathbf{Z}/d\mathbf{Z}} \mathbf{e}\bigl(a(dy+cz)^2/b\bigr) \\ &= \sum_{y\in\mathbf{Z}/c\mathbf{Z}} \mathbf{e}(ady^2/c) \sum_{z\in\mathbf{Z}/d\mathbf{Z}} \mathbf{e}(acz^2/d) = G(ad,\,c)G(ac,\,d), \end{aligned}$$

which proves (2.10). Now suppose (2.9) is true for $G(*,\, c)$ and $G(*,\, d)$ (with positive c and d). Then

$$G(a,\, b) = \left(\frac{ad}{c}\right)\left(\frac{ac}{d}\right)\varepsilon_c\varepsilon_d\sqrt{cd} = \left(\frac{a}{b}\right)\eta\sqrt{b}$$

with $\eta = \left(\frac{d}{c}\right)\left(\frac{c}{d}\right)\varepsilon_c\varepsilon_d$. If $c \equiv d \equiv 3 \pmod 4$, then by (2.6), $\eta = -i^2 = 1$, and $b-1 \in 4\mathbf{Z}$. If $c-1 \in 4\mathbf{Z}$, then $b-d \in 4\mathbf{Z}$ and $\eta = \varepsilon_d = \varepsilon_b$; similarly $\eta = \varepsilon_b$ if $d-1 \in 4\mathbf{Z}$. Thus we obtain (2.9) for $b = cd$. This means that it is sufficient to prove (2.9) when $b = p^m$ with an odd prime number p and $0 < m \in \mathbf{Z}$. Suppose $m > 1$; since the map $\bigl(y \pmod{p^{m-1}},\, z \pmod p\bigr) \mapsto y + p^{m-1}z \pmod{p^m}$ gives a bijection of $(\mathbf{Z}/p^{m-1}\mathbf{Z}) \times (\mathbf{Z}/p\mathbf{Z})$ onto $\mathbf{Z}/p^m\mathbf{Z}$, we have $G(a,\, p^m) = \sum_{y=1}^{p^{m-1}} \mathbf{e}(ay^2/p^m)\sum_{z=1}^{p} \mathbf{e}(2ayz/p)$. The last sum over $1 \le z \le p$ equals p if $y \in p\mathbf{Z}$. For $y \notin p\mathbf{Z}$ we have $\sum_{z=1}^{p} \mathbf{e}(2ayz/p) = \sum_{w=1}^{p} \mathbf{e}(w/p) = 0$. Therefore

$$G(a,\, p^m) = p\sum_{u=1}^{p^{m-2}} \mathbf{e}(ap^2u^2/p^m) = pG(a,\, p^{m-2})$$

for $m > 1$. If $m = 2$ in particular, we see that $G(a,\, p^2) = p$. Thus for $0 < n \in \mathbf{Z}$ we obtain $G(a,\, p^{2n}) = p^n$ and $G(a,\, p^{2n+1}) = p^nG(a,\, p)$. Therefore our problem can be reduced to the formula

$$G(a,\, p) = \left(\frac{a}{p}\right)\varepsilon_p\sqrt{p}. \tag{2.11}$$

To show this, put $\chi(y) = \left(\frac{y}{p}\right)$. Dividing the set $\{y \in \mathbf{Z} \mid 0 < y < p\}$ into the set of quadratic residues modulo p and that of nonresidues, we see that

$$\begin{aligned} G(a,p) &= \sum_{x=1}^{p} \mathbf{e}(ax^2/p) = \sum_{y=1}^{p} \chi(y)\mathbf{e}(ay/p) + \sum_{y=1}^{p} \mathbf{e}(ay/p) \\ &= \sum_{u=1}^{p} \chi(u)\mathbf{e}(au/p) = \chi(a)G(\chi), \end{aligned}$$

and so (2.11) follows from (2.4b). This completes the proof.

Remark. We can determine $G(a,\, b)$ even for odd negative b as follows. We first take $c = -1$ in (2.10) and obtain

$$G(a,\, -d) = G(-a,\, d), \tag{2.12}$$

since $G(x,\, -1) = 1$ for every $x \in \mathbf{Z}$. Therefore, if b is odd and negative, then $G(a,\, b) = G(a,\, -|b|) = G(-a,\, |b|)$ by (2.12), and so by (2.9) we obtain

$$G(a,\, b) = \left(\frac{-a}{|b|}\right)\varepsilon_{|b|}\sqrt{|b|} = \left(\frac{a}{|b|}\right)\left(\frac{-1}{|b|}\right)\varepsilon_{|b|}\sqrt{|b|}. \tag{2.13}$$

Applying (2.7) to the factor $\left(\frac{-1}{|b|}\right)$, we can give another formula. Thus, for $0 \ge b + 1 \in 2\mathbf{Z}$ we have

$$G(a,\, b) = \left(\frac{a}{|b|}\right) \varepsilon'(b)\sqrt{|b|}, \tag{2.14}$$

where $\varepsilon'(b) = 1$ if $b - 3 \in 4\mathbf{Z}$ and $\varepsilon'(b) = -i$ if $b - 1 \in 4\mathbf{Z}$.

2.7. Let χ_0 be the principal character as defined in §2.4. We define its Gauss sum $G(\chi_0)$ by

$$G(\chi_0) = 1. \tag{2.15}$$

Let χ be a character modulo an integer $r > 1$ that is not primitive. We then call χ **imprimitive**. If χ is nontrivial, then we can find a character χ' modulo a proper divisor c of r, > 1, such that $\chi(a) = \chi'(a)$ for every a prime to r. Moreover, among such characters χ' there is a unique one that is primitive. We then call χ' the **primitive character associated with** χ, and call c the **conductor of** χ. If χ is trivial, we call the principal character χ_0 the **primitive character associated with** χ, and define the conductor of χ to be 1. In both cases, given χ, we take the primitive character χ' associated with χ, and define the Gauss sum $G(\chi)$ by $G(\chi) = G(\chi')$.

For $1 \le i \le m$ let χ_i be a Dirichlet character modulo r_i, and let r be the least common multiple of the r_i. Then we denote by $\chi_1 \cdots \chi_m$ the character modulo r defined by $(\chi_1 \cdots \chi_m)(a) = \chi_1(a) \cdots \chi_m(a)$ for a prime to r.

Let $\mathrm{Aut}(\mathbf{C})$ denote the group of all ring-automorphisms of the field $\mathbf{C}$. For $\sigma \in \mathrm{Aut}(\mathbf{C})$ we can define a character χ_i^σ modulo r_i by $\chi_i^\sigma(a) = \chi_i(a)^\sigma$ for every a. Then we have:

Lemma 2.8. *Put* $q(\chi_1, \ldots, \chi_m) = G(\chi_1) \cdots G(\chi_m)/G(\chi_1 \cdots \chi_m)$. *Let* K *be the field generated over* $\mathbf{Q}$ *by the values* $\chi_i(a)$ *for all* i *and all* a. *Then* $q(\chi_1, \ldots, \chi_m)$ *belongs to* K, *and for every automorphism* σ *of* $\overline{\mathbf{Q}}$ *we have*

$$q(\chi_1, \ldots, \chi_m)^\sigma = q(\chi_1^\sigma, \ldots, \chi_m^\sigma). \tag{2.16}$$

Proof. Let $\zeta = \mathbf{e}(1/N)$ with a multiple N of $r_1 \cdots r_m$. We take N so that $\chi_i(a) \in \mathbf{Q}(\zeta)$ for every i and a. Given σ, we can find an integer t prime to N such that $\mathbf{e}(a/N)^\sigma = \mathbf{e}(ta/N)$. Then from (2.3) we obtain

$$\chi(t)^\sigma G(\chi)^\sigma = G(\chi^\sigma) \tag{2.17}$$

for every character χ modulo N. Formula (2.16) is an immediate consequence of this fact. Suppose σ is the identity map on K; then $\chi_i^\sigma = \chi_i$ for every i, and so $q(\chi_1, \ldots, \chi_m)^\sigma = q(\chi_1, \ldots, \chi_m)$. This shows that $q(\chi_1, \ldots, \chi_m) \in K$, and our proof is complete.

To state an easy application of (2.17), for every primitive or imprimitive character modulo N and every $q \in \mathbf{Z}$, put

$$[\chi, q] = G(\chi)^{-1} \sum_{n=1}^{N} \chi(n)\mathbf{e}(nq/N). \tag{2.17a}$$

Then from (2.17) we can easily derive

(2.17b) $[\chi, q] \in \mathbf{Q}_{\rm ab}$ *and* $[\chi, q]^\sigma = [\chi^\sigma, q]$ *for every* $\sigma \in \mathrm{Gal}(\mathbf{Q}_{\rm ab}/\mathbf{Q})$.

In fact, we will give an explicit form of $[\chi, q]$ in Lemma A3.1 of the Appendix, which implies (2.17b).

2.9. Given a primitive or an imprimitive Dirichlet character χ, we put

$$L(s, \chi) = \sum_{n=1}^{\infty} \chi(n)n^{-s}, \tag{2.18a}$$

$$L_N(s, \chi) = L(s, \chi)\prod_{p|N}\left[1 - \chi(p)p^{-s}\right] = \sum_{(n, N)=1} \chi(n)n^{-s}, \tag{2.18b}$$

where $0 < N \in \mathbf{Z}$. The function of (2.18a) is called the **Dirichlet L-function** of χ. We will state some of its analytic properties in §8.3. For the moment we just note that $L(m, \chi)$ and $L(1-m, \chi)$ for a positive integer m such that $\chi(-1) = (-1)^m$ is meaningful. Then we put

$$P_N(m, \chi) = G(\chi)^{-1}(\pi i)^{-m}L_N(m, \chi). \tag{2.19}$$

Lemma 2.10. *Let χ and m be as above. Then $P_N(m, \chi) \in \mathbf{Q}_{\rm ab}$ and $L(1-m, \chi) \in \mathbf{Q}_{\rm ab}$. Moreover, for every $\sigma \in \mathrm{Gal}(\mathbf{Q}_{\rm ab}/\mathbf{Q})$ we have*

$$P_N(m, \chi)^\sigma = P_N(m, \chi^\sigma), \tag{2.20}$$

$$L(1-m, \chi)^\sigma = L(1-m, \chi^\sigma). \tag{2.21}$$

PROOF. We can easily reduce the problem to the case where χ is primitive and $N = 1$. Then

$$2 \cdot m!(2\pi i)^{-m}G(\overline{\chi})L(m, \chi) = md^{1-m}L(1-m, \overline{\chi}) = -\sum_{a=1}^{d} \overline{\chi}(a)B_m(a/d),$$

where d is the conductor of χ and B_m the mth Bernoulli polynomial. If χ is principal, this means

$$2 \cdot n!(2\pi i)^{-n}\zeta(n) = n\zeta(1-n) = -B_n$$

for $0 < n \in 2\mathbf{Z}$, where B_n is the nth Bernoulli number. This formula is ancient. The preceding one was first noted by Hecke in 1940. For the proof and other formulas for $L(m, \chi)$ and $L(1-m, \chi)$ we refer the reader to Section 4 of [S07]. Our lemma now follows from these formulas. Notice that the complex conjugate of χ^σ is $(\overline{\chi})^\sigma$, and $G(\overline{\chi}) = d\chi(-1)G(\chi)^{-1}$ by (2.3b).

Lemma 2.11. *Let V be a finite-dimensional vector space over a field F and let $W = V \otimes_F K$ with a finite or an infinite extension K of F. Let $\mathrm{End}(V)$ denote the algebra of all F-linear transformations of V. Given a subset A of $\mathrm{End}(V)$, for each $\alpha \in A$ denote by $\tilde{\alpha}$ the K-linear extension of α to W. Then $\bigcap_{\alpha \in A} \mathrm{Ker}(\tilde{\alpha}) = \bigl\{\bigcap_{\alpha \in A} \mathrm{Ker}(\alpha)\bigr\} \otimes_F K$.*

PROOF. Take a basis $\{y_i\}_{i\in I}$ of V over F and also a basis B of K over F. For $\alpha \in A$ put $\alpha y_i = \sum_{j\in I} p_{ij}^{\alpha} y_j$ with $p_{ij}^{\alpha} \in F$. Let $u = \sum_{i\in I} c_i y_i \in \bigcap_{\alpha\in A} \mathrm{Ker}(\tilde{\alpha})$ with $c_i \in K$. Then there is a finite subset E of B such that $c_i = \sum_{e\in E} a_{ie} e$ with $a_{ie} \in F$ for every $i \in I$. Then for $\alpha \in A$ we have $0 = \tilde{\alpha} u = \sum_{e\in E} e \sum_{i,j\in I} a_{ie} p_{ij}^{\alpha} y_j$, and so $\sum_{e\in E} e \sum_{i\in I} a_{ie} p_{ij}^{\alpha} = 0$ for every j. Thus $\sum_{i\in I} a_{ie} p_{ij}^{\alpha} = 0$ for every j and every $e \in E$. Now $u = \sum_{e\in E} e v_e$ with $v_e = \sum_{i\in I} a_{ie} y_i$. We have $v_e \in V$ and $\alpha v_e = \sum_{i,j\in I} a_{ie} p_{ij}^{\alpha} y_j = 0$, and so $v_e \in \bigcap_{\alpha\in A} \mathrm{Ker}(\alpha)$, which proves our lemma.

Lemma 2.12. *Let V, F, and $\mathrm{End}(V)$ be as in Lemma 2.11. Suppose that $V \neq \{0\}$ and F is algebraically closed. Let R be a subspace of $\mathrm{End}(V)$, $\neq \{0\}$, whose elements are mutually commutative. Then V has a nonzero element u that is an eigenvector of every element of R.*

PROOF. We prove this by induction on the dimension of R. Let $0 \neq \alpha \in R$. We can take an eigenvector v of α and put $\alpha v = cv$ with $c \in F$. (This settles the case where $\dim(R) = 1$.) Let $W = \{x \in V \mid \alpha x = cx\}$. Then $W \neq \{0\}$, since $v \in W$. Assuming that $\dim(R) > 1$, take a subspace S of R so that $R = F\alpha \oplus S$. Since the elements of S commute with α, we easily see that W is stable under S. By the induction assumption we can find an element u of W that is an eigenvector of every element of S. Since $\alpha u = cu$, this proves our lemma.

Lemma 2.13. *Let K be a finite Galois extension of a field F and W a vector space over K of finite dimension; let $G = \mathrm{Gal}(K/F)$. Suppose there is an action of G on W, written $(x, \sigma) \mapsto x^{\sigma}$ for $x \in W$ and $\sigma \in G$, such that $(ax)^{\sigma} = a^{\sigma} x^{\sigma}$ for $a \in K$ and $x \in W$. Then $W = V \otimes_F K$ with $V = \{y \in W \mid y^{\sigma} = y$ for every $\sigma \in G\}$.*

PROOF. Let B be a basis of K over F, and Y a finite set of elements of V that are linearly independent over F. Suppose $\sum_{b\in B,\, y\in Y} a_{by} b y = 0$ with $a_{by} \in F$. Then $\sum_{b\in B,\, y\in Y} a_{by} b^{\sigma} y = 0$ for every $\sigma \in G$. Since $\det[b^{\sigma}]_{b,\sigma} \neq 0$, we obtain $\sum_{y\in Y} a_{by} y = 0$ for every $b \in B$, and so $a_{by} = 0$ for every b and y. If W has dimension n over K, then W is a vector space of dimension $n \cdot \#B$ over F, and so $\#Y \leq n$. Given any $x \in W$, put $z_b = \sum_{\sigma\in G} b^{\sigma} x^{\sigma}$ for $b \in B$. Then $z_b \in V$. Since $\det[b^{\sigma}]_{b,\sigma} \neq 0$, x is a K-linear combination of the z_b. This shows that V has dimension n over F, and we can find Y such that $\#Y = n$. This proves our lemma.

Lemma 2.14. *Given a positive integer M and $\kappa = 0$ or 1, there exists a primitive character χ such that M is prime to the conductor of χ, $\chi(-1) = (-1)^{\kappa}$, and χ^2 is nontrivial.*

PROOF. Since there exist infinitely many prime numbers p such that $p - 1 \in 8\mathbf{Z}$, we can find such a p prime to M. Let ψ be a character modulo p

of order $p-1$. Take $\chi = \psi$ if $\kappa = 1$ and $\chi = \psi^2$ if $\kappa = 0$. Then χ has the desired properties.

3. Modular forms of integral weight

3.1. In this book we assume that the reader has some notion of modular forms of integral weight with respect to a congruence subgroup of $SL_2(\mathbf{Z})$, though we try to make our exposition as self-contained as possible. As to their well-known properties, sometimes we will only state them, dispensing with the proof. To define modular forms on $\mathfrak{H}_n$, we first put $\mathbf{e}(c) = \exp(2\pi i c)$ for $c \in \mathbf{C}$, as we did in (0.3). Next, for $k \in \mathbf{Z}$, $\alpha \in Gp^+(n, \mathbf{R})$, and a function $f : \mathfrak{H}_n \to \mathbf{C}$ we define $f\|_k\alpha : \mathfrak{H}_n \to \mathbf{C}$ by

$$(3.1) \qquad (f\|_k\alpha)(z) = j_\alpha(z)^{-k} f(\alpha z) \qquad (z \in \mathfrak{H}_n).$$

From this definition and (1.7) we obtain

$$(3.2a) \qquad f\|_k(c1_{2n}) = c^{-nk} f \qquad (c \in \mathbf{R}^\times),$$

$$(3.2b) \qquad (f\|_k\,\alpha\beta) = (f\|_k\,\alpha)\|_k\,\beta \quad \text{if} \quad \alpha,\, \beta \in Gp^+(n,\, \mathbf{R}).$$

For a positive integer N put

$$(3.3) \qquad \Gamma(N) = \{\gamma \in Sp(n,\, \mathbf{Z}) \mid \gamma - 1_{2n} \prec N\mathbf{Z}\}.$$

(For the symbol $\prec$ see §0.1.) By a **congruence subgroup** of $Sp(n, \mathbf{Q})$, or simply, a congruence subgroup, we mean a subgroup Γ of $Sp(n, \mathbf{Q})$ that contains $\Gamma(N)$ as a subgroup of finite index for some N. If Γ and Γ' are congruence subgroups and $\Gamma \subset \Gamma'$, we call Γ a congruence subgroup of Γ'. Given a congruence subgroup Γ and $k \in \mathbf{Z}$, we denote by $\mathscr{M}_k(\Gamma)$ the set of all functions $f : \mathfrak{H}_n \to \mathbf{C}$ satisfying the following conditions:

(3.4a) *f is holomorphic;*
(3.4b) *$f\|_k\gamma = f$ for every $\gamma \in \Gamma$;*
(3.4c) *f is holomorphic at every cusp.*

An element of $\mathscr{M}_k(\Gamma)$ is called a (holomorphic) **modular form of weight** k with respect to Γ. Condition (3.4c) is necessary only when $n = 1$, in which case it means the following condition:

(3.4d) *For every $\alpha \in SL_2(\mathbf{Q})$ we have* $j_\alpha(z)^{-k} f(\alpha(z)) = \sum_{m=0}^{\infty} c_\alpha(m)\mathbf{e}(mz/r_\alpha)$

with $0 < r_\alpha \in \mathbf{Z}$ and $c_\alpha(m) \in \mathbf{C}$.

In fact, if this is satisfied for every $\alpha \in SL_2(\mathbf{Z})$, then it holds for every $\alpha \in SL_2(\mathbf{Q})$. Indeed, by Lemma 2.2(iii) every element α of $SL_2(\mathbf{Q})$ can be written $\alpha = \beta\gamma$ with $\beta \in SL_2(\mathbf{Z})$ and γ of the form $\gamma = \begin{bmatrix} r & s \\ 0 & t \end{bmatrix}$. Therefore (3.4d) for $\alpha = \beta\gamma$ can easily be derived from the case with β in place of α.

For $\beta \in Gp(n, \mathbf{Q})$ we easily see that $\beta^{-1}\Gamma\beta$ is a congruence subgroup, and $\mathscr{M}_k(\Gamma)\|_k\beta = \mathscr{M}_k(\beta^{-1}\Gamma\beta)$ provided $\nu(\beta) > 0$. We put

$$\mathscr{M}_k = \bigcup_{N=1}^{\infty} \mathscr{M}_k\bigl(\Gamma(N)\bigr). \tag{3.5}$$

3.2. Let us now explain why (3.4c) is unnecessary if $n > 1$. Let Γ be a congruence subgroup of $\Gamma(1)$. Then $\Gamma(N) \subset \Gamma$ for some $N \in \mathbf{Z}, > 0$, and so there is a $\mathbf{Z}$-lattice M in $S_n(\mathbf{Z})$ and also a subgroup U of $SL_n(\mathbf{Z})$ of finite index such that

$$\left\{\begin{bmatrix} 1_n & b \\ 0 & 1_n \end{bmatrix} \middle| \, b \in M\right\} \subset \Gamma, \qquad \left\{\begin{bmatrix} a & 0 \\ 0 & {}^t a^{-1} \end{bmatrix} \middle| \, a \in U\right\} \subset \Gamma.$$

Therefore if $f \in \mathscr{M}_k(\Gamma)$, then by (1.17b) and (1.17c) we have

$$f(z+b) = f(z) \quad \text{for every } b \in M, \tag{3.6a}$$

$$f(az \cdot {}^t a) = f(z) \quad \text{for every } a \in U. \tag{3.6b}$$

Let $S = S_n(\mathbf{Q})$ and $L = \{h \in S \,|\, \mathrm{tr}(hM) \subset \mathbf{Z}\}$. Then L is a $\mathbf{Z}$-lattice in S and (3.6a) guarantees an expansion of the form

$$f(z) = \sum_{h \in L} c(h)\mathbf{e}\bigl(\mathrm{tr}(hz)\bigr) \tag{3.7a}$$

with $c(h) \in \mathbf{C}$. This will be proven in §A1.1 of the Appendix. We often put

$$f(z) = \sum_{h \in S} c(h)\mathbf{e}\bigl(\mathrm{tr}(hz)\bigr) \tag{3.7b}$$

by defining $c(h)$ to be 0 for $h \in S, \notin L$. Usually we call the right-hand side of (3.7a) or (3.7b) the **Fourier expansion of** f, and call the $c(h)$ the **Fourier coefficients of** f.

Lemma 3.3. *Suppose $n > 1$; let f be a holomorphic function on $\mathfrak{H}_n$ of the form (3.7a) satisfying (3.6b) with a subgroup U of $SL_n(\mathbf{Z})$ of finite index. Then we have (3.7a) with $c(h) \neq 0$ only if h is nonnegative.*

The proof will be given in §A1.2 of the Appendix. Similar results can be proved for the Fourier expansion of an automorphic form of a more general type; see [S97, Propositions A4.2 and A4.5] and [S00, Proposition 5.7]. In fact, we do not need Lemma 3.3 in this book, since the modular forms in our later treatment in the case $n > 1$ will always be given explicitly, and so they have expansions of type (3.7a) with $c(h) \neq 0$ only for nonnegative h.

CHAPTER II

THETA FUNCTIONS AND FACTORS OF AUTOMORPHY

4. Classical theta functions

4.1. We define the classical theta function θ and its modification φ by

$$\theta(u,\, z;\, r,\, s) = \sum_{g-r\in\mathbf{Z}^n} \mathbf{e}\big(2^{-1}\cdot {}^tgzg + {}^tg(u+s)\big), \tag{4.1}$$

$$\varphi(u,\, z;\, r,\, s) = \mathbf{e}\big(2^{-1}\cdot {}^tu(z-\overline{z})^{-1}u\big)\theta(u,\, z;\, r,\, s). \tag{4.2}$$

Here $u \in \mathbf{C}^n$, $z \in \mathfrak{H}_n$, and $r,\, s \in \mathbf{R}^n$. To prove the convergence of the infinite series of (4.1), put $g = h + r$ with $h \in \mathbf{Z}^n$ and $y = \mathrm{Im}(z)$; take compact subsets U of $\mathbf{C}^n$ and Z of $\mathfrak{H}_n$. Then for fixed $r,\, s \in \mathbf{R}^n$, $u \in U$, and $z \in Z$ we easily see that

$$\mathrm{Re}\big(\pi i\cdot {}^tgzg + 2\pi i\cdot {}^tg(u+s)\big) = -\pi\cdot {}^thyh + {}^thv + w$$

with $v \in \mathbf{R}^n$ and $w \in \mathbf{R}$ in some compact sets depending on r, s, U, and Z. Let λ be the smallest eigenvalue of πy. Then $\pi\cdot {}^thyh \geq \lambda\cdot {}^thh$, and so the sum of (4.1) is majorized by

$$\sum_{h\in\mathbf{Z}^n} \exp(-\lambda\cdot {}^thh + {}^thv + w) = e^w \prod_{j=1}^{n}\sum_{k\in\mathbf{Z}} \exp(-\lambda k^2 + kv_j),$$

where v_j is the jth component of v. Therefore we see that the right-hand side of (4.1) is locally uniformly convergent on $\mathbf{C}^n \times \mathfrak{H}_n$, and so defines a holomorphic function in $(u,\, z)$.

The function $\theta(u,\, z;\, r,\, s)$ is called **Riemann's theta function**. By purely formal calculations we can easily verify

$$\begin{aligned}\theta(u&+za+b,\, z;\, r,\, s)\\ &= \mathbf{e}\big(-2^{-1}\cdot {}^taza - {}^ta(u+b+s)\big)\theta(u,\, z;\, r+a,\, s+b) \qquad (a,\, b \in \mathbf{R}^n),\end{aligned} \tag{4.3}$$

$$\begin{aligned}\theta(u&+za+b,\, z;\, r,\, s)\\ &= \mathbf{e}\big(-2^{-1}\cdot {}^taza - {}^tau + {}^trb - {}^tsa\big)\theta(u,\, z;\, r,\, s) \qquad (a,\, b \in \mathbf{Z}^n),\end{aligned} \tag{4.4}$$

$$\begin{aligned}\theta(u&,\, z;\, r+a,\, s+b)\\ &= \mathbf{e}({}^trb)\theta(u,\, z;\, r,\, s) = \theta(u+b,\, z;\, r,\, s) \qquad (a,\, b \in \mathbf{Z}^n).\end{aligned} \tag{4.5}$$

G. Shimura, *Modular Forms: Basics and Beyond*, Springer Monographs in Mathematics,
DOI 10.1007/978-1-4614-2125-2_2, © Springer Science+Business Media, LLC 2012

The function $\theta(u, z; r, s)$ was introduced by Riemann for the purpose of studying abelian integrals on an algebraic curve. In fact, these theta functions are essential in the geometric investigation of abelian varieties. In the present book, however, we merely employ them as a technical tool for studying automorphic forms of several types. The reader who is interested in their geometric and other aspects may be referred to [S98] and earlier articles by various authors cited there.

4.2. Let us now put

$$\Gamma_n^\theta = \Gamma^\theta = \{\gamma \in \Gamma(1) \mid \{a_\gamma \cdot {}^t b_\gamma\} \equiv \{c_\gamma \cdot {}^t d_\gamma\} \equiv 0 \pmod{2\mathbf{Z}^n}\}, \tag{4.6}$$

where $\{s\}$ is the column vector consisting of the diagonal elements of s. Notice that $a_\gamma \cdot {}^t b_\gamma$ and $c_\gamma \cdot {}^t d_\gamma$ belong to $S_n(\mathbf{Z})$ by (1.2a). We note an easy fact:

(4.7) *For* $s \in S_n(\mathbf{Z})$ *we have*

$$\begin{aligned} {}^t xsx \in 2\mathbf{Z} \text{ for every } x \in \mathbf{Z}^n &\iff \{s\} \in 2\mathbf{Z}^n \\ &\Longrightarrow \{{}^t ysy\} \in 2\mathbf{Z}^n \text{ for every } y \in \mathbf{Z}_n^n. \end{aligned}$$

Clearly $\Gamma(2) \subset \Gamma^\theta$. Put $F(x, y) = x \cdot {}^t y$ for $x, y \in \mathbf{Z}_n^1$. By (1.2c) we have

$$F\big((x, y)\gamma\big) - F(x, y) = xa_\gamma \cdot {}^t b_\gamma \cdot {}^t x + yc_\gamma \cdot {}^t d_\gamma \cdot {}^t y + 2xb_\gamma \cdot {}^t c_\gamma \cdot {}^t y$$

for $\gamma \in \Gamma(1)$, and so we see that

$$\Gamma^\theta = \{\gamma \in \Gamma(1) \mid F\big((x, y)\gamma\big) - F(x, y) \in 2\mathbf{Z} \text{ for every } x, y \in \mathbf{Z}_n^1\}. \tag{4.8a}$$

This shows that Γ^θ is a subgroup of $\Gamma(1)$. Taking γ^{-1} in place of γ, from (1.2b) we see that

$$\Gamma^\theta = \{\gamma \in \Gamma(1) \mid \{{}^t a_\gamma c_\gamma\} \equiv \{{}^t b_\gamma d_\gamma\} \equiv 0 \pmod{2\mathbf{Z}^n}\}. \tag{4.8b}$$

We can let $Sp(n, \mathbf{R})$ act on $\mathbf{C}^n \times \mathfrak{H}_n$ by the rule

$$\gamma(u, z) = \big({}^t \mu_\gamma(z)^{-1} u, \gamma z\big) \quad \text{for} \quad \gamma \in Sp(n, \mathbf{R}),\ u \in \mathbf{C}^n, \text{ and } z \in \mathfrak{H}_n. \tag{4.9}$$

From (1.7) we obtain $(\beta\gamma)(u, z) = \beta\big(\gamma(u, z)\big)$ for $\beta, \gamma \in Sp(n, \mathbf{R})$; also 1_{2n} gives the identity map on $\mathbf{C}^n \times \mathfrak{H}_n$.

Lemma 4.3. (i) $\mu_\gamma(z)^{-1} c_\gamma \in S_n(\mathbf{C})$ *for every* $\gamma \in Sp(n, \mathbf{R})$ *and* $z \in \mathfrak{H}_n$.
(ii) *Let* $\kappa(u, z) = {}^t u(z - \bar{z})^{-1} u$ *for* $u \in \mathbf{C}^n$ *and* $z \in \mathfrak{H}_n$. *Then*

$$\kappa\big(\gamma(u, z)\big) - \kappa(u, z) = -{}^t u \mu_\gamma(z)^{-1} c_\gamma u \tag{4.10}$$

for every $\gamma \in Sp(n, \mathbf{R})$.

PROOF. Put $p_z(u, v) = {}^t u(z - \bar{z})^{-1} v$ for $u, v \in \mathbf{C}^n$. Then $p_z(u, v) = p_z(v, u)$ and $\kappa(u, z) = p_z(u, u)$. From (1.16) we obtain

$$(\gamma z - \gamma\bar{z})^{-1} = \mu_\gamma(z)(z - \bar{z})^{-1} \cdot {}^t \mu_\gamma(\bar{z}) \tag{4.11}$$

for every $\gamma \in Sp(n, \mathbf{R})$. Therefore

$$p_{\gamma z}\big({}^t\mu_\gamma(z)^{-1}u,\ {}^t\mu_\gamma(z)^{-1}v\big) = {}^tu\mu_\gamma(z)^{-1}(\gamma z - \gamma\bar{z})^{-1}\cdot {}^t\mu_\gamma(z)^{-1}v$$
$$= {}^tu(z-\bar{z})^{-1}\cdot {}^t\mu_\gamma(\bar{z})\cdot {}^t\mu_\gamma(z)^{-1}v.$$

Since $\mu_\gamma(\bar{z}) = \mu_\gamma(z) - c_\gamma(z-\bar{z})$, we have ${}^t\mu_\gamma(\bar{z})\cdot {}^t\mu_\gamma(z)^{-1} = 1_n - (z-\bar{z})\cdot {}^tc_\gamma\cdot {}^t\mu_\gamma(z)^{-1}$. Thus

$$p_{\gamma z}\big({}^t\mu_\gamma(z)^{-1}u,\ {}^t\mu_\gamma(z)^{-1}v\big) - p_z(u,\ v) = -{}^tu\cdot {}^tc_\gamma\cdot {}^t\mu_\gamma(z)^{-1}v.$$

Since the left-hand side is symmetric in $(u,\ v)$, we see that ${}^tc_\gamma\cdot {}^t\mu_\gamma(z)^{-1}\in S_n(\mathbf{C})$, which proves (i). Putting $u=v$, we obtain (ii). Here we add a more direct proof of (i). We have $c_\gamma\cdot {}^t\mu_\gamma(z) = c_\gamma z\cdot {}^tc_\gamma + c_\gamma\cdot {}^td_\gamma\in S_n(\mathbf{C})$, as can be seen from (1.2a). Thus $\mu_\gamma(z)^{-1}c_\gamma = \mu_\gamma(z)^{-1}c_\gamma\cdot {}^t\mu_\gamma(z)\cdot {}^t\mu_\gamma(z)^{-1}\in S_n(\mathbf{C})$ as expected.

Theorem 4.4. (i) *For every* $\gamma\in\Gamma(1)$ *we have*

$$(4.12)\qquad \theta\big(\gamma(u,\ z);\ r,\ s\big) = \zeta\cdot j_\gamma(z)^{1/2}\mathbf{e}\big(2^{-1}\cdot {}^tu\mu_\gamma(z)^{-1}c_\gamma u\big)\theta(u,\ z;\ r'',\ s'')$$

with a constant $\zeta\in\mathbf{T}$ *depending on* $r,\ s,\ \gamma$, *and a suitable choice of a branch of* $j_\gamma(z)^{1/2}$, *and*

$$\begin{bmatrix} r'' \\ s''\end{bmatrix} = {}^t\gamma\begin{bmatrix} r \\ s\end{bmatrix} + \frac{1}{2}\begin{bmatrix}\{{}^tac\} \\ \{{}^tbd\}\end{bmatrix},$$

where the symbol $\{*\}$ *is defined in* §4.2.

(ii) *For every* $\gamma\in\Gamma^\theta$ *there is a holomorphic function* $h_\gamma(z)$ *in* $z\in\mathfrak{H}_n$, *written also* $h(\gamma,\ z)$, *such that* $h_\gamma(z)^2 = \zeta\cdot j_\gamma(z)$ *with a constant* $\zeta\in\mathbf{T}$ *and*

$$(4.13)\qquad \theta\big(\gamma(u,\ z);\ r,\ s\big)$$
$$= \mathbf{e}\big(2^{-1}({}^trs - {}^tr's')\big)h_\gamma(z)\mathbf{e}\big(2^{-1}\cdot {}^tu\mu_\gamma(z)^{-1}c_\gamma u\big)\theta(u,\ z;\ r',\ s')$$

with $\begin{bmatrix} r' \\ s'\end{bmatrix} = {}^t\gamma\begin{bmatrix} r \\ s\end{bmatrix}.$

(iii) $h_\gamma(z)^4 = j_\gamma(z)^2$ *if* $\gamma\in\Gamma(2)$.

These formulas are classical (see [KP92], for example). We will give a shorter proof by proving (4.12) for some special γ in §4.9, and the remaining statements in Section A1 of the Appendix.

Putting $u=0$, from (4.12) and (4.13) we obtain

$$(4.14)\qquad \theta(0,\ \gamma z;\ r,\ s) = \zeta\cdot j_\gamma(z)^{1/2}\theta(0,\ z;\ r'',\ s'')\quad\text{if}\quad \gamma\in\Gamma(1),$$
$$(4.15)\qquad \theta(0,\ \gamma z;\ r,\ s) = \mathbf{e}\big(2^{-1}({}^trs - {}^tr's')\big)h_\gamma(z)\theta(0,\ z;\ r',\ s')\quad\text{if}\quad \gamma\in\Gamma^\theta,$$

where $(r'',\ s'')$ is as in (4.12) and $(r',\ s')$ is as in (4.13).

4.5. Put $q(u,\ z) = \mathbf{e}\big(2^{-1}\cdot {}^tu(z-\bar{z})^{-1}u\big)$. Then from (4.10) we obtain

$$(4.16)\qquad q\big(\gamma(u,\ z)\big) = q(u,\ z)\mathbf{e}\big(-2^{-1}\cdot {}^tu\mu_\gamma(z)^{-1}c_\gamma u\big).$$

We now consider φ of (4.2). Since $\varphi(u,\ z;\ r,\ s) = q(u,\ z)\theta(u,\ z;\ r,\ s)$, we easily see that (4.12) and (4.13) are equivalent to

$$\varphi\big(\gamma(u,\,z);\,r,\,s\big) = \zeta\cdot j_\gamma(z)^{1/2}\varphi(u,\,z;\,r'',\,s''), \tag{4.17}$$

$$\varphi\big(\gamma(u,\,z);\,r,\,s\big) = \mathbf{e}\big(2^{-1}({}^t rs - {}^t r's')\big)h_\gamma(z)\varphi(u,\,z;\,r',\,s'). \tag{4.18}$$

Thus the transformation formulas for φ under $(u,\,z)\mapsto\gamma(u,\,z)$ are simpler than those for θ. It is mainly for this reason that we consider φ in addition to θ.

4.6. Let us now put

$$\theta(z) = \sum_{g\in\mathbf{Z}^n}\mathbf{e}\big(2^{-1}\cdot{}^t gzg\big) \qquad (z\in\mathfrak{H}_n). \tag{4.19}$$

Then $\theta(z) = \theta(0,\,z;\,0,\,0) = \varphi(0,\,z;\,0,\,0)$, and so from (4.13) we obtain

$$\theta(\gamma z) = h_\gamma(z)\theta(z) \qquad (\gamma\in\Gamma^\theta), \tag{4.20}$$

and consequently

$$h(\beta\gamma,\,z) = h_\beta(\gamma z)h_\gamma(z) \qquad (\beta,\,\gamma\in\Gamma^\theta). \tag{4.21}$$

Clearly $\theta(az\cdot{}^t a) = \theta(z+b) = \theta(z)$ for $a\in GL_n(\mathbf{Z})$ and $b\in 2S_n(\mathbf{Z})$, and so

$$h_\alpha = 1 \ \text{ for } \alpha = \operatorname{diag}[a,\,{}^t a^{-1}],\ \ a\in GL_n(\mathbf{Z}), \tag{4.22a}$$

$$h_\beta = 1 \ \text{ for } \beta = \begin{bmatrix}1 & b\\ 0 & 1\end{bmatrix},\ \ b\in 2S_n(\mathbf{Z}), \tag{4.22b}$$

$$h_{-\gamma} = h_\gamma \ \text{ for every } \ \gamma\in\Gamma^\theta. \tag{4.22c}$$

In the following theorem we will see that h_γ^2 coincides with j_γ if $\gamma\in\Gamma(4)$. For every congruence subgroup Γ of Γ^θ we denote by $\mathscr{M}_{1/2}(\Gamma)$ the set of all holomorphic functions f on $\mathfrak{H}_n$ such that $f^2\in\mathscr{M}_1$ and

$$f(\gamma z) = h_\gamma(z)f(z) \quad \textit{for every} \quad \gamma\in\Gamma. \tag{4.23}$$

In fact, the condition $f^2\in\mathscr{M}_1$ follows from (4.23) if $n>1$. Indeed, if (4.23) is satisfied, then $f(\gamma z)^2 = j_\gamma(z)f(z)^2$ for every $\gamma\in\Gamma\cap\Gamma(4)$, which implies that $f^2\in\mathscr{M}_1$. If $f\in\mathscr{M}_{1/2}(\Gamma)$, then f has expansion (3.7a) with $c(h)\neq 0$ only for nonnegative h. Indeed, from (4.22a, b) we see that f satisfies (3.6a, b) for suitable M and U, and so we have at least an expansion of type (3.7a). If $n>1$, Lemma 3.3 gives the desired result. If $n=1$, the condition $f^2\in\mathscr{M}_1$ implies condition (3.4d), as will be explained in §5.1. In fact, we will introduce in §5.1 modular forms of weight k for an arbitrary $k\in 2^{-1}\mathbf{Z}$ and will discuss (3.4d) in that context. We put

$$\mathscr{M}_{1/2} = \bigcup_{N=1}^{\infty}\mathscr{M}_{1/2}\big(\Gamma(2N)\big). \tag{4.24}$$

Theorem 4.7. (1) *If* $\gamma\in\Gamma^\theta$ *and* $\det(d_\gamma)\neq 0$, *then*

$$\lim_{z\to 0}h_\gamma(z) = \sum_{x\in A}\mathbf{e}\big(-{}^t xd_\gamma^{-1}c_\gamma x/2\big) = \sum_{x\in B}\mathbf{e}\big({}^t xb_\gamma d_\gamma^{-1}x/2\big), \tag{4.25}$$

where $A = \mathbf{Z}^n/{}^t d_\gamma \mathbf{Z}^n$ and $B = \mathbf{Z}^n/d_\gamma \mathbf{Z}^n$.

(2) *If* $\gamma \in \Gamma^\theta$ *and* $\det(d_\gamma)$ *is odd, then* $h_\gamma(z)^2 = \left(\frac{-1}{\det(d_\gamma)}\right) j_\gamma(z)$. *In particular,* $h_\gamma(z)^2 = j_\gamma(z)$ *for every* $\gamma \in \Gamma(4)$.

(3) $h_\iota(z) = \det(-iz)^{1/2}$ *with the branch of the square root such that* $h_\iota(z) > 0$ *for* $\mathrm{Re}(z) = 0$.

(4) *Let* $\alpha \in Gp^+(n, \mathbf{Q})$ *and* $r(z) = j_\alpha(z)^{1/2}$ *with any choice of a branch. Then there is a congruence subgroup* Δ *of* Γ^θ *such that*

$$h(\alpha\gamma\alpha^{-1}, \alpha z) = r(\gamma z)h(\gamma, z)r(z)^{-1}$$

for every $\gamma \in \Delta$.

We will prove the first two assertions in Section A1 of the Appendix; (3) will be proven in §4.9 and (4) in §5.3. Clearly (2) implies Theorem 4.4(iii). Once (2) of the above theorem is established, we see from (4.14) and (4.15) that $\theta(0, z; r, s)$ as a function of z belongs to $\mathscr{M}_{1/2}$ if $r, s \in \mathbf{Q}^n$. If $n = 1$, condition (3.4d) for $\theta(0, z; r, s)^2$ follows from (4.14).

We will often consider $\det(-iz)^{\pm 1/2}$ in our later treatment. We use the convention that it always means the function as in (3) above and its inverse.

4.8. Let us now recall some elementary facts on Fourier analysis. For $f \in L^1(\mathbf{R}^n)$ we define its **Fourier transform** $\hat{f}$ by

$$\hat{f}(x) = \int_{\mathbf{R}^n} f(y)\mathbf{e}(-{}^t xy)dy \qquad (x \in \mathbf{R}^n), \tag{4.26}$$

where we consider x and y column vectors, so that ${}^t xy = \sum_{\nu=1}^n x_\nu y_\nu$, and dy is the standard volume element of $\mathbf{R}^n$. Then $\hat{f}$ is a continuous function. If f is continuous, then we have

$$\sum_{g \in \mathbf{Z}^n} f(r+g) = \sum_{h \in \mathbf{Z}^n} \hat{f}(h)\mathbf{e}({}^t hr) \qquad (r \in \mathbf{R}^n), \tag{4.27}$$

provided both sides converge absolutely and uniformly. This is called the **Poisson summation formula** (see [S07, Theorem 2.3], for example). If we exchange f for $\hat{f}$, then we obtain

$$\sum_{g \in \mathbf{Z}^n} \hat{f}(r+g) = \sum_{h \in \mathbf{Z}^n} f(h)\mathbf{e}(-{}^t hr) \qquad (r \in \mathbf{R}^n). \tag{4.27a}$$

It is well known that the function $\exp(-\pi x^2)$ is its own Fourier transform, that is,

$$\int_{\mathbf{R}} \exp(-\pi x^2)\mathbf{e}(-xt)dx = \exp(-\pi t^2). \tag{4.28}$$

For a short proof, see [S07, pp. 14–15]. An n-dimensional version of (4.28) can be given as follows:

$$\int_{\mathbf{R}^n} \exp\big(-\pi h[x]\big)\mathbf{e}(-{}^t vx)dx = \det(h)^{-1/2}\exp\big(-\pi h^{-1}[v]\big) \qquad (v \in \mathbf{R}^n,\ 0 < h \in S_n(\mathbf{R})), \tag{4.29}$$

where $h[x] = {}^t xhx$. This can be proved by taking a real matrix α such that ${}^t\alpha h\alpha = 1_n$ and replacing x by $\alpha^{-1}x$, which reduces the problem to (4.28). From this we obtain

$$\int_{\mathbf{R}^n} \mathbf{e}(2^{-1}z[x])\mathbf{e}(-{}^t vx)dx = \det(-iz)^{-1/2}\mathbf{e}\big(-2^{-1}z^{-1}[v]\big) \qquad (v \in \mathbf{R}^n,\ z \in \mathfrak{H}_n). \tag{4.30}$$

Indeed, if $z = ih$ with $0 < h \in S_n(\mathbf{R})$, this is exactly (4.29). Now the left-hand side of (4.30) is convergent and defines a holomorphic function in z; the right-hand side is clearly holomorphic in z. Since they coincide on "the imaginary axis" of $\mathfrak{H}_n$, we obtain (4.30) on the whole $\mathfrak{H}_n$.

4.9. We now prove formula (4.12) for the elements γ of $\Gamma(1)$ of the forms

$$\begin{bmatrix} a & 0 \\ 0 & d \end{bmatrix}, \quad \begin{bmatrix} 1 & b \\ 0 & 1 \end{bmatrix}, \quad \begin{bmatrix} 0 & -1 \\ 1 & 0 \end{bmatrix}. \tag{4.31}$$

If $\gamma = \mathrm{diag}[a,\ d] \in \Gamma(1)$, then $d = {}^t a^{-1} \in GL_n(\mathbf{Z})$ and $\gamma(z) = az \cdot {}^t a$, and so we easily see that

$$\theta(au,\ az \cdot {}^t a;\ r,\ s) = \theta(u,\ z;\ {}^t ar,\ a^{-1}s). \tag{4.32}$$

Next, if $\gamma = \begin{bmatrix} 1 & b \\ 0 & 1 \end{bmatrix} \in \Gamma(1)$, then ${}^t b = b$ and $\gamma(z) = z + b$. Observing that ${}^t xbx/2 \equiv {}^t x\{b\}/2 \pmod{\mathbf{Z}}$ if $x \in \mathbf{Z}^n$ (with $\{*\}$ defined as in §4.2), we obtain

$$\theta(u,\ z+b;\ r,\ s) = \mathbf{e}\big(-2^{-1}({}^t rbr + {}^t r\{b\})\big)\theta\big(u,\ z;\ r,\ s + br + 2^{-1}\{b\}\big), \tag{4.33}$$

which is (4.12) for the present γ. Finally, if $\gamma = \begin{bmatrix} 0 & -1 \\ 1 & 0 \end{bmatrix}$, then $\gamma(z) = -z^{-1}$. To discuss this case, we first put $v = y - u - s$ in (4.30) with real vectors $y,\ u,\ s$, and consider the Fourier transform $\hat{f}$ of the function $f(x) = \mathbf{e}\big(2^{-1}z[x] + {}^t x(u+s)\big)$. Then (4.30) shows that

$$\hat{f}(y) = \det(-iz)^{-1/2}\mathbf{e}\big(-2^{-1}z^{-1}[y-u-s]\big).$$

Applying the Poisson summation formula (4.27) to the present case we obtain

$$\det(-iz)^{1/2}\mathbf{e}\big(2^{-1}z^{-1}[u]\big)\theta(u,\ z;\ r,\ s) = \mathbf{e}({}^t rs)\theta(z^{-1}u,\ -z^{-1};\ -s,\ r) \tag{4.34}$$

for real u, and so for all $u \in \mathbf{C}^n$, since both sides are holomorphic in u. Consequently, we know that (4.12) holds for γ of the forms of (4.31). Taking $u = r = s = 0$ in (4.34), we obtain

$$\det(-iz)^{1/2}\theta(z) = \theta(-z^{-1}), \tag{4.34a}$$

which gives Theorem 4.7(3).

4.10. Let us now look more closely at the special case $n = 1$. In this case our function takes the form

(4.35) $$\theta(u,\, z;\, r,\, s) = \sum_{m\in\mathbf{Z}} \mathbf{e}\bigl(2^{-1}(m+r)^2 z + (m+r)(u+s)\bigr),$$

where $u \in \mathbf{C}$, $z \in \mathfrak{H}_1$, and $r,\, s \in \mathbf{R}$; in particular,

(4.36) $$\theta(z) = \theta(0,\, z;\, 0,\, 0) = \sum_{m\in\mathbf{Z}} \mathbf{e}(m^2 z/2).$$

The function of (4.35) satisfies the differential equation

(4.37) $$\frac{\partial^2\theta(u,\, z;\, r,\, s)}{\partial u^2} = 4\pi i \cdot \frac{\partial\theta(u,\, z;\, r,\, s)}{\partial z}.$$

Also, we have

(4.38) $$\Gamma^\theta = \Gamma(2) \cup \Gamma(2)\iota, \quad \iota = \begin{bmatrix} 0 & -1 \\ 1 & 0 \end{bmatrix}.$$

Indeed, that $\Gamma(2)\cup\Gamma(2)\iota \subset \Gamma^\theta$ can easily be seen. Suppose $\gamma = \begin{bmatrix} a & b \\ c & d \end{bmatrix} \in \Gamma^\theta$ and $a \notin 2\mathbf{Z}$. Then $c \in 2\mathbf{Z}$ and $ad - 1 \in 2\mathbf{Z}$. Thus $d \notin 2\mathbf{Z}$ and $b \in 2\mathbf{Z}$, and so $\gamma \in \Gamma(2)$. Suppose $\gamma \in \Gamma^\theta$ and $\gamma \notin \Gamma(2)$. Then $a \in 2\mathbf{Z}$, and so $b \notin 2\mathbf{Z}$. Since $b = a_{\gamma\iota}$, we see that $\gamma\iota \in \Gamma(2)$. This proves (4.38).

From Theorem 4.7(3) we obtain

(4.39) $$h_\iota(z) = (-iz)^{1/2},$$

where the branch of $(-iz)^{1/2}$ is chosen so that it is a positive real number if $z = iy$ with $0 < y \in \mathbf{R}$. We use this convention throughout the present book.

Let us now show that

(4.40) $$h_\gamma(z) = \varepsilon_d^{-1}\left(\frac{2c}{d}\right)(cz+d)^{1/2} \quad \text{if} \quad \gamma = \begin{bmatrix} a & b \\ c & d \end{bmatrix} \in \Gamma(2).$$

Here the branch of $(cz+d)^{1/2}$ is taken so that $-\pi/2 < \arg(cz+d)^{1/2} \le \pi/2$, which means that

(4.41a) $$\lim_{z\to 0}(cz+d)^{1/2} = \begin{cases} \sqrt{d} & \text{if } d > 0, \\ i\sqrt{|d|} & \text{if } d < 0 \text{ and } c \ge 0, \\ -i\sqrt{|d|} & \text{if } d < 0 \text{ and } c < 0; \end{cases}$$

$\left(\dfrac{2c}{d}\right)$ is the symbol defined in (0.4) and (0.5), and ε_d is defined by (as already done in (0.6))

(4.41b) $$\varepsilon_d = \begin{cases} 1 & \text{if } d - 1 \in 4\mathbf{Z}, \\ i & \text{if } d + 1 \in 4\mathbf{Z}. \end{cases}$$

To prove (4.40), let us assume $c \ne 0$, since the case $c = 0$ is easy. Take the branch of $(cz+d)^{1/2}$ as specified, and let $q = \lim_{z\to 0}(cz+d)^{1/2}$. By Theorem 4.7(2), $h_\gamma(z)^2 = \pm j_\gamma(z)$, and so we can put $h_\gamma(z) = \eta(cz+d)^{1/2}$ with η such that $\eta^2 = \pm 1$. By (4.25) we obtain

$$\eta q = \sum_{x\in \mathbf{Z}/d\mathbf{Z}} \mathbf{e}\big(-cx^2/(2d)\big) = \sum_{y=1}^{|d|} \mathbf{e}\big(-2cy^2/d\big) = G(-2c,\, d)$$

with $G(*.*)$ of (2.5). Suppose $d > 0$; then by (2.9), $\eta d^{1/2} = \left(\frac{-2c}{d}\right)\varepsilon_d d^{1/2}$, and so $\eta = \varepsilon_d^{-1}\left(\frac{2c}{d}\right)$, which proves (4.40) when $d > 0$. If $d < 0$, employing (2.13), we obtain $q\eta = \left(\frac{2c}{-d}\right)\varepsilon_{-d}\sqrt{|d|}$. Comparing this with (4.41a), we obtain the desired result. We can also use the fact $h_\gamma = h_{-\gamma}$. However, for all practical purposes we need the formula only when $d > 0$; in fact, it is not advisable to use it when $d < 0$.

By (4.5), $\theta(u,\, z;\, r,\, s)$ up to a factor of $\mathbf{T}$ depends only on r, s modulo $\mathbf{Z}$. In particular, there are four functions determined by $r,\, s \in \{0,\, 1/2\}$. We can replace $1/2$ by $-1/2$ by multiplying the function by a root of unity. Now the explicit forms of these four functions are given as follows:

(4.42a) $$\theta(u,\, z;\, 0,\, 0) = \sum_{m\in\mathbf{Z}} \mathbf{e}(m^2 z/2)\mathbf{e}(mu),$$

(4.42b) $$\theta(u,\, z;\, 0,\, 1/2) = \sum_{m\in\mathbf{Z}} (-1)^m \mathbf{e}(m^2 z/2)\mathbf{e}(mu),$$

(4.42c) $$\theta(u,\, z;\, 1/2,\, 0) = \sum_{m\in\mathbf{Z}} \mathbf{e}\big((2m+1)^2 z/8\big)\mathbf{e}\big((2m+1)u/2\big),$$

(4.42d) $$\theta(u,\, z;\, 1/2,\, -1/2) = -i\sum_{m\in\mathbf{Z}} (-1)^m \mathbf{e}\big((2m+1)^2 z/8\big)\mathbf{e}\big((2m+1)u/2\big).$$

These were introduced by Jacobi, and so it is natural to call them **Jacobi's theta functions**. They are traditionally denoted by $\vartheta_\nu(u,\, z)$ with $0 \le \nu \le 3$, but the numbering depends on the author. Formulas (4.3), (4.4), (4.5), and (4.13) are of course valid for these, and the explicit transformation formulas for ϑ_ν are given in any textbook on elliptic functions. We do not need these ϑ_ν in this book, but we note them here for the purpose of giving a perspective that *the functions* ϑ_ν *and their transformation formulas are merely special cases obtained by taking* n *to be* 1 *and substituting* 0 *or* $\pm 1/2$ *for* r *or* s, and this point is worthy of emphasis.

We add here a few facts about $\partial\theta/\partial u$. Namely, put

(4.43) $$\theta^*(z;\, r,\, s) = (2\pi i)^{-1}(\partial\theta/\partial u)(0,\, z;\, r,\, s).$$

Then

(4.44) $$\theta^*(z;\, r,\, s) = \sum_{m\in\mathbf{Z}} (m+r)\mathbf{e}\big(2^{-1}(m+r)^2 z + (m+r)s\big),$$

and from (4.13) we immediately obtain

(4.45) $$\theta^*(\gamma z;\, r,\, s) = \mathbf{e}\big((rs - r's')/2\big) j_\gamma(z) h_\gamma(z) \theta^*(z;\, r',\, s') \quad \text{if} \quad \gamma \in \Gamma^\theta,$$

where $[r' \quad s'] = [r \quad s]\gamma$. In particular, put $\theta^*(z) = \theta^*(z;\, 0,\, 0)$; then

$$\theta^*(z) = \sum_{m\in\mathbf{Z}} m\mathbf{e}(m^2z/2), \tag{4.46}$$

$$\theta^*(\gamma z) = j_\gamma(z)h_\gamma(z)\theta^*(z) \quad \text{if} \quad \gamma \in \Gamma^\theta. \tag{4.47}$$

4.11. Returning to the case with $n \geq 1$, we consider the functions of the forms

$$\varphi(u,\, z;\, \lambda) = \mathbf{e}\big(2^{-1}\cdot {}^t u(z-\bar{z})^{-1}u\big)\theta(u,\, z;\, \lambda), \tag{4.48a}$$

$$\theta(u,\, z;\, \lambda) = \sum_{\xi\in\mathbf{Q}^n} \lambda(\xi)\mathbf{e}\big(2^{-1}\cdot {}^t\xi z\xi + {}^t\xi u\big). \tag{4.48b}$$

Here $u \in \mathbf{C}^n$ and $z \in \mathfrak{H}_n$ as before; λ is an element of the set $\mathscr{L}(\mathbf{Q}^n)$ defined in §0.3. By our explanation in §0.3, $\theta(u,\, z;\, \lambda)$ is a finite $\mathbf{C}$-linear combination of $\theta(u,\, z;\, r,\, s)$ with $r,\, s \in \mathbf{Q}^n$, and such a $\theta(u,\, z;\, r,\, s)$ is a special case of (4.48b). We now reformulate Theorems 4.4 and 4.7 as follows.

Theorem 4.12. *Let P be the group defined in Lemma 2.2(iii). For $\alpha \in P\Gamma^\theta$ and $\lambda \in \mathscr{L}(\mathbf{Q}^n)$ we can define a holomorphic function $h_\alpha(z)$ on $\mathfrak{H}_n$ and an element λ^α of $\mathscr{L}(\mathbf{Q}^n)$ with the following properties:*

(4.49a) $\varphi\big(\alpha(u,\, z);\, \lambda\big) = h_\alpha(z)\varphi(u,\, z;\, \lambda^\alpha)$.

(4.49b) $h_\alpha(z)^2 = \zeta j_\alpha(z)$ *with* $\zeta \in \mathbf{T}$.

(4.49c) $h_\alpha(z) = |\det(d_\alpha)|^{1/2}$ *if* $\alpha \in P$.

(4.49d) $h_{\rho\alpha\tau}(z) = h_\rho(z)h_\alpha(\tau z)h_\tau(z)$ *if* $\rho \in P$ *and* $\tau \in \Gamma^\theta$.

(4.49e) $\lambda^{\rho\alpha\tau} = \big((\lambda^\rho)^\alpha\big)^\tau$ *if* $\rho \in P$ *and* $\tau \in \Gamma^\theta$.

(4.49f) *For each λ the set $\{\alpha \in \Gamma^\theta \mid \lambda^\alpha = \lambda\}$ is a congruence subgroup.*

(4.49g) *Let $\alpha \in Sp(n,\, \mathbf{Q})$ and $r(z) = j_\alpha(z)^{1/2}$ with any choice of a branch. Then for every $\lambda \in \mathscr{L}(\mathbf{Q}^n)$ there exists an element μ of $\mathscr{L}(\mathbf{Q}^n)$ such that $\varphi\big(\alpha(u,\, z);\, \lambda\big) = r(z)\varphi(u,\, z;\, \mu)$.*

PROOF. We already have $h_\gamma(z)$ for $\gamma \in \Gamma^\theta$. Formula (4.18) (or rather its $\mathbf{C}$-linear combination) means that λ^γ can be determined by (4.49a). For $\beta = \begin{bmatrix} a & b \\ 0 & d \end{bmatrix} \in P$ we have $\beta(u,\, z) = ({}^t d^{-1}u,\, az\cdot {}^t a + bd^{-1})$ and $q\big(\beta(u,\, z)\big) = q(u,\, z)$ by (4.16), and so $\varphi\big(\beta(u,\, z),\, \lambda\big) = \varphi(u,\, z;\, \lambda')$ with $\lambda'(\xi) = \mathbf{e}(2^{-1}\cdot {}^t\xi\cdot {}^t db\xi)\lambda(d\xi)$. Thus putting $h_\beta(z) = |\det(d)|^{1/2}$ and $\lambda^\beta = |\det(d)|^{-1/2}\mathbf{e}(2^{-1}\cdot {}^t\xi\cdot {}^t db\xi)\lambda(d\xi)$, we have (4.49a) with β in place of α. We can easily verify that $h_{\beta'\beta}(z) = h_{\beta'}(\beta z)h_\beta(z)$ and $\lambda^{\beta'\beta} = (\lambda^{\beta'})^\beta$ for $\beta', \beta \in P$. If $\gamma \in \Gamma^\theta \cap P$, then $h_\gamma = 1$ by (4.22a, b), and so it coincides with h_γ defined for γ as an element of P. The same is true for λ^γ, as it is determined by (4.49a). Now given $\alpha = \beta\gamma$ with $\beta \in P$ and $\gamma \in \Gamma^\theta$, we define h_α and λ^α by $h_\alpha(z) = h_\beta(\gamma z)h_\gamma(z)$ and $\lambda^\alpha = (\lambda^\beta)^\gamma$. These are well defined, since we have shown the consistency on

$\Gamma^\theta \cap P$. Then formulas (4.49a, b, c, d, e) can be verified in a straightforward way. To prove (4.49f), it is sufficient to show that given $r, s \in \mathbf{Q}^n$, there is a congruence subgroup Γ such that $\varphi\bigl(\gamma(u, z); r, s\bigr) = h_\gamma(z)\varphi(u, z; r, s)$ for every $\gamma \in \Gamma$. For that purpose take a positive integer m so that $mr, ms \in \mathbf{Z}^n$. Then for $\gamma \in \Gamma(2m^2)$ we have ${}^t rs - {}^t r's' \in 2\mathbf{Z}$ in (4.13), and so (4.18) gives the desired result. As for (4.49g), since $Sp(n, \mathbf{Q})$ is generated by P and ι as noted in Lemma 2.2(iv), successive applications of (4.49a) with elements of P and ι as α establish (4.49g). This completes the proof.

Putting $u = 0$ in (4.48a, b) and (4.49a), we obtain

(4.50a) $$\varphi(0, z; \lambda) = \theta(0, z; \lambda) = \sum_{\xi \in \mathbf{Q}^n} \lambda(\xi)\mathbf{e}(2^{-1} \cdot {}^t\xi z \xi),$$

(4.50b) $$\theta(0, \alpha z; \lambda) = h_\alpha(z)\theta(0, z; \lambda^\alpha) \quad \text{for every} \quad \alpha \in P\Gamma^\theta.$$

The factors of automorphy $\mu_\gamma(z)$ and $j_\gamma(z)$ are defined for every $\gamma \in Sp(n, \mathbf{Q})$ and the "associativity" of the type (1.7) holds on the whole group $Sp(n, \mathbf{Q})$, but in the case of $h_\alpha(z)$ it is defined only for $\alpha \in P\Gamma^\theta$ and the associativity holds only to the extent given by (4.49d, e). However, these are sufficient for practical purposes, though we always have to be careful about our calculation involving $h_\alpha(z)$.

4.13 As generalizations of (4.43) and (4.47) when $n = 1$, we put

(4.51) $$\theta^*(z, \lambda) = \sum_{\xi \in \mathbf{Q}} \lambda(\xi)\xi\mathbf{e}(\xi^2 z/2) \qquad \bigl(z \in \mathfrak{H}_1,\ \lambda \in \mathscr{L}(\mathbf{Q})\bigr).$$

Since $\theta^*(z, \lambda) = (2\pi i)^{-1}(\partial\varphi/\partial u)(0, z; \lambda)$, from (4.49a) we obtain

(4.52) $$\theta^*(\alpha z, \lambda) = h_\alpha(z) j_\alpha(z)\theta^*(z, \lambda^\alpha) \quad \text{for every} \quad \alpha \in P\Gamma^\theta.$$

5. Modular forms of half-integral weight

5.1. By a **weight** (of a modular form) we mean an element of $2^{-1}\mathbf{Z}$; we call it an **integral weight** if it belongs to $\mathbf{Z}$ and a **half-integral weight** otherwise. To make our formulas short, for a weight k and $\alpha \in Sp(n, \mathbf{Q})$ we hereafter put

(5.1a) $$[k] = \begin{cases} k & \text{if } k \in \mathbf{Z}, \\ k - 1/2 & \text{if } k \notin \mathbf{Z}, \end{cases}$$

(5.1b) $$j_\alpha^k(z) = \begin{cases} j_\alpha(z)^k & \text{if } k \in \mathbf{Z}, \\ h_\alpha(z) j_\alpha(z)^{[k]} & \text{if } k \notin \mathbf{Z}. \end{cases}$$

Here we assume that $\alpha \in P\Gamma^\theta$ when $k \notin \mathbf{Z}$. We also put

(5.2) $$Gp^+(n, \mathbf{Q}) = \bigl\{\alpha \in Gp(n, \mathbf{Q}) \bigm| \nu(\alpha) > 0\bigr\}.$$

For k, α as in (5.1b) and a function f on $\mathfrak{H}_n$ we define a function $f\|_k \alpha$ on $\mathfrak{H}_n$ by

(5.3a) $$(f\|_k\,\alpha)(z) = j_\alpha^k(z)^{-1} f\big(\alpha(z)\big).$$

This is the same as (3.1) if $k \in \mathbf{Z}$ and $\alpha \in Sp(n,\,\mathbf{Q})$. Notice that

(5.3b) $$f\|_k(-1) = (-1)^{n[k]} f.$$

Also, from Theorem 4.7(2) we see that

(5.3c) $$j_\gamma^{-k}(z) = \left(\frac{-1}{\det(d_\gamma)}\right) j_\gamma^k(z)^{-1} \quad \text{if } k \notin \mathbf{Z},\ \gamma \in \Gamma^\theta \text{ and } \det(d_\gamma) \notin 2\mathbf{Z}.$$

As for the analogue of formula (3.2b), from (1.7) and (4.21) we obtain

(5.4) $$f\|_k(\gamma\delta) = (f\|_k\,\gamma)\|_k\,\delta \quad \text{if } \gamma, \delta \in \Gamma^\theta \quad \text{and } k \notin \mathbf{Z}.$$

In fact, the last formula can be extended to the cases covered by (4.49d, e).

Suppose $k \notin \mathbf{Z}$. The symbols $j_\alpha^k(z)$ and $f\|_k\,\alpha$ are defined with no ambiguity if $\alpha \in P\Gamma^\theta$. However, we will have to consider an arbitrary $\alpha \in Gp^+(n,\,\mathbf{Q})$. For that purpose we make the following convention: *Whenever we write $j_\alpha(z)^{-k} f(\alpha z)$ for $\alpha \in Gp^+(n,\,\mathbf{Q})$, the symbol $j_\alpha(z)^{-k}$ means any branch of the function.* There is no danger in doing so, since the statement in each case is valid with any choice of a branch.

Let Γ be a congruence subgroup of $Sp(n,\,\mathbf{Q})$. For an integral weight k we defined $\mathscr{M}_k(\Gamma)$ in §3.1. Suppose now k is half-integral and $\Gamma \subset \Gamma^\theta$. We then define $\mathscr{M}_k(\Gamma)$ to be the set of all holomorphic functions $f \in \mathfrak{H}_n$ such that $f\|_k\,\gamma = f$ for every $\gamma \in \Gamma$ and $f^2 \in \mathscr{M}_{2k}$. The last condition is automatically satisfied if $n > 1$. Indeed, by Theorem 4.7(2), $f^2\|_{2k}\,\gamma = f^2$ for every $\gamma \in \Gamma \cap \Gamma(4)$, and so $f^2 \in \mathscr{M}_{2k}$, since condition (3.4c) is unnecessary if $n > 1$. We call an element of $\mathscr{M}_k(\Gamma)$ a **modular form of weight** k with respect to Γ.

Suppose $n = 1$ and k is half-integral. Let f be a holomorphic function on $\mathfrak{H}_1$ such that $f\|_k\,\gamma = f$ for every $\gamma \in \Gamma$. Then $f^2 \in \mathscr{M}_{2k}$ if and only if (3.4d) is satisfied with any choice of a branch of $(cz+d)^{-k}$. Indeed, for such an f satisfying (3.4d) we have $f^2\|_{2k}\,\gamma = f^2$ for every $\gamma \in \Gamma \cap \Gamma(4)$, since $\big(h_\gamma(z) j_\gamma(z)^{k-1/2}\big)^2 = j_\gamma(z)^{2k}$ for $\gamma \in \Gamma(4)$, and (3.4d) is satisfied with $(f^2,\,2k)$ in place of $(f,\,k)$, and so $f^2 \in \mathscr{M}_{2k}$.

Conversely, suppose $f^2 \in \mathscr{M}_{2k}$ and $f\|_k\,\gamma = f$ for every $\gamma \in \Gamma$. Let $\alpha \in SL_2(\mathbf{Q})$ and let $r(z)$ be a branch of $j_\alpha(z)^k$. Let $\gamma = \begin{bmatrix} 1 & m \\ 0 & 1 \end{bmatrix}$ with $0 < m \in 4\mathbf{Z}$. We can find m such that $\gamma \in \Gamma \cap \alpha^{-1}\Gamma\alpha \cap \alpha^{-1}\Gamma(4)\alpha$. Put $\delta = \alpha\gamma\alpha^{-1}$ and $\kappa = 2k$. Then $\kappa \in \mathbf{Z}$, $\delta\alpha = \alpha\gamma$, and $j_\delta(\alpha z)^\kappa j_\alpha(z)^\kappa = j_{\delta\alpha}(z)^\kappa = j_{\alpha\gamma}(z)^\kappa = j_\alpha(\gamma z)^\kappa j_\gamma(z)^\kappa$, which means that $h_\delta(\alpha z)^\kappa r(z) = \pm r(\gamma z) h_\gamma(z)^\kappa$. Put $g(z) = r(z)^{-1} f(\alpha z)$. Since $\delta \in \Gamma \cap \Gamma(4)$ and $h_\gamma = 1$, we have $g(\gamma z) = r(\gamma z)^{-1} f\big(\alpha\gamma(z)\big) = r(\gamma z)^{-1} f\big(\delta\alpha(z)\big) = r(\gamma z)^{-1} h_\delta(\alpha z)^\kappa f(\alpha z) = \pm r(z)^{-1} \cdot f(\alpha z) = \pm g(z)$. Thus $g(z+m) = \pm g(z)$, and so $g(z+2m) = g(z)$. Consequently we can put $g(z) = \sum_{\nu\in\mathbf{Z}} c(\nu) q(z)^\nu$ with $c(\nu) \in \mathbf{C}$ and $q(z) =$

$\mathbf{e}(z/(2m))$. If $f^2 \in \mathcal{M}_\kappa$, then (3.4d) applied to f^2 implies that g^2 as a function of $q(z)$ has no singularity at $q(z) = 0$. Then the same must be true for $g(z)$, and so $c(\nu) = 0$ for $\nu < 0$. Thus f satisfies (3.4d).

We now put

(5.5)
$$\mathcal{M}_k = \bigcup_{N=1}^{\infty} \mathcal{M}_k\bigl(\Gamma(2N)\bigr).$$

Let k and ℓ be weights. Then for $f \in \mathcal{M}_k$ and $g \in \mathcal{M}_\ell$, we easily see that $fg \in \mathcal{M}_{k+\ell}$.

Also, $\theta(0,\, z;\, \lambda) \in \mathcal{M}_{1/2}$ for every $\lambda \in \mathcal{L}(\mathbf{Q}^n)$. This follows from (4.49a) and (4.49f) if $n > 1$. Now suppose $n = 1$. Putting $u = 0$ in (4.49g), we find that (3.4d) is satisfied with $k = 1/2$ and $f(z) = \theta(0,\, z;\, \lambda)$ as expected.

If $n = 1$, $\theta^*(z,\, \lambda)$ belongs to $\mathcal{M}_{3/2}$ for every $\lambda \in \mathcal{L}(\mathbf{Q}^n)$. To show this, we first note that $\theta^*(\gamma z,\, \lambda) = j_\gamma^{3/2}(z)\theta^*(z,\, \lambda)$ for γ in a suitable congruence subgroup, which follows from (4.52) and (4.49f). Since $\theta^*(z,\, \lambda) = (2\pi i)^{-1} \cdot(\partial\varphi/\partial u)(0,\, z;\, \lambda)$, from (4.49g) we obtain $\theta^*(\alpha z,\, \lambda) = j_\alpha(z)^{3/2}\theta^*(z,\, \mu)$, and so $\theta^*(z,\, \lambda)$ satisfies (3.4d) with $k = 3/2$. Thus $\theta^*(z,\, \lambda) \in \mathcal{M}_{3/2}$.

Let $f \in \mathcal{M}_k$ with half-integral k and $n > 1$. From (4.22a, b) we see that (3.6a) and (3.6b) hold for f with suitable M and U. Therefore we have an expansion of f of the form (3.7a) or (3.7b). Thus we can speak of the Fourier expansion of f in the case of half-integral k. If $n = 1$, we have shown that f satisfies (3.4d).

5.2. For $k \in \mathbf{Z}$ our definitions of $f\|_k\,\alpha$ and $\mathcal{M}_k(\Gamma)$ in Section 3 are standard, but the case of half-integral k is not so clear-cut. Indeed, for half-integral k we could have defined $f\|_k\,\alpha$ by $(f\|_k\,\alpha)(z) = h_\alpha(z)^{-2k}f(\alpha z)$ instead of (5.3a). The reason for adopting (5.3a) is that its natural generalization to the Hilbert modular case is the best definition. This of course requires a clarification, but without going into details here we refer the reader to Section 17 of the present book and [S87].

The factor of automorphy h_γ of (4.40) is different from what we introduced in [S73a], which has been employed by many researchers and which is given by

(5.6)
$$h'_\delta(z) = \varepsilon_d^{-1}\left(\frac{c}{d}\right)(cz + d)^{1/2}$$

for $\delta = \begin{bmatrix} a & b \\ c & d \end{bmatrix} \in \Gamma(1)$ with $c \in 4\mathbf{Z}$. For such a δ put $\gamma = \begin{bmatrix} a & 2b \\ c/2 & d \end{bmatrix}$. Then $\gamma \in \Gamma(2)$ and $h_\gamma(2z) = h'_\delta(z)$. In the present book for various reasons, we develop the theory by using h_γ of (4.40). The formulation in terms of h'_δ of (5.6) can easily be obtained by means of the equality $h_\gamma(2z) = h'_\delta(z)$.

5.3. Let us now prove Theorem 4.7(4). Let α and $r(z)$ be as in that statement. We first assume that $\alpha \in Sp(n,\, \mathbf{Q})$. Put $\Delta_\lambda = \{\gamma \in \Gamma^\theta \mid \lambda^\gamma = \lambda\}$

for $\lambda \in \mathscr{L}(\mathbf{Q}^n)$. By (4.49f), Δ_λ is a congruence subgroup of $Sp(n, \mathbf{Q})$. Now take λ to be the characteristic function of $\mathbf{Z}^n$ and take also μ as in (4.49g). Let $\gamma \in \Delta_\mu \cap \alpha^{-1}\Delta_\lambda\alpha$ and $\delta = \alpha\gamma\alpha^{-1}$. Then $\delta\alpha = \alpha\gamma$, $\lambda^\delta = \lambda$, $\mu^\gamma = \mu$, and $\varphi\big(\delta\alpha(u, z); \lambda\big) = h_\delta(\alpha z)\varphi\big(\alpha(u, z); \lambda\big) = h_\delta(\alpha z)r(z)\varphi(u, z; \mu)$. This equals $\varphi\big(\alpha\gamma(u, z); \lambda\big) = r(\gamma z)\varphi\big(\gamma(u, z); \mu\big) = r(\gamma z)h_\gamma(z)\varphi(u, z; \mu)$. Since $\varphi(u, z; \mu)$ is a nonzero function, we obtain $h_\delta(\alpha z)r(z) = r(\gamma z)h_\gamma(z)$. This proves Theorem 4.7(4) for $\alpha \in Sp(n, \mathbf{Q})$.

By (1.3) every element of $Gp^+(n, \mathbf{Q})$ is the product of an element of $Sp(n, \mathbf{Q})$ and an element β of the form $\beta = \mathrm{diag}[e1_n, 1_n]$ with $0 < e \in \mathbf{Q}$. Therefore we easily see that it is sufficient to prove Theorem 4.7(4) when α is such a β; we may even assume that $e \in \mathbf{Z}$. Thus our task is to show that $h(\beta\gamma\beta^{-1}, \beta z) = h_\gamma(z)$ for γ in a suitable congruence subgroup. Suppose $n = 1$ and $\gamma = \begin{bmatrix} a & b \\ c & d \end{bmatrix} \in \Gamma(2e)$; then $\beta\gamma\beta^{-1} = \begin{bmatrix} a & eb \\ e^{-1}c & d \end{bmatrix} \in \Gamma(2)$, and from (4.40) we see that $h(\beta\gamma\beta^{-1}, \beta z) = \big(\frac{e}{d}\big)h_\gamma(z)$, and so $h(\beta\gamma\beta^{-1}, \beta z) = h_\gamma(z)$ if $d - 1 \in 4e\mathbf{Z}$, which gives the desired result.

If $n > 1$, the proof is more involved and requires some facts on the Gauss sum of a quadratic form as in (4.25). We refer the reader to [S00, Theorems 6.8 and 6.9], whose proof is given in [S00, §A2.9]. Howeve,r we need Theorem 4.7(4) only for the proof of the following lemma, which we employ only when $n = 1$ in the present book.

Lemma 5.4. *Let k be an integral or a half-integral weight. Given $\alpha \in Gp^+(n, \mathbf{Q})$ and $f \in \mathscr{M}_k$, put $g(z) = j_\alpha(z)^{-k}f(\alpha z)$. Then $g \in \mathscr{M}_k$.*

For integral k this was already noted in §3.1. If k is half-integral, this is an immediate consequence of Theorem 4.7(4), and so the proof may be left to the reader.

Lemma 5.5. *Let ψ be a primitive or an imprimitive character modulo r, and let $\psi(-1) = (-1)^\mu$ with $\mu = 0$ or 1. Put $k = (2\mu + 1)/2$ and*

(5.7) $$\theta_\psi(z) = 2^{-1}\sum_{m\in\mathbf{Z}} \psi(m)m^\mu\mathbf{e}(m^2 z/2).$$

Then $\theta_\psi \in \mathscr{M}_k$ and

(5.8) $$\theta_\psi\|_k\,\gamma = \psi(d_\gamma)\theta_\psi \quad \textit{for every } \gamma \in \Gamma(2) \textit{ such that } c_\gamma \in 2r^2\mathbf{Z},$$

(5.9) $$\theta_\psi\|_k \begin{bmatrix} 0 & -r^{-1} \\ r & 0 \end{bmatrix} = \psi(-1)r^{-1/2}G(\psi)\theta_{\bar\psi} \quad \textit{if } \psi \textit{ is primitive.}$$

PROOF. Notice that $2^{-1}\sum_{m\in\mathbf{Z}}$ of (5.7) can be replaced by $\sum_{m=1}^\infty$ if ψ is not the principal character. If $r = 1$, then ψ is the principal character and $\theta_\psi = 2^{-1}\theta$ with θ of (4.36), and so (5.8) and (5.9) in that case follow from (4.20). Here we prove (5.8) when ψ is primitive; the general case will be settled after the proof of Lemma 7.13. Assuming that $r > 1$ and $\mu = 0$, for

$s \in \mathbf{Z}$ put $f_s(z) = \sum_{m\in\mathbf{Z}} \mathbf{e}(ms/r)\mathbf{e}(m^2z/2)$. Then $f_s(z) = \theta(0,\, z;\, 0,\, s/r)$. Let $\gamma = \begin{bmatrix} a & b \\ c & d \end{bmatrix} \in \Gamma(2)$ with $c \in 2r^2\mathbf{Z}$. By (4.15) we have

$$\text{(5.10)} \qquad \begin{aligned} f_s\|_{1/2}\gamma &= \mathbf{e}\big(-cds^2/(2r^2)\big)\theta(0,\, z;\, cs/r,\, ds/r) \\ &= \theta(0,\, z;\, 0,\, ds/r) = f_{ds}. \end{aligned}$$

Now by (2.3a) we have

$$\sum_{s=1}^{r} \bar{\psi}(s) f_s(z) = G(\bar{\psi}) \sum_{m\in\mathbf{Z}} \psi(m)\mathbf{e}(m^2z/2) = 2G(\bar{\psi})\theta_\psi(z).$$

This combined with (5.10) shows that

$$2G(\bar{\psi})\theta_\psi\|_{1/2}\gamma = \sum_{s=1}^{r} \bar{\psi}(s) f_s\|_{1/2}\gamma = \sum_{s=1}^{r} \bar{\psi}(s) f_{ds} = 2G(\bar{\psi})\psi(d)\theta_\psi.$$

This proves (5.8) when $\mu = 0$, since $G(\bar{\psi}) \neq 0$ as can be seen from (2.3c). Next, suppose $\mu = 1$; let $g_s(z) = \theta^*(z;\, 0,\, s/r)$ with θ^* of (4.44). Then $k = 3/2$ and $\sum_{s=1}^{r} \bar{\psi}(s) g_s = 2G(\bar{\psi})\theta_\psi$. Also, we see from (4.45) that $g_s\|_k\gamma = g_{ds}$. Therefore we obtain (5.8) when $\mu = 1$ in the same manner as in the case $\mu = 0$.

As for (5.9) when $\mu = 1$, from (4.45) we obtain $g_s(\iota z) = j_\iota^k(z)\theta^*(z;\, s/r,\, 0)$, and so

$$2G(\bar{\psi})\theta_\psi(-1/r^2z) = j_\iota^k(r^2z) \sum_{s=1}^{r} \bar{\psi}(s)\theta^*(r^2z;\, s/r,\, 0)$$

$$= r^{-1} j_\iota^k(r^2z) \sum_{s=1}^{r} \sum_{m\in\mathbf{Z}} \bar{\psi}(s)(rm+s)\mathbf{e}\big((rm+s)^2z/2\big) = 2r^{2k-1} j_\iota^k(z)\theta_{\bar{\psi}}(z).$$

For $\alpha = \begin{bmatrix} 0 & -r^{-1} \\ r & 0 \end{bmatrix}$ we have $j_\alpha^k(z) = r^k j_\iota^k(z)$ and so $\theta_\psi\|_k\alpha = r^{1/2}G(\bar{\psi})^{-1}\theta_{\bar{\psi}} = \psi(-1)r^{-1/2}G(\psi)\theta_{\bar{\psi}}$ by (2.3b, c), which gives (5.9) for $\mu = 1$. The case $\mu = 0$ can be proved by employing (4.15) instead of (4.45).

5.6. We note here an interesting special case. Take $\varphi(m) = \left(\dfrac{3}{m}\right)$ and put

$$\text{(5.11)} \qquad \eta(z) = \theta_\varphi(z/12) = \sum_{m=1}^{\infty} \left(\frac{3}{m}\right)\mathbf{e}(m^2z/24).$$

In this case $r = 12$ and $G(\varphi) = 2\sqrt{3}$ by (2.4a), and so from (5.9) we obtain $\eta\|_{1/2}\iota = \eta$. Also, $m^2 - 1 \in 24\mathbf{Z}$ for every integer m prime to 6, and so $\eta(z+1) = \mathbf{e}(1/24)\eta(z)$. Employing these relations we can easily prove

$$\text{(5.12)} \qquad \eta(z) = \Delta(z)^{1/24} = \mathbf{e}(z/24)\prod_{n=1}^{\infty}\big(1 - \mathbf{e}(nz)\big),$$

where $\Delta(z) = \mathbf{e}(z)\prod_{n=1}^{\infty}\big(1 - \mathbf{e}(nz)\big)^{24}$. We leave the proof to the reader. In fact, it was explained in [S07, p. 19]. We also note an easy fact:

(5.12a) $\eta\|_{1/2}\gamma = \eta$ for every $\gamma \in \Gamma(1)$ such that $b_\gamma,\ c_\gamma \in 24\mathbf{Z}$.

Lemma 5.7. *For every congruence subgroup Γ of Γ^θ and $k \in 2^{-1}\mathbf{Z}$ we have $\mathscr{M}_k(\Gamma) = \{0\}$ if $k < 0$ and $\mathscr{M}_0(\Gamma) = \mathbf{C}$.*

PROOF. Since $f^2 \in \mathscr{M}_{2k}$ for $f \in \mathscr{M}_k$, it is sufficient to treat the case $k \in \mathbf{Z}$. If $n = 1$, our assertions are well known, and so we assume that $n > 1$. We do not need this lemma for our later treatment, but we prove here that $\mathscr{M}_k(\Gamma) = \{0\}$ if $k < 0$. For $\alpha \in SL_2(\mathbf{Q})$ put $\sigma_\alpha = \begin{bmatrix} a_\alpha 1_n & b_\alpha 1_n \\ c_\alpha 1_n & d_\alpha 1_n \end{bmatrix}$; put also $p(z) = z1_n$ for $z \in \mathfrak{H}_1$. Then $\sigma_\alpha \in Sp(n,\ \mathbf{Q})$, $p(z) \in \mathfrak{H}_n$, $\sigma_\alpha\big(p(z)\big) = p\big(\alpha(z)\big)$, and $j\big(\sigma_\alpha,\ p(z)\big) = j_\alpha(z)^n$. Let $f \in \mathscr{M}_k(\Gamma)$ with a congruence subgroup Γ of $Sp(n,\ \mathbf{Q})$ and $0 > k \in \mathbf{Z}$. Put $\Gamma_1 = \big\{\alpha \in SL_2(\mathbf{Z}) \,\big|\, \sigma_\alpha \in \Gamma\big\}$ and $g(z) = f\big(p(z)\big)$ for $z \in \mathfrak{H}_1$. Then Γ_1 is a congruence subgroup of $SL_2(\mathbf{Q})$ and $(g\|_{nk}\,\alpha)\big(p(z)\big) = (f\|_k\,\sigma_\alpha)\big(p(z)\big)$, and so $g\|_{nk}\,\delta = g$ for $\delta \in \Gamma_1$. Moreover, since $f\|_k\,\sigma_\alpha \in \mathscr{M}_k$, we see that g satisfies (3.4d). Therefore $g \in \mathscr{M}_k(\Gamma_1)$, and so $g = 0$. This means that $f\big(p(z)\big) = 0$ for every $z \in \mathfrak{H}_1$. Given $w \in \mathfrak{H}_n$, as shown at the end of Section 1, we can find an element of the form $\beta = \begin{bmatrix} 1 & s \\ 0 & 1 \end{bmatrix}\begin{bmatrix} a & 0 \\ 0 & {}^ta^{-1} \end{bmatrix}$ with $a \in GL_n(\mathbf{R})$ and $s \in S_n(\mathbf{R})$ such that $\beta(i1_n) = w$. Since $GL_n(\mathbf{Q})$ and $S_n(\mathbf{Q})$ are dense in $GL_n(\mathbf{R})$ and $S_n(\mathbf{R})$, respectively, we can find an element δ of P that is in any small neighborhood of β. In other words, the set $\big\{\delta(i1_n) \,\big|\, \delta \in P\big\}$ is dense in $\mathfrak{H}_n$. Put $f_1 = f\|_k\,\delta$. Then $f_1 \in \mathscr{M}_k$, and so $f_1\big(p(z)\big) = 0$, which means that $f\big(\delta(i1_n)\big) = 0$. Since this is so for every $\delta \in P$, we obtain $f = 0$ as expected.

5.8. Let $f \in \mathscr{M}_k$ with $0 \le k \in 2^{-1}\mathbf{Z}$. Given $\alpha \in Sp(n,\ \mathbf{Q})$, by Lemmas 5.4 and 3.3 we have

$$j_\alpha(z)^{-k} f(\alpha z) = \sum_{0 \le h \in S} c_\alpha(h)\mathbf{e}\big(\mathrm{tr}(hz)\big), \qquad S = S_n(\mathbf{Q}),$$

with $c_\alpha(h) \in \mathbf{C}$. We call f a **cusp form** if $c_\alpha(h) \ne 0$ only when $h > 0$ for every α, and we denote by $\mathscr{S}_k(\Gamma)$ the set of all cusp forms contained in $\mathscr{M}_k(\Gamma)$. We put

$$\mathscr{S}_k = \bigcup_{N=1}^{\infty} \mathscr{S}_k\big(\Gamma(2N)\big). \tag{5.13}$$

From our definition and Lemma 5.4 we easily see that if $f \in \mathscr{S}_k$, then $j_\alpha(z)^{-k} f(\alpha z) \in \mathscr{S}_k$ for every $\alpha \in Gp^+(n,\ \mathbf{Q})$. Clearly $\mathscr{S}_0 = \{0\}$.

We note that $\theta^*(z,\ \lambda) \in \mathscr{S}_{3/2}$ for every $\lambda \in \mathscr{L}(\mathbf{Q})$. Indeed, since $\Gamma(1)$ is generated by ι and $\Gamma(1) \cap P_1$, from (4.52) we see that $j_\alpha(z)^{-3/2}\theta^*(\alpha z,\ \lambda) = \theta^*(z,\ \mu)$ with some $\mu \in \mathscr{L}(\mathbf{Q})$ for every $\alpha \in SL_2(\mathbf{Q})$. Clearly $\theta^*(z,\ \mu)$ has 0 as its constant term, and so $\theta^*(z,\ \lambda) \in \mathscr{S}_{3/2}$.

Lemma 5.9. *Let $f(z) = \sum_{h \in S} c(h)\mathbf{e}\big(\mathrm{tr}(hz)\big) \in \mathscr{M}_k(\Gamma)$ with $k \in 2^{-1}\mathbf{Z}$ and a congruence subgroup Γ, where $S = S_n(\mathbf{Q})$. Put*

$$f_\rho(z) = \sum_{h \in S} \overline{c(h)} \mathbf{e}\big(\mathrm{tr}(hz)\big). \tag{5.14}$$

Then $f_\rho(z) = \overline{f(-\bar{z})}$ *and* $f_\rho \in \mathscr{M}_k(\varepsilon \Gamma \varepsilon^{-1})$ *with* $\varepsilon = \mathrm{diag}[-1_n, 1_n]$. *Moreover,*

$$\overline{j_\alpha^k(z)} = j_{\alpha'}^k(-\bar{z}), \tag{5.15}$$

$$f_\rho \|_k \alpha = (f\|_k \alpha')_\rho \tag{5.16}$$

for $\alpha \in Sp(n, \mathbf{Q})$, *whenever* $j_\alpha^k(z)$ *is defined, where* $\alpha' = \varepsilon\alpha\varepsilon^{-1}$.

PROOF. Clearly, $f_\rho(z) = \overline{f(-\bar{z})}$ and $\alpha' \in Sp(n, \mathbf{Q})$; also, $\alpha' \in P\Gamma^\theta$ if $\alpha \in P\Gamma^\theta$. Moreover, $-\overline{\alpha(z)} = \alpha'(-\bar{z})$, $\overline{j_\alpha(z)} = j_{\alpha'}(-\bar{z})$, and

$$f_\rho(\alpha z) = \overline{f\big(-\overline{\alpha z}\big)} = \overline{f\big(\alpha'(-\bar{z})\big)}. \tag{5.17}$$

Take θ as f. Then $\theta_\rho = \theta$ and (5.17) shows that if $\alpha \in \Gamma^\theta$, then $\overline{h_\alpha(z)\theta(z)} = h_{\alpha'}(-\bar{z})\theta(z)$, and so $\overline{h_\alpha(z)} = h_{\alpha'}(-\bar{z})$, which is also true for $\alpha \in P$ because of (4.49c). Thus $\overline{h_\alpha(z)} = h_{\alpha'}(-\bar{z})$ for every $\alpha \in P\Gamma^\theta$ by (4.49d), and we obtain (5.15). Then (5.16) follows immediately from (5.15). Take α in $\varepsilon\Gamma\varepsilon^{-1}$ in (5.17). Then $\alpha' \in \Gamma$ and we obtain $f_\rho(\alpha z) = \overline{j_{\alpha'}^k(-\bar{z}) f(-\bar{z})} = j_\alpha^k(z) f_\rho(z)$, and so $f_\rho \in \mathscr{M}_k(\varepsilon\Gamma\varepsilon^{-1})$ as expected. This completes the proof.

6. Holomorphic and nonholomorphic modular forms on $\mathfrak{H}_1$

6.1. Hereafter we will mainly be concerned with functions on $\mathfrak{H}_1$, and so we write simply $\mathfrak{H}$ for $\mathfrak{H}_1$. We also put

$$GL_2^+(\mathbf{R}) = \{\alpha \in GL_2(\mathbf{R}) \mid \det(\alpha) > 0\}, \tag{6.1a}$$

$$GL_2^+(\mathbf{Q}) = GL_2^+(\mathbf{R}) \cap GL_2(\mathbf{Q}), \tag{6.1b}$$

$$P = \{\alpha \in SL_2(\mathbf{Q}) \mid c_\alpha = 0\}. \tag{6.1c}$$

Hereafter we work with $GL_2(\mathbf{R})$ and $\mathfrak{H}_1$; the group P is P_n (with $n = 1$) of Lemma 2.2(iii).

Formulas (1.10) and (1.16) in the case $n = 1$ take the forms:

$$d(\alpha z) = j_\alpha(z)^{-2} dz \quad \text{for every} \quad \alpha \in SL_2(\mathbf{R}), \tag{6.2}$$

$$\mathrm{Im}(\alpha z) = |j_\alpha(z)|^{-2} \mathrm{Im}(z) \quad \text{for every} \quad \alpha \in SL_2(\mathbf{R}). \tag{6.3}$$

Writing simply $y = \mathrm{Im}(z)$, from (6.2) and (6.3) we see that $y^{-2} dz \wedge d\bar{z}$ is invariant under $SL_2(\mathbf{R})$. Since $dz \wedge d\bar{z} = -2i dx \wedge dy$, we obtain

$$(y^{-2} dx \wedge dy) \circ \alpha = y^{-2} dx \wedge dy \quad \text{for every} \quad \alpha \in SL_2(\mathbf{R}). \tag{6.4}$$

Thus the form $y^{-2} dx \wedge dy$ gives a measure on $\mathfrak{H}$ invariant under $SL_2(\mathbf{R})$. To make our formulas short, we put

$$\mathbf{d}z = y^{-2} dx dy, \tag{6.5}$$

and define a measure μ on $\mathfrak{H}$ by

(6.5a) $$\mu(A) = \int_A \mathbf{d}z$$

for $A \subset \mathfrak{H}$. It is well known that $\mu(\Gamma(1)\backslash\mathfrak{H}) = \pi/3$, and so

(6.6) $$\mu(\Gamma\backslash\mathfrak{H}) = \bigl[\Gamma(1) : \{\pm 1\}\Gamma\bigr]\pi/3 \quad \text{if} \quad \Gamma \subset \Gamma(1).$$

Lemma 6.2. *Let $f \in \mathscr{M}_k$ and let $f(z) = \sum_{m=0}^{\infty} c_m \mathbf{e}(mz/N)$ with $c_m \in \mathbf{C}$ and $0 < N \in \mathbf{Z}$. Then the following assertions hold:*

(i) *There exists a positive number A such that $|f(z)| \le A(1 + y^{-k})$ on the whole $\mathfrak{H}$, where $y = \mathrm{Im}(z)$. Moreover, if f is a cusp form, then A can be chosen so that $|f(z)| \le Ay^{-k/2}$ on the whole $\mathfrak{H}$.*

(ii) $c_m = O(m^k)$.

(iii) $c_m = O(m^{k/2})$ *if f is a cusp form.*

PROOF. For the proof of these facts when $k \in \mathbf{Z}$ the reader is referred to [S07, Lemmas 1.7 and 1.8]. To prove (i) when $k \notin \mathbf{Z}$, take $(f^2, 2k)$ as (f, k) in (i). Then we obtain the desired fact for (f, k). To prove (ii) and (iii), we first note, for $0 \le m \in \mathbf{Z}$,

$$Nc_m \exp(-2\pi my/N) = \int_0^N f(z)\mathbf{e}(-mx/N)dx.$$

Take $y = 1/m$. Then (ii) and (iii) follow from (i).

Results of the same type as the above lemma hold in the case $n > 1$. The case of cusp form is easy, but there are some nontrivial technical problems in the case of non-cusp forms; see [S00, Proposition A6.4 and formula (A6.7)] and the proof given there.

6.3. The notion of a cusp can be defined for a certain class of discrete subgroups of $SL_2(\mathbf{R})$ that includes congruence subgroups of $SL_2(\mathbf{Q})$. Since we deal only with such congruence subgroups in this book, a **cusp** is a point of $\mathbf{Q} \cup \{\infty\}$, and vice versa. Then the map $\alpha \mapsto \alpha(\infty)$ with $\alpha \in SL_2(\mathbf{Q})$ is a bijection of $SL_2(\mathbf{Q})/P$ onto $\mathbf{Q} \cup \{\infty\}$, and so $\Gamma\backslash(\mathbf{Q} \cup \{\infty\})$ is the set of Γ-equivalence classes of cusps, which is in one-to-one correspondence with $\Gamma\backslash SL_2(\mathbf{Q})/P$. We often use $P\backslash SL_2(\mathbf{Q})/\Gamma$ instead, by considering the inverse map.

For a congruence subgroup Γ of $SL_2(\mathbf{Z})$ and a weight k we denote by $C_k(\Gamma)$ the set of all C^∞ functions $f(z)$ on $\mathfrak{H}$ such that

(6.7) $$f\|_k \gamma = f \quad \text{for every } \gamma \in \Gamma;$$

we assume that $\Gamma \subset \Gamma^\theta$ if $k \notin \mathbf{Z}$. Then from (6.3) we see that

(6.8) $$|y^{k/2} f(z)| \text{ is a } \Gamma\text{-invariant function on } \mathfrak{H}.$$

Therefore we consider more generally a C^∞ function $f(z)$ satisfying (6.8) for some Γ and k. We say that such an f is **slowly increasing** or **rapidly decreasing at every cusp** according as the following condition (6.9a) or (6.9b) is satisfied:

(6.9a) *For every* $\alpha \in SL_2(\mathbf{Q})$ *there exist positive constants* A, B, *and* c *depending on* f *and* α *such that*

$$|\mathrm{Im}(\alpha z)^{k/2} f(\alpha z)| < Ay^c \quad \text{if } y = \mathrm{Im}(z) > B.$$

(6.9b) *For every* $\alpha \in SL_2(\mathbf{Q})$ *and* $c \in \mathbf{R}, > 0$, *there exist positive constants* A *and* B *depending on* f, α, *and* c *such that*

$$|\mathrm{Im}(\alpha z)^{k/2} f(\alpha z)| < Ay^{-c} \quad \text{if } y = \mathrm{Im}(z) > B.$$

Notice that these are conditions on $|f|$, rather than on f. Since $SL_2(\mathbf{Q}) = \Gamma(1)P$ by Lemma 2.2(iii), we can find a finite set X such that

$$SL_2(\mathbf{Q}) = \bigsqcup_{\xi \in X} \Gamma \xi P, \qquad X \subset \Gamma(1). \tag{6.10}$$

Then we easily see that condition (6.9a) or (6.9b) is satisfied for every $\alpha \in SL_2(\mathbf{Q})$ if it is satisfied for every $\alpha \in X$. Also, if f satisfies (6.9a) or (6.9b), so does $j_\beta(z)^{-k} f(\beta z)$ for every $\beta \in GL_2^+(\mathbf{Q})$.

Since $|\mathrm{Im}(\alpha z)^{k/2} f(\alpha z)|$ is invariant under $z \mapsto z+m$ with a positive integer m and the set $\{x+iy \mid 0 \le x \le m,\ B' \le y \le B\}$ is compact for every $B' < B$, we easily see that changing A in (6.9a, b) suitably, we can replace B in (6.9a, b) by an arbitrary positive number.

We will be considering a function $f(z, s)$ of $(z, s) \in \mathfrak{H} \times D$, also written $f_s(z)$, with a domain D in $\mathbf{C}$ such that $f_s(z)$ for each fixed s as a function of z satisfies (6.8) with the same Γ and k. If for every compact subset K of D, f_s satisfies condition (6.9a) for $s \in K$ with A, B, and c depending only on K and α, then we say that f is **slowly increasing at every cusp locally uniformly on** D (or locally uniformly in s).

Lemma 6.4. (i) *Let* f *be a holomorphic function on* $\mathfrak{H}$ *belonging to* $C_k(\Gamma)$. *Then* $f \in \mathscr{M}_k$ *if and only if* f *is slowly increasing at every cusp.*

(ii) *For* f *as in* (i), $f \in \mathscr{S}_k$ *if and only if* f *is rapidly decreasing at every cusp.*

(iii) *An element of* $\mathscr{M}_k(\Gamma)$ *is a cusp form if the constant term of the Fourier expansion of* $j_\xi(z)^{-k} f(\xi z)$ *is* 0 *for every* ξ *in the set* X *of* (6.10).

(iv) *Let* f *be an element of* $C_k(\Gamma)$ *that is slowly increasing at every cusp. Then there exist two positive constants* A *and* c *such that*

$$|y^{k/2} f(x+iy)| \le A(y^c + y^{-c}) \quad \text{for every } x+iy \in \mathfrak{H}.$$

(v) *If an element* f *of* $C_k(\Gamma)$ *is rapidly decreasing at every cusp, then* $|\mathrm{Im}(z)^{k/2} f(z)|$ *is bounded on the whole* $\mathfrak{H}$.

PROOF. Given a holomorphic f satisfying (6.7), for $\alpha \in SL_2(\mathbf{Q})$ put $g(z) = j_\alpha(z)^{-k} f(\alpha z)$. Then $g \circ \gamma = \pm g$ for every $\gamma \in \alpha^{-1}\Gamma\alpha \cap P$. Now $(\alpha^{-1}\Gamma\alpha \cap P)\{\pm 1\}/\{\pm 1\}$ is a cyclic group generated by an element of the form $\begin{bmatrix} 1 & t_\alpha \\ 0 & 1 \end{bmatrix}$ with $0 < t_\alpha \in \mathbf{Q}$. Therefore $g(z + 2t_\alpha) = g(z)$, and so $g(z) = \sum_{m \in \mathbf{Z}} c_\alpha(m)\mathbf{e}(mz/r_\alpha)$ with $r_\alpha = 2t_\alpha$ and $c_\alpha(m) \in \mathbf{C}$. Suppose f is slowly increasing at every cusp. Then $|g(z)| \le Ay^c$ for $y > B$ as in (6.9a). Put $q = \mathbf{e}(z/r_\alpha)$ and $h(q) = \sum_{m \in \mathbf{Z}} c_\alpha(m) q^m$. Since $|q| = \exp(-2\pi y/r_\alpha)$, we see that $\lim_{q \to 0} qh(q) = 0$, which means that $c_\alpha(m) = 0$ for $m \le 0$. Thus f satisfies (3.4d), and so $f \in \mathscr{M}_k$. Suppose f is rapidly decreasing at every cusp. Then $\lim_{q \to 0} h(q) = 0$, and so $c_\alpha(0) = 0$. Thus $f \in \mathscr{S}_k$.

Conversely, given $f \in \mathscr{M}_k$, let $j_\alpha(z)^{-k} f(\alpha z) = \sum_{m=0}^{\infty} c_\alpha(m)\mathbf{e}(mz/r_\alpha)$ as in (3.4d). Put $q = \mathbf{e}(z/r_\alpha)$. Then $j_\alpha^{-k} f(\alpha z) = h(q)$ with $h(q) = \sum_{m=0}^{\infty} c_\alpha(m) q^m$. Since h is a holomorphic function for $|q| < 1$, there is a constant A such that $|h(q)| \le A$ for $|q| < 1/2$. Thus f is slowly increasing at every cusp. Next suppose $f \in \mathscr{S}_k$; then $c_\alpha(0) = 0$, and so $|h(q)| \le A'|q|$ for $|q| < 1/2$ with a positive constant A'. Consequently, $|\mathrm{Im}(\alpha z)^{k/2} f(\alpha z)| < Ay^{k/2}\exp(-2\pi y/r_\alpha)$ if $y > B$ with a suitable B, and so f satisfies (6.9b). Assertion (iii) is clear.

To prove (iv), put

(6.10a) $$T = \{x + iy \in \mathbf{C} \mid |x| \le 1/2,\ y > 1/2\}.$$

Since T contains a fundamental domain for $\Gamma(1)\backslash\mathfrak{H}$, we can find a finite subset E of $\Gamma(1)$ such that

(6.10b) $$\mathfrak{H} = \bigcup_{\varepsilon \in E} \Gamma\varepsilon T.$$

Then $|\mathrm{Im}(\varepsilon z)^{k/2} f(\varepsilon z)| \le A \cdot \mathrm{Im}(z)^c$ for every $\varepsilon \in E$ and $z \in T$ with some A and c. Given $z \in \mathfrak{H}$, we can put $z = \gamma\varepsilon w$ with some $\gamma \in \Gamma$, $\varepsilon \in E$, and $w \in T$. Then $|\mathrm{Im}(z)^{k/2} f(z)| = |\mathrm{Im}(\gamma\varepsilon w)^{k/2} f(\gamma\varepsilon w)| = |\mathrm{Im}(\varepsilon w)^{k/2} f(\varepsilon w)| \le A \cdot \mathrm{Im}(w)^c$. Let $(\gamma\varepsilon)^{-1} = \begin{bmatrix} * & * \\ r & s \end{bmatrix}$. If $r = 0$, then $\mathrm{Im}(w) = \mathrm{Im}(z)$ since $(\gamma\varepsilon)^{-1} \in \Gamma(1) \cap P$. If $r \ne 0$, then $\mathrm{Im}(w) = \mathrm{Im}(z)|rz + s|^{-2} \le \mathrm{Im}(z)^{-1}$. Thus $|\mathrm{Im}(z)^{k/2} f(z)| \le A\big(\mathrm{Im}(z)^c + \mathrm{Im}(z)^{-c}\big)$, which proves (iv).

If f is as in (v), then $|\mathrm{Im}(\varepsilon z)^{k/2} f(\varepsilon z)| < Ay^{-c}$ for every $\varepsilon \in E$ and $z \in T$ with some A and $c > 0$. Then we easily see that $|\mathrm{Im}(\varepsilon z)^{k/2} f(\varepsilon z)|$ is bounded on T. Since $|\mathrm{Im}(z)^{k/2} f(z)|$ is a Γ-invariant function on $\mathfrak{H}$ and $\mathfrak{H} = \bigcup_{\varepsilon \in E} \Gamma\varepsilon T$, we obtain (v). This completes the proof.

6.5. Given $f,\ g \in C_k(\Gamma)$, we put

(6.11) $$\langle f,\ g\rangle = \mu(\Phi)^{-1}\int_{\Phi} \bar{f} g y^k \mathbf{d}z, \qquad \Phi = \Gamma\backslash\mathfrak{H}.$$

The integral over Φ is formally meaningful, since $\bar{f}gy^k$ is Γ-invariant, as can be seen from (6.3). We call $\langle f, g\rangle$ the **inner product** of f and g if the integral is convergent, in which case we easily see that the quantity of (6.11) is independent of the choice of Γ.

The integral is convergent if fg is rapidly decreasing at every cusp. Indeed, by Lemma 6.4(v), $|y^k fg|$ is bounded on the whole $\mathfrak{H}$, which implies that the integral of (6.11) is convergent, since $\int_\Phi y^{-2}dxdy < \infty$. We easily see that

(6.12) $$\langle f\|_k\,\alpha,\ g\|_k\,\alpha\rangle = \langle f,\ g\rangle$$

for every $\alpha \in SL_2(\mathbf{Q})$. Here, if $k \notin \mathbf{Z}$, then we either assume that $\alpha \in P\Gamma^\theta$, or put $f\|_k\,\alpha = \kappa j_\alpha(z)^{-k} f(\alpha z)$ with $\kappa \in \mathbf{T}$ and any choice of a branch of $j_\alpha(z)^{-k}$. We also note an easy fact:

(6.12a) $$t^k\langle f(tz),\ g(tz)\rangle = \langle f,\ g\rangle \quad \text{if} \quad 0 < t \in \mathbf{Q}.$$

Lemma 6.6. *The inner product $\langle f, g\rangle$ is meaningful for every $f, g \in \mathscr{M}_{1/2}$, even when neither f nor g is a cusp form.*

PROOF. In this proof we put $k = 1/2$. Take $\Gamma \subset \Gamma(2)$ so that $f, g \in \mathscr{M}_k(\Gamma)$, and take T and E as in (6.10a, b). For each $\varepsilon \in E$ put $f_\varepsilon(z) = j_\varepsilon(z)^{-k}f(\varepsilon z)$ and $g_\varepsilon(z) = j_\varepsilon(z)^{-k}g(\varepsilon z)$ with any choice of a branch of $j_\varepsilon(z)^{-k}$. Then $f_\varepsilon, g_\varepsilon \in \mathscr{M}_k$ by Lemma 5.4, and so by Lemma 6.2(i), $|f_\varepsilon g_\varepsilon| \le A_\varepsilon$ on T with a constant A_ε. Since $\Gamma\backslash\mathfrak{H}$ is covered by $\bigcup_{\varepsilon\in E}\varepsilon T$, we have

$$|\langle f,\ g\rangle| \le \sum_{\varepsilon\in E}\int_{\varepsilon T}|fg|y^k\mathbf{d}z = \sum_{\varepsilon\in E}\int_T(|fgy^k|\circ\varepsilon)\mathbf{d}z.$$

Now $|fgy^k|\circ\varepsilon = y^k|f_\varepsilon g_\varepsilon| \le A_\varepsilon y^k$ on T, and so

$$|\langle f,\ g\rangle| \le \sum_{\varepsilon\in E}A_\varepsilon\int_T y^{k-2}dxdy = \sum_{\varepsilon\in E}A_\varepsilon\int_{1/2}^\infty y^{-3/2}dy < \infty.$$

This proves our lemma.

6.7. We put $y = \mathrm{Im}(z)$ and view it as a real-valued function on $\mathfrak{H}$ as we have done in previous sections. For $k \in \mathbf{R}$ we define differential operators ε, δ_k, and L_k acting on C^∞ functions f on $\mathfrak{H}$ by

(6.13a) $$\varepsilon f = -y^2\partial f/\partial\bar{z},$$

(6.13b) $$\delta_k f = y^{-k}(\partial/\partial z)(y^k f) = \frac{kf}{2iy} + \frac{\partial f}{\partial z},$$

(6.13c) $$L_k = 4\delta_{k-2}\varepsilon = -y^2\left(\frac{\partial^2}{\partial x^2} + \frac{\partial^2}{\partial y^2}\right) + 2iky\frac{\partial}{\partial\bar{z}} = 4\varepsilon\delta_k - k,$$

and define also δ_k^p for $0 \le p \in \mathbf{Z}$ inductively by

(6.13d) $$\delta_k^{p+1} = \delta_{k+2p}\delta_k^p, \quad \delta_k^1 = \delta_k, \quad \delta_k^0 = 1.$$

For every $\alpha \in GL_2^+(\mathbf{R})$ these operators satisfy

(6.14a) $$\varepsilon(f\|_k\,\alpha) = (\varepsilon f)\|_{k-2}\,\alpha,$$

(6.14b) $$\delta_k(f\|_k\,\alpha) = (\delta_k f)\|_{k+2}\,\alpha,$$

(6.14c) $$\delta_k^p(f\|_k\,\alpha) = (\delta_k^p f)\|_{k+2p}\,\alpha,$$

(6.14d) $$L_k(f\|_k\,\alpha) = (L_k f)\|_k\,\alpha.$$

Here $f\|_k\,\alpha$ for $k \notin \mathbf{Z}$ can be defined by $(f\|_k\,\alpha)(z) = j_\alpha(z)^{-k}f(\alpha z)$ with any choice of a branch of $j_\alpha(z)^{-k}$. Then we define $(f\|_{k+2p}\,\alpha)(z)$ with $j_\alpha(z)^{-k-2p} = j_\alpha(z)^{-k}j_\alpha(z)^{-2p}$ for every $p \in \mathbf{Z}$. From (6.13c) we easily obtain

(6.14e) $$L_{k-2}\varepsilon = \varepsilon(L_k - k + 2), \qquad L_{k+2}\delta_k = \delta_k(L_k + k).$$

To prove the above formulas, we first note that

(6.15) $$\frac{\partial}{\partial z}f(\alpha z) = \frac{\partial f}{\partial z}(\alpha z)j_\alpha(z)^{-2}, \qquad \frac{\partial}{\partial \bar{z}}f(\alpha z) = \frac{\partial f}{\partial \bar{z}}(\alpha z)\overline{j_\alpha(z)}^{-2}.$$

Employing (6.3) and (6.15), we have

$$\begin{aligned}
&\delta_k(f\|_k\,\alpha) = y^{-k}(\partial/\partial z)\big(y^k j_\alpha^{-k}f(\alpha z)\big) = y^{-k}(\partial/\partial z)\big(y(\alpha z)^k\overline{j_\alpha}^{-k}f(\alpha z)\big)\\
&= y^{-k}\overline{j_\alpha}^{-k}(\partial/\partial z)\big((y^k f)\circ\alpha\big) = y^{-k}\overline{j_\alpha}^{-k}j_\alpha^{-2}\{(\partial/\partial z)(y^k f)\}(\alpha z)\\
&= j_\alpha^{-k-2}y(\alpha z)^{-k}\{(\partial/\partial z)(y^k f)\}(\alpha z) = (\delta_k f)\|_{k+2}\alpha,
\end{aligned}$$

which gives (6.14b). Formula (6.14a) can be proved in a similar way. Then (6.14c) and (6.14d) follow immediately from (6.14a, b).

From (6.2) we easily see that $y^{-2}(dx^2 + dy^2)$ is a Riemannian metric on $\mathfrak{H}$ invariant under $SL_2(\mathbf{R})$. Therefore $-L_0$ is exactly the **Laplace-Beltrami operator** with respect to this metric.

Theorem 6.8. *Let $f \in C_k(\Gamma)$ and $h \in C_{k-2}(\Gamma)$ with a congruence subgroup Γ. Then*

(6.16) $$\langle f,\, \delta_{k-2}h\rangle = \langle \varepsilon f,\, h\rangle,$$

provided fh, $f\cdot\delta_{k-2}h$ and $(\varepsilon f)h$ are rapidly decreasing at every cusp.

PROOF. Replacing Γ by $\Gamma\cap\Gamma(4)$ if necessary, we may assume that $\Gamma\subset\Gamma(4)$. Then Γ has no elements of finite order other than 1. Throughout the proof, we put $y = \mathrm{Im}(z)$. For $0 < r \in \mathbf{R}$ put

(6.17) $$T_r = \{z \in \mathfrak{H}\,|\, y > r\}, \qquad M_r = \{z \in \mathfrak{H}\,|\, y = r\}.$$

Take a finite subset X of $\Gamma(1)$ as in (6.10). Then $\Gamma\backslash(\mathbf{Q}\cup\{\infty\})$ is represented by $\{\xi(\infty)\,|\,\xi \in X\}$. Put $Q_\xi = \xi^{-1}\Gamma\xi \cap P$ for each $\xi \in X$. Then $\xi Q_\xi \xi^{-1} = \{\gamma \in \Gamma\,|\,\gamma\xi(\infty) = \xi(\infty)\}$. Now $\Gamma\backslash(\mathfrak{H}\cup\mathbf{Q}\cup\{\infty\})$ can be viewed as a compact Riemann surface. We can find a sufficiently large r such that the set $\xi(Q_\xi\backslash T_r)$ for $\xi \in X$ can be embedded in $\Gamma\backslash\mathfrak{H}$ without overlap. Let K be the complement of $\bigcup_{\xi\in X}\xi(Q_\xi\backslash T_r)$ in $\Gamma\backslash\mathfrak{H}$. Then K is a compact manifold with boundary, and

$$\partial K = \sum_{\xi \in X} \xi(B_\xi), \quad B_\xi = Q_\xi \backslash M_r.$$

Let φ be a Γ-invariant C^∞ 1-form on $\mathfrak{H}$. Then

$$\int_K d\varphi = \int_{\partial K} \varphi = \sum_{\xi \in X} \int_{B_\xi} \varphi \circ \xi.$$

Given f and h as in our theorem, take $\varphi = \bar{f} h y^{k-2} d\bar{z}$. Then

$$\begin{aligned} d\varphi &= (\partial/\partial z)(\bar{f} h y^{k-2}) dz \wedge d\bar{z} \\ &= \{(\partial \bar{f}/\partial z) h y^{k-2} + \bar{f}(\partial h/\partial z) y^{k-2} + (-i/2)(k-2)\bar{f} h y^{k-3}\} dz \wedge d\bar{z}. \end{aligned}$$

Since $dz \wedge d\bar{z} = -2i dx \wedge dy$ and $(\partial \bar{f}/\partial z) = \overline{(\partial f/\partial \bar{z})}$, we see that

(6.18) $$(2i)^{-1} d\varphi = \overline{\varepsilon f} \cdot h y^{k-4} dx \wedge dy - \bar{f} \cdot (\delta_{k-2} h) y^{k-2} dx \wedge dy.$$

Thus

$$\int_K \overline{\varepsilon f} \cdot h y^{k-4} dx \wedge dy - \int_K \bar{f} \cdot (\delta_{k-2} h) y^{k-2} dx \wedge dy = (2i)^{-1} \sum_{\xi \in X} \int_{B_\xi} \varphi \circ \xi.$$

We now take the limit when $r \to \infty$. By our assumption the left-hand side converges to $\mu(\Gamma \backslash \mathfrak{H})\{\langle \varepsilon f, h \rangle - \langle f, \delta_{k-2} h \rangle\}$. We have $\varphi \circ \xi = p_\xi(z) d\bar{z}$ with $p_\xi(z) = \overline{j_\xi(z)}^{-2} (\bar{f} h y^{k-2}) \circ \xi$. Notice that $y|p_\xi(z)| = |\bar{f} h y^{k-1}| \circ \xi$. Now Q_ξ is generated by a matrix of the form $\begin{bmatrix} 1 & t_\xi \\ 0 & 1 \end{bmatrix}$ with $0 < t_\xi \in \mathbf{Q}$, and so $Q_\xi \backslash M_r$ can be identified with the line segment $\{x + ir \mid 0 \le x \le t_\xi\}$. Thus

$$\left| \int_{B_\xi} \varphi \circ \xi \right| \le \int_0^{t_\xi} |p_\xi(x + ir)| dx.$$

Suppose fh is rapidly decreasing at every cusp; then $|p_\xi(x + iy)| \le A_\xi y^{-c}$ with positive constants A_ξ and c for sufficiently large y. Therefore we obtain our theorem.

Corollary 6.9. (i) *Let Γ be a congruence subgroup and let $f \in \mathscr{S}_k$. Then $\langle f, \delta_{k-2} h \rangle = 0$ for every $h \in C_{k-2}(\Gamma)$ such that both h and $\delta_{k-2} h$ are slowly increasing at every cusp.*

(ii) *Let $f \in C_k(\Gamma)$. Suppose both f and εf are rapidly decreasing at every cusp and $L_k f = 0$. Then $f \in \mathscr{S}_k$.*

(iii) *Let $f, g \in C_k(\Gamma)$. Then under a suitable condition (see the following* PROOF*) we have*

(6.19) $$\langle f, L_k g \rangle = \langle L_k f, g \rangle,$$

(6.20) $$\langle f, L_k f \rangle \ge 0.$$

PROOF. Let f and h be as in (i). Then, by Lemma 6.4(ii), both fh and $f\delta_{k-2} h$ are rapidly decreasing at every cusp. Since f is holomorphic, we have $\varepsilon f = 0$, and so $\langle f, \delta_{k-2} h \rangle = 0$ by (6.16). This proves (i). Next let f be as in (ii). Then $\delta_{k-2}\varepsilon f = 4^{-1}L_k f = 0$, and so $\langle \varepsilon f, \varepsilon f\rangle = \langle f, \delta_{k-2}\varepsilon f\rangle = 0$ by (6.16). Thus $\varepsilon f = 0$, which means that f is holomorphic, and so $f \in \mathscr{S}_k$ by Lemma 6.4(ii). As for (iii), we have, by (6.16), $\langle f, \delta_{k-2}\varepsilon g\rangle = \langle \varepsilon f, \varepsilon g\rangle = \langle \delta_{k-2}\varepsilon f, g\rangle$, which gives (6.19). To justify the last sequence of equalities, we need some conditions on f and g as stated in Theorem 6.8, which are often easy to verify, and so we leave the precise statements to the reader. If we take $f = g$, then $\langle f, L_k f\rangle = 4\langle \varepsilon f, \varepsilon f\rangle \geq 0$, which proves (6.20). This completes the proof.

Remark. In Corollary 6.9(ii) the condition on εf is unnecessary. This will be explained after the proof of Lemma 9.3.

Lemma 6.10. *If $f \in \mathscr{M}_k(\Gamma)$, then $\delta_k^p f$ for every $p \in \mathbf{Z}, \geq 0$, belongs to $C_{k+2p}(\Gamma)$, and is slowly increasing at every cusp. Moreover, it is rapidly decreasing at every cusp if $f \in \mathscr{S}_k$.*

PROOF. That $\delta_k^p f \in C_{k+2p}(\Gamma)$ follows from (6.14c). Let $\alpha \in \Gamma(1)$ and $g(z) = j_\alpha(z)^{-k} f(\alpha z) = \sum_{m\in\mathbf{Z}} c_\alpha(m)\mathbf{e}(mz/r_\alpha)$ as in (3.4d). By (6.14c) we have $j_\alpha(z)^{-k-2p}(\delta_k^p f)(\alpha z) = \delta_k^p g(z)$, and by induction on p we easily see that $\delta_k^p g = \sum_{\nu=0}^{p} a_\nu y^{-\nu}(\partial/\partial z)^{p-\nu} g$ with $a_\nu \in \mathbf{C}$, and so

$$\delta_k^p g(z) = \sum_{\nu=0}^{p} a_\nu y^{-\nu} \sum_{m=0}^{\infty} c_\alpha(m)(2\pi i m/r_\alpha)^{p-\nu}\mathbf{e}(mz/r_\alpha).$$

Since $c_\alpha(m) = O(m^k)$ by Lemma 6.2(iii), we can easily verify the inequality of (6.9a) for $\delta_k^p g$. If $f \in \mathscr{S}_k$, then $c_\alpha(0) = 0$, and so $\delta_k^p g$ satisfies (6.9b). Thus we obtain our lemma.

6.11. We have been discussing modular forms as functions on $\mathfrak{H}$. Instead, we can treat them as functions on $SL_2(\mathbf{R})$ or its covering. Let us explain the idea in the case $k \in \mathbf{Z}$ for simplicity. Put $K = \{\alpha \in SL_2(\mathbf{R}) \,|\, {}^t\alpha\alpha = 1_2\}$ and $\rho(\xi) = j_\xi^k(i)^{-1}$ for $\xi \in K$. Then ρ is a continuous homomorphism of K into $\mathbf{T}$. Given $f \in C_k(\Gamma)$, define a function $\tilde{f}$ on $SL_2(\mathbf{R})$ by $\tilde{f}(\alpha) = (f\|_k\,\alpha)(i)$ for $\alpha \in SL_2(\mathbf{R})$. Then $\tilde{f}(\gamma\alpha) = \tilde{f}(\alpha)$ for every $\gamma \in \Gamma$ and $\tilde{f}(\alpha\xi) = \rho(\xi)\tilde{f}(\alpha)$ for every $\xi \in K$. In this way we can associate with f a function on $\Gamma\backslash G$ belonging to a representation ρ of K.

One consequence of this association is that the action of the differential operators ε and δ_k^p corresponds to that of some elements of the universal enveloping algebra $\mathfrak{U}$ of the Lie algebra of $SL_2(\mathbf{R})$. This approach gives a certain conceptual perspective, and even clarifies some technical points, especially when we deal with higher-dimensional Lie groups and symmetric spaces. In this book, however, we stay within the traditional framework of functions on $\mathfrak{H}$. The reader who is interested in the higher-dimensional cases and also in the Lie-theoretical treatment of this topic is referred to the author's articles as follows: the operators in the higher-dimensional cases are discussed in [S94],

[S00, Sections 12 and A8], and [S04, Section A1]; the connection with $\mathfrak{U}$ is explained in [S90, Section 7], and [S02, vol. IV, pp. 739–740].

In particular, (6.16) formulated on a Lie group G can be given in the form

$$\int_{\Gamma\backslash G} Xf \cdot h\, d\mu = -\int_{\Gamma\backslash G} f \cdot Xh\, d\mu, \tag{6.21}$$

where X is an element of the Lie algebra of G. However, in almost all papers and textbooks this is proved under the condition that f and h have compact support, or under a similar strong condition, which makes its application impractical. The fact under a much weaker condition is given in [S00, p. 287, Lemma A8.3].

CHAPTER III

THE RATIONALITY AND EISENSTEIN SERIES

7. The rationality of modular forms

7.1. We employ the symbols $\overline{\mathbf{Q}}$ and $\mathbf{Q}_{\mathrm{ab}}$ defined in §0.1, which are the algebraic closure of $\mathbf{Q}$ in $\mathbf{C}$ and the maximal abelian extension of $\mathbf{Q}$ in $\overline{\mathbf{Q}}$. We also denote by $\mathrm{Aut}(\mathbf{C})$ the group of all ring-automorphisms of $\mathbf{C}$, and for $\sigma \in \mathrm{Aut}(\mathbf{C})$ and $x \in \mathbf{C}$ we denote by x^σ the image of x under σ. Thus, for $\tau \in \mathrm{Aut}(\mathbf{C})$ the product $\sigma\tau$ is defined by $x^{\sigma\tau} = (x^\sigma)^\tau$. These are consistent with what we said in §0.4. Given two subfields K and L of a field M, we denote by KL their composite, that is, the subfield of M generated by K and L.

Putting $\mathfrak{H}^* = \mathfrak{H} \cup \mathbf{Q} \cup \{\infty\}$, we recall the basic fact that for a congruence subgroup Γ of $SL_2(\mathbf{Q})$ the orbit space $\Gamma\backslash\mathfrak{H}^*$ has a structure of compact Riemann surface, which can naturally be viewed as an algebraic curve defined over $\mathbf{C}$; for this the reader is referred to [S71]. (In the present book we mean by an algebraic curve a nonsingular projective curve.) We denote by $\mathscr{A}_0(\Gamma)$ the field of all Γ-invariant meromorphic functions on $\mathfrak{H}$ which can be viewed as meromorphic functions on the compact Riemann surface $\Gamma\backslash\mathfrak{H}^*$, and put

$$\mathscr{A}_0 = \bigcup_{N=1}^{\infty} \mathscr{A}_0\bigl(\Gamma(N)\bigr). \tag{7.0}$$

If $f \in \mathscr{A}_0$, we can put

$$f(z) = \sum_{n_0 \le n \in \mathbf{Z}} a_n \mathbf{e}(nz/t) \tag{7.1}$$

with $n_0 \in \mathbf{Z}$, $0 < t \in \mathbf{Z}$ and $a_n \in \mathbf{C}$. For a subfield Φ of $\mathbf{C}$ we denote by $\mathscr{A}_0(\Phi)$ the set of all $f \in \mathscr{A}_0$ such that $a_n \in \Phi$ for every n, and put $\mathscr{A}_0(\Gamma, \Phi) = \mathscr{A}_0(\Phi) \cap \mathscr{A}_0(\Gamma)$. If $\mathscr{A}_0(\Gamma) = \mathbf{C}\mathscr{A}_0(\Gamma, \Phi)$, then the curve $\Gamma\backslash\mathfrak{H}^*$ has a model over Φ. Now we have:

Theorem 7.2. (i) $\mathscr{A}_0\bigl(\Gamma(N)\bigr) = \mathbf{C}\mathscr{A}_0\bigl(\Gamma(N), \mathbf{Q}\bigr)$ *for every* $N \in \mathbf{Z}, > 0$.

(ii) *Given* $\sigma \in \mathrm{Aut}(\mathbf{C})$ *and* $f \in \mathscr{A}_0$ *as in (7.1), there exists an element* f^σ *of* $\mathscr{A}_0$ *such that*

$$f^\sigma(z) = \sum_{n_0 \le n \in \mathbf{Z}} a_n^\sigma \mathbf{e}(nz/t). \tag{7.2}$$

G. Shimura, *Modular Forms: Basics and Beyond*, Springer Monographs in Mathematics,
DOI 10.1007/978-1-4614-2125-2_3,

(iii) $\mathscr{A}_0(\Gamma) = \mathbf{C}\mathscr{A}_0(\Gamma, \mathbf{Q}_{ab})$ *for every congruence subgroup* Γ *of* $SL_2(\mathbf{Q})$.

PROOF. The first assertion is proved in [S71, Proposition 6.9]. To prove (ii), let $\mathscr{A}$ denote the set of all formal infinite sums of the form (7.1). We easily see that $\mathscr{A}$ is a field containing $\mathscr{A}_0$. For $f \in \mathscr{A}$ as in (7.1) and $\sigma \in \mathrm{Aut}(\mathbf{C})$ define $f^\sigma \in \mathscr{A}$ by (7.2). Clearly $f \mapsto f^\sigma$ is an automorphism of $\mathscr{A}$. Let $f \in \mathscr{A}_0(\Gamma(N))$. By (i) we can put $f = (\sum_\mu c_\mu g_\mu)/(\sum_\nu d_\nu h_\nu)$ with finitely many elements $c_\mu, d_\nu \in \mathbf{C}$ and $g_\mu, h_\nu \in \mathscr{A}_0(\Gamma(N), \mathbf{Q})$. Viewing this expression of f as an equality in $\mathscr{A}_0$ and applying σ to it, we obtain $f^\sigma = (\sum_\mu c_\mu^\sigma g_\mu)/(\sum_\nu d_\nu^\sigma h_\nu)$, since $g_\mu^\sigma = g_\mu$ and $h_\nu^\sigma = h_\nu$. This shows that $f^\sigma \in \mathscr{A}_0$, which proves (ii). As for (iii), a detailed discussion about the fields of definition for $\Gamma\backslash\mathfrak{H}^*$ is given in [S71, Section 6.7], which includes (iii); see especially Propositions 6.27 and 6.30 of the book.

7.3. We next consider $\mathscr{M}_k$ with $0 \le k \in 2^{-1}\mathbf{Z}$ and define f^σ for $f \in \mathscr{M}_k$. In fact, we do this for a more general class of functions. Namely we consider a weight $k \ge 0$ and a function f on $\mathfrak{H}$ such that

(7.3a) $$f(z) = \sum_{0 \le \xi \in \mathbf{Q}} \sum_{a=0}^{m} c_a(\xi)(\pi y)^{-a}\mathbf{e}(\xi z), \quad y = \mathrm{Im}(z),$$

(7.3b) $$f\|_k\gamma = f \quad \text{for every } \gamma \in \Gamma,$$

where $0 \le m \in \mathbf{Z}$, $c_a(\xi) \in \mathbf{C}$, and Γ is a congruence subgroup of $SL_2(\mathbf{Q})$, which is contained in Γ^θ if $k \notin \mathbf{Z}$. One more condition:

(7.3c) *For every* $\alpha \in SL_2(\mathbf{Q})$ *the function* $j_\alpha(z)^{-k}f(\alpha z)$ *can be written in the form (7.3a).*

Since (7.3b) implies that $f(z + t) = f(z)$ for some $t \in \mathbf{Z}, > 0$, we have $c_a(\xi) \ne 0$ in (7.3a) only for $t\xi \in \mathbf{Z}$. Then we denote by $\mathscr{N}_k^m(\Gamma)$ the set of all f satisfying (7.3a, b, c), and put

(7.3d) $$\mathscr{N}_k^m = \bigcup_{N=1}^\infty \mathscr{N}_k^m(\Gamma(2N)), \qquad \mathscr{N}_k = \bigcup_{m=0}^\infty \mathscr{N}_k^m.$$

For a subfield Φ of $\mathbf{C}$ we denote by $\mathscr{N}_k^m(\Phi)$ the set of all $f \in \mathscr{N}_k^m$ such that $c_a(\xi)$ in (7.3a) belongs to Φ for every a and ξ; we then put $\mathscr{N}_k^m(\Gamma, \Phi) = \mathscr{N}_k^m(\Gamma) \cap \mathscr{N}_k^m(\Phi)$. We also use these symbols with $\mathscr{N}_k$ in place of $\mathscr{N}_k^m$, when m is not specified, or rather $\mathscr{N}_k$ is understood in the sense of (7.3d). Clearly $\mathscr{M}_k = \mathscr{N}_k^0$; notice that (7.3c) implies (3.4d) if $f \in \mathscr{M}_k$. We call an element of $\mathscr{N}_k$ a **nearly holomorphic modular form of weight** k. If an element of $\mathscr{N}_k$ belongs to $\mathscr{N}_k(\Phi)$, then we call it **Φ-rational.** We put $\mathscr{M}_k(\Phi) = \mathscr{M}_k \cap \mathscr{N}_k(\Phi)$, $\mathscr{M}_k(\Gamma, \Phi) = \mathscr{M}_k(\Gamma) \cap \mathscr{M}_k(\Phi)$, $\mathscr{S}_k(\Phi) = \mathscr{S}_k \cap \mathscr{M}_k(\Phi)$, and $\mathscr{S}_k(\Gamma, \Phi) = \mathscr{S}_k(\Gamma) \cap \mathscr{M}_k(\Phi)$.

7.4. Let Γ be a congruence subgroup of $\Gamma(1)$ such that $\mathbf{C}\mathscr{A}_0(\Gamma, \mathbf{Q}) = \mathscr{A}_0(\Gamma)$. Then the curve $\Gamma\backslash\mathfrak{H}^*$ has a $\mathbf{Q}$-rational model V whose function-field over $\mathbf{Q}$ can be identified with $\mathscr{A}_0(\Gamma, \mathbf{Q})$. We then denote by $V_{\mathbf{C}}$ the algebraic curve $\Gamma\backslash\mathfrak{H}^*$ over $\mathbf{C}$ in which V is embedded. By a **divisor** on $V_{\mathbf{C}}$ we mean as usual a formal finite sum $A = \sum_{P\in V_{\mathbf{C}}} c_P P$ with $c_P \in \mathbf{Z}$. We write $A \succ A'$ if $A' = \sum_{P\in V_{\mathbf{C}}} c'_P P$ and $c_P \geq c'_P$ for every P. We also consider a sum $B = \sum_{P\in V_{\mathbf{C}}} d_P P$ with $d_P \in \mathbf{Q}$; we call it a **fractional divisor** on $V_{\mathbf{C}}$. Given a divisor A we put

$$L(A) = \{0\} \cup \{0 \neq g \in \mathscr{A}_0(\Gamma, \mathbf{C}) \mid \mathrm{div}(g) \succ -A\}, \tag{7.4}$$

$$L(A, \Phi) = L(A) \cap \mathscr{A}_0(\Gamma, \Phi), \tag{7.4a}$$

where $\mathrm{div}(g)$ is a divisor of g and Φ is a subfield of $\mathbf{C}$. It is well known that if A is Φ-rational, then

$$L(A) = L(A, \Phi) \otimes_{\Phi} \mathbf{C}. \tag{7.4b}$$

By Theorem 7.2 we can take $\Gamma(N)$ as Γ; then we denote V by V_N. Let p be the natural projection map of V_N onto V_1. Then p is defined over $\mathbf{Q}$, since it corresponds to the injection $\mathscr{A}_0(\Gamma(1), \mathbf{Q}) \to \mathscr{A}_0(\Gamma(N), \mathbf{Q})$. Let C_N be the divisor that is the sum of all inequivalent cusps of Γ_N viewed as points of $(V_N)_{\mathbf{C}}$. Then C_1 is the point on V_1 represented by ∞, which is $\mathbf{Q}$-rational. We have $p^{-1}(C_1) = \mu_N C_N$ with a positive integer μ_N, and so C_N is a $\mathbf{Q}$-rational divisor on V_N. Now, for $0 \neq f \in \mathscr{M}_k(\Gamma(N))$ with $k \in \mathbf{Z}$ we can define the divisor of f, written $\mathrm{div}(f)$, as a fractional divisor on $(V_N)_{\mathbf{C}}$; see [S71, §2.4]. We can even define $\mathrm{div}(f)$ for $f \in \mathscr{M}_k(\Gamma(N))$ with $N \in 2\mathbf{Z}$ and $k \notin \mathbf{Z}$. We merely put $\mathrm{div}(f) = (1/4)\mathrm{div}(f^4)$, which is well defined, since $f^4 \in \mathscr{M}_{4k}(\Gamma(N))$ as can be seen from Theorem 4.4(iii).

Theorem 7.5. *Let k be a weight ≥ 0, and let $\mathscr{X}$ denote any of the three symbols $\mathscr{M}$, $\mathscr{N}$, and $\mathscr{S}$. For $\sigma \in \mathrm{Aut}(\mathbf{C})$ and $f \in \mathscr{N}_k$ given by (7.3a) define f^σ as a formal infinite series by*

$$f^\sigma(z) = \sum_{0\leq\xi\in\mathbf{Q}} \sum_{a=0}^{m} c_a(\xi)^\sigma (\pi y)^{-a} \mathbf{e}(\xi z). \tag{7.5}$$

Then the following assertions hold:

(i) *We have $\mathscr{X}_k = \mathscr{X}_k(\mathbf{Q}) \otimes_{\mathbf{Q}} \mathbf{C}$, and consequently for every $f \in \mathscr{X}_k$ and $\sigma \in \mathrm{Aut}(\mathbf{C})$ the series of (7.5) is convergent and defines an element of $\mathscr{X}_k$.*

(ii) *For two positive integers M and N put*

$$\Gamma(M, N) = \{\gamma \in \Gamma(1) \mid b_\gamma \in M\mathbf{Z},\ c_\gamma \in N\mathbf{Z}\}.$$

(This is clearly a congruence subgroup of $\Gamma(1)$.) Given a character ψ modulo MN, assuming that $M, N \in 2\mathbf{Z}$ if $k \notin \mathbf{Z}$, put

$$\mathscr{X}_k(M, N; \psi) = \{f \in \mathscr{X}_k \mid f\|_k \gamma = \psi(d_\gamma) f \text{ for every } \gamma \in \Gamma(M, N)\}. \tag{7.6}$$

Then (iia) *$\mathscr{X}_k(M, N; \psi)^\sigma = \mathscr{X}_k(M, N; \psi^\sigma)$;* (iib) *$\mathscr{X}_k(M, N; \psi)$ is spanned by its $\mathbf{Q}_{\mathrm{ab}}$-rational elements.*

(iii) *For every subfield* Φ *of* $\mathbf{C}$ *containing* $\mathbf{Q}_{\mathrm{ab}}$, *the set* $\mathscr{X}_k(\Phi)$ *is stable under* $f \mapsto (c_\alpha z + d_\alpha)^{-k} f(\alpha z)$ *for every* $\alpha \in GL_2^+(\mathbf{Q})$.

(iv) $\mathscr{X}_k$ *is spanned by* $\mathscr{X}_k(M, N; \psi)$ *for all combinations of* (M, N, ψ).

(v) *If* $f(z) = \sum_{\xi \in \mathbf{Q}} c(\xi)\mathbf{e}(\xi z) \in \mathscr{M}_k$, *then the coefficients* $c(\xi)$ *belong to a field that is finitely generated over* $\mathbf{Q}$.

PROOF. Before starting our proof, we note that $\mathscr{X}_k(M, N; \psi) \neq \{0\}$ only if $\psi(-1) = (-1)^{[k]}$, since $f\|_k(-1) = (-1)^{[k]}f$. Now we prove the statements of our theorem first for $\mathscr{M}_k$ and $\mathscr{S}_k$. The case of $\mathscr{N}_k$ will be treated in §7.9. We take η and Δ of (5.12) and assume that $24|N$. By Theorem 4.7(2) and (5.12a), $\eta^{2k} \in \mathscr{M}_k(\Gamma(N), \mathbf{Q})$ for every weight $k > 0$. Clearly Δ has no zeros in $\mathfrak{H}$, and so $\mathrm{div}(\Delta)$ considered on V_1 is C_1. Since $\mathbf{e}(z/N)$ is the local parameter at ∞ on $(V_N)_{\mathbf{C}}$ and $(V_N)_{\mathbf{C}}$ is a Galois covering of $(V_1)_{\mathbf{C}}$, we see that $\mathrm{div}(\Delta)$ considered on V_N is NC_N. Thus $\mathrm{div}(\eta)$ considered on V_N is $(N/24)C_N$. Suppose $f \in \mathscr{M}_k(\Gamma(N))$. Put $g = f/\eta^{2k}$ and $D_N = (kN/12)C_N$. Then $g \in \mathscr{A}_0(\Gamma(N))$ and $\mathrm{div}(g) = \mathrm{div}(f) - \mathrm{div}(\eta^{2k}) \succ -D_N$, and so $g \in L(D_N)$. Conversely, given $g \in L(D_N)$, we easily see that $g\eta^{2k} \in \mathscr{M}_k(\Gamma(N))$. Moreover, $f \in \mathscr{M}_k(\mathbf{Q})$ if and only if $g \in \mathscr{A}_0(\mathbf{Q})$, and so $\mathscr{M}_k(\Gamma(N), \mathbf{Q}) = \{g\eta^{2k} \mid g \in L(D_N, \mathbf{Q})\}$. Thus from (7.4b) we obtain

$$\mathscr{M}_k(\Gamma(N)) = \mathscr{M}_k(\Gamma(N), \mathbf{Q}) \otimes_{\mathbf{Q}} \mathbf{C} \tag{7.7}$$

for every weight k at least when $24|N$. Therefore $\mathscr{M}_k = \mathscr{M}_k(\mathbf{Q}) \otimes_{\mathbf{Q}} \mathbf{C}$, and so every element f of $\mathscr{M}_k$ is a sum $f = \sum_{a \in A} aq_a$ with a finite subset A of $\mathbf{C}$ and $q_a \in \mathscr{M}_k(\mathbf{Q})$. Then for $\sigma \in \mathrm{Aut}(\mathbf{C})$ we see that f^σ defined by (7.5) equals $\sum_{a \in A} a^\sigma q_a$, which belongs to $\mathscr{M}_k$. This proves (i) for $\mathscr{M}_k$. Also, if f is as in (v), then the $c(\xi)$ are contained in the field generated by all $a \in A$ over $\mathbf{Q}$. This proves (v).

To discuss $\mathscr{S}_k$, we have to be careful about the contribution of cusps to the divisor in question. To simplify the matter, put $t_\alpha(z) = \eta(\alpha z)/\eta(z)$ for $\alpha \in \Gamma(1)$. Then $t_\alpha(z)^2 = \zeta j_\alpha(z)$ with a root of unity ζ, $t_{\alpha\beta}(z) = t_\alpha(\beta z)t_\beta(z)$ for $\alpha, \beta \in \Gamma(1)$, and by (5.12a), $t_\alpha(z) = h_\alpha(z)$ if $\alpha \in \Gamma(24)$. Put

$$g|_k\alpha = t_\alpha(z)^{-2k} g(\alpha z) \tag{7.7a}$$

for a function g on $\mathfrak{H}$. Let $f \in \mathscr{S}_k(\Gamma(N))$ with $N \in 24\mathbf{Z}$. Then for every $\alpha \in \Gamma(1)$, we see that $f|_k\alpha \in \mathscr{S}_k(\Gamma(N))$ and $f|_k\alpha = \sum_{n=1}^{\infty} c_\alpha(n)\mathbf{e}(nz/N)$ with $c_\alpha(n) \in \mathbf{C}$, and so we see that $\mathrm{div}(f) \succ C_N$, where $\mathrm{div}(f)$ is considered on V_N. The above argument about $g = f/\eta^{2k}$ with f restricted to $\mathscr{S}_k(\Gamma(N))$ shows that $\mathscr{S}_k(\Gamma(N)) = \{g\eta^{2k} \mid g \in L(D_N - C_N)\}$ and $\mathscr{S}_k(\Gamma(N), \mathbf{Q}) = \{g\eta^{2k} \mid g \in L(D_N - C_N, \mathbf{Q})\}$. Thus from (7.4b) we obtain

$$\mathscr{S}_k(\Gamma(N)) = \mathscr{S}_k(\Gamma(N), \mathbf{Q}) \otimes_{\mathbf{Q}} \mathbf{C} \tag{7.7b}$$

if $24|N$. This proves (i) for $\mathscr{S}_k$.

As for (iii), we note that $GL_2^+(\mathbf{Q})$ is generated by ι, P, and the elements $\mathrm{diag}[e, 1]$ with $0 < e \in \mathbf{Q}$, by Lemma 2.2(iv) and (1.3). Thus it is sufficient to prove (iii) for α of three such types. The cases with the latter two types are easy. We also note that $\mathscr{A}_0(\mathbf{Q}_{\mathrm{ab}})$ is stable under ι. (In fact $\mathscr{A}_0(\mathbf{Q}_{\mathrm{ab}})$ is stable under $GL_2^+(\mathbf{Q})$; see [S71, Theorem 6.23].) Now let $f \in \mathscr{M}_k(\mathbf{Q}_{\mathrm{ab}})$. Then $f|_k \iota \in \mathscr{M}_k$ by Lemma 5.4, $f/\eta^{2k} \in \mathscr{A}_0(\mathbf{Q}_{\mathrm{ab}})$, and $(f|_k \iota)/(\eta^{2k}|_k \iota) = (f/\eta^{2k}) \circ \iota \in \mathscr{A}_0(\mathbf{Q}_{\mathrm{ab}})$. Since $\eta^{2k}|_k \iota = \eta^{2k}$, we see that $f|_k\iota \in \mathscr{M}_k(\mathbf{Q}_{\mathrm{ab}})$. Thus we obtain (iii) for $\mathscr{M}_k$. Since $\mathscr{S}_k(\mathbf{Q}_{\mathrm{ab}}) = \mathscr{M}_k(\mathbf{Q}_{\mathrm{ab}}) \cap \mathscr{S}_k$ and $\mathscr{S}_k$ is stable under $f \mapsto f|_k \iota$ as noted at the end of §5.8, we obtain (iii) for $\mathscr{S}_k$.

To prove (iv), we observe that $f \mapsto f\|_k\gamma$ for $\gamma \in \Gamma(N, N)$ for even N defines an action of $\Gamma(N, N)/\Gamma(N)$ on $\mathscr{X}_k(\Gamma(N))$ with $\mathscr{X} = \mathscr{M}$ or $\mathscr{S}$. (This is so even for $\mathscr{N}_k$, provided (iii) is established for $\mathscr{N}_k$.) Now the map $\gamma \mapsto a_\gamma$ is a homomorphism of $\Gamma(N, N)$ into $(\mathbf{Z}/N\mathbf{Z})^\times$ whose kernel is $\Gamma(N)$, and so $\mathscr{X}_k(\Gamma(N))$ as a representation space of $\Gamma(N, N)/\Gamma(N)$, or rather of $(\mathbf{Z}/N\mathbf{Z})^\times$, can be decomposed as the direct sum of $\mathscr{X}_k(N, N; \psi)$ with the characters ψ modulo N such that $\psi(-1) = (-1)^{[k]}$, from which we can easily derive (iv). As for (ii), we first prove:

Lemma 7.6. (i) *Given* $\sigma \in \mathrm{Aut}(\mathbf{C})$, *three positive integers* M, N, K *such that* $K \in M\mathbf{Z} \cap N\mathbf{Z}$, *and* $\gamma \in \Gamma(M, N)$, *there exists an element* $\beta \in \Gamma(M, N)$ *such that* $d_\beta - d_\gamma \in K\mathbf{Z}$ *and* $f^\sigma\|_k \gamma = (f\|_k \beta)^\sigma$ *for every* $f \in \mathscr{M}_k(\Gamma(K))$, *where we assume that both* M *and* N *are even if* $k \notin \mathbf{Z}$.

(ii) *For* $f \in \mathscr{M}_k(M, N; \psi) \cap \mathscr{M}_k(\mathbf{Q}_{\mathrm{ab}})$ *with* $k \notin \mathbf{Z}$ *and even* M, N, *define* f^X *by* $f^X(z) = (-iNz)^{-k} f(-(Nz)^{-1})$. *Then* $f^X \in \mathscr{M}_k(2, MN; \bar{\psi}\rho) \cap \mathscr{M}_k(\mathbf{Q}_{\mathrm{ab}})$, *with* $\rho(d) = \left(\dfrac{N}{d}\right)$, *and for every* $\sigma \in \mathrm{Gal}(\mathbf{Q}_{\mathrm{ab}}/\mathbf{Q})$ *we have* $(f^\sigma)^X = \chi(s)^\sigma (f^X)^\sigma$, *where* $\chi(d) = \psi(d)\left(\dfrac{-1}{d}\right)^{[k]}$ *and* s *is an integer prime to* MN *such that* $\mathbf{e}(1/MN)^\sigma = \mathbf{e}(s/MN)$.

PROOF. We put $G = GL_2(\mathbf{Q})$ and define the adelization $G_{\mathbf{A}}$ of G as usual; see [S71, §6.4] or [S97, Section 8]. We also put $G_{\mathbf{a}+} = GL_2^+(\mathbf{R})$ and define subgroups $G_{\mathbf{A}+}$, U, and U_N of $G_{\mathbf{A}}$ by

$$G_{\mathbf{A}+} = \{x \in G_{\mathbf{A}} \mid x_{\mathbf{a}} \in G_{\mathbf{a}+}\}, \quad U = G_{\mathbf{a}+} \textstyle\prod_p GL_2(\mathbf{Z}_p), \tag{7.7c}$$

$$U_N = \{x \in U \mid x_p - 1 \prec N\mathbf{Z}_p \text{ for every } p\}, \tag{7.7d}$$

where $x_{\mathbf{a}}$ resp. x_p denotes the archimedean component resp. p-component of x, and $\prod_p$ is the product over all prime numbers p. Notice that $\Gamma(1) = SL_2(\mathbf{Q}) \cap U$ and $\Gamma(N) = SL_2(\mathbf{Q}) \cap U(N)$. We can let $G_{\mathbf{A}+}$ act on $\mathscr{A}_0(\mathbf{Q}_{\mathrm{ab}})$ as a group of automorphisms; see [S71, Theorem 6.23]. This action can be extended to the graded algebra $\sum_{k=0}^{\infty} \mathscr{A}_k(\mathbf{Q}_{\mathrm{ab}})$, where $\mathscr{A}_k(\mathbf{Q}_{\mathrm{ab}})$ is the $\mathscr{A}_0(\mathbf{Q}_{\mathrm{ab}})$-linear span of $\mathscr{M}_k(\mathbf{Q}_{\mathrm{ab}})$. This is explained in [S07, Section A5]. Namely, there is an action of $G_{\mathbf{A}+}$ on $\sum_{k=0}^{\infty} \mathscr{A}_k(\mathbf{Q}_{\mathrm{ab}})$, written $(x, f) \mapsto f^{[x]}$ for $x \in G_{\mathbf{A}+}$ and $f \in \mathscr{A}_k(\mathbf{Q}_{\mathrm{ab}})$, with the following properties:

(7.8a) *The map* $f \mapsto f^{[x]}$ *gives a* $\mathbf{Q}$*-linear map of* $\mathscr{A}_k(\mathbf{Q}_{\mathrm{ab}})$ *onto itself.*

(7.8b) $(f^{[x]})^{[y]} = f^{[xy]}$, $(fg)^{[x]} = f^{[x]}g^{[x]}$.

(7.8c) $f^{[\alpha]} = (c_\alpha z + d_\alpha)^{-k} f(\alpha z)$ *if* $\alpha \in GL_2^+(\mathbf{Q})$.

(7.8d) $f^{[u]} = f^{\{t\}}$ *if* $u = \mathrm{diag}[1, t^{-1}]$ *with* $t \in \prod_p \mathbf{Z}_p^\times$, *where* $\{t\}$ *is the image of* t *under the canonical homomorphism of* $\mathbf{Q}_\mathbf{A}^\times$ *onto* $\mathrm{Gal}(\mathbf{Q}_{\mathrm{ab}}/\mathbf{Q})$ *and* $f^{\{t\}}$ *is understood in the sense of (7.5).*

(7.8e) *For* $f, g \in \mathscr{A}_k(\mathbf{Q}_{\mathrm{ab}})$, $g \neq 0$, *we have* $f^{[x]}/g^{[x]} = (f/g)^{\tau(x)}$ *with* $\tau(x)$ *defined in* [S71, §6.6].

(7.8f) *Given* $f \in \mathscr{A}_k(\mathbf{Q}_{\mathrm{ab}})$, *there exists* $M \in \mathbf{Z}, > 0$, *such that* $f^{[v]} = f$ *for every* $v \in U_M$.

(7.8g) *If* $t \in \prod_p \mathbf{Z}_p^\times$, $s \in \mathbf{Z}$, $0 < N \in \mathbf{Z}$, *and* $st_p - 1 \in N\mathbf{Z}_p$ *for all prime numbers* p, *then* $\mathbf{e}(1/N)^{\{t\}} = \mathbf{e}(s/N)$.

We put $\mathscr{A}_k\big(\Gamma(N), \mathbf{Q}_{\mathrm{ab}}\big) = \big\{f \in \mathscr{A}_k(\mathbf{Q}_{\mathrm{ab}}) \,\big|\, f\|_k \gamma = f$ for every $\gamma \in \Gamma(N)\big\}$ and prove

(7.9) $f^{[w]} = f$ *for every* $f \in \mathscr{A}_k\big(\Gamma(N), \mathbf{Q}_{\mathrm{ab}}\big)$ *and* $w \in U_N \cap SL_2(\mathbf{Q})_\mathbf{A}$.

To prove this, given $f \in \mathscr{M}_k\big(\Gamma(N), \mathbf{Q}_{\mathrm{ab}}\big)$, take M as in (7.8f). We may assume that $N|M$. Put $W_N = U_N \cap SL_2(\mathbf{Q})_\mathbf{A}$. By strong approximation (see [S71, Lemma 6.15]), $SL_2(\mathbf{Q})_\mathbf{A} = W_M SL_2(\mathbf{Q})$, and so if $w \in W_N$, then $w = y\gamma$ with $y \in W_M$ and $\gamma \in SL_2(\mathbf{Q})$. We see that $\gamma \in U_N \cap SL_2(\mathbf{Q}) = \Gamma(N)$ and $f^{[y]} = f$ by (7.8f). Thus $f^{[w]} = f^{[y][\gamma]} = f^{[\gamma]} = f\|_k \gamma = f$, which proves (7.9).

In view of (7.7) it is sufficient to prove (i) of our lemma when $f \in \mathscr{M}_k\big(\Gamma(K), \mathbf{Q}\big)$. Let $\sigma \in \mathrm{Aut}(\mathbf{C})$ and $\gamma \in \Gamma(M, N)$. We can find $t \in \prod_p \mathbf{Z}_p^\times$ such that $\sigma = \{t\}$ on $\mathbf{Q}_{\mathrm{ab}}$. Put $u = \mathrm{diag}[1, t^{-1}]$. Again by strong approximation, $u\gamma u^{-1} = x\beta$ with $x \in W_K$ and $\beta \in SL_2(\mathbf{Q})$. Then $d_\beta - d_\gamma \in K\mathbf{Z}$, $\beta \in \Gamma(M, N)$, and $f^{[u\gamma]} = f^{[x\beta u]}$. By (7.8c, d), $f^{[u\gamma]} = f^\sigma\|_k \gamma$ and by (7.9), $f^{[x\beta u]} = f^{[\beta u]} = (f\|_k \beta)^{[u]} = (f\|_k \beta)^{\{t\}} = (f\|_k \beta)^\sigma$. This proves (i) for integral k.

Suppose $k \notin \mathbf{Z}$ and $M, N \in 2\mathbf{Z}$; let $f \in \mathscr{M}_k\big(\Gamma(K), \mathbf{Q}\big)$. Put $m = k - 1/2$. Then $\theta^{-1}f \in \mathscr{A}_m\big(\Gamma(K), \mathbf{Q}\big)$, and the above argument with f replaced by $\theta^{-1}f$ shows that $(\theta^{-1}f)^\sigma\|_m \gamma = \big((\theta^{-1}f)\|_m \beta\big)^\sigma$. We have $(\theta^{-1}f)^\sigma\|_m \gamma = (\theta\|_{1/2}\gamma)^{-1}(f^\sigma\|_k \gamma)$, $(\theta^{-1}f)\|_m \beta = (\theta\|_{1/2}\beta)^{-1}(f\|_k \beta)$, and $\theta\|_{1/2}\gamma = \theta\|_{1/2}\beta = \theta$, and so we obtain our lemma when $k \notin \mathbf{Z}$. This completes the proof of (i).

As for (ii), $f^X \in \mathscr{M}_k(\mathbf{Q}_{\mathrm{ab}})$ by Theorem 7.5(iii). Now take $t \in \prod_p \mathbf{Z}_p^\times$ and $s \in \mathbf{Z}$ so that $\sigma = \{t\}$ on $\mathbf{Q}_{\mathrm{ab}}$ and $st_p - 1 \in MN\mathbf{Z}_p$ for all prime numbers p. Then $\mathbf{e}(1/MN)^\sigma = \mathbf{e}(s/MN)$ by (7.8g). Put $\alpha = \begin{bmatrix} 0 & -1 \\ N & 0 \end{bmatrix}$, $u =$

$\mathrm{diag}[1,\, t^{-1}]$, and $v = \mathrm{diag}[t^{-1},\, t]$. Take $\gamma \in \Gamma(1)$ so that $\gamma - v_p \prec MN\mathbf{Z}_p$ for all prime numbers p. Then $\alpha u = vu\alpha$. Put $g = \theta^{-2k} f$. Then $g \in \mathscr{A}_0(\mathbf{Q}_{\mathrm{ab}})$. Since $\theta^{2k} \in \mathscr{M}_k(2,\, 2;\, \varphi)$ with $\varphi(d) = \left(\dfrac{-1}{d}\right)^{[k]}$, we see that $g \circ \varepsilon = \chi(d_\varepsilon) g$ for every $\varepsilon \in \Gamma(M,\, N)$, where $\chi = \psi\varphi$. By (7.9), $g \circ \gamma = g^{[v]}$, and so $(g \circ \alpha)^\sigma = g^{[\alpha u]} = g^{[vu\alpha]} = (g \circ \gamma)^\sigma \circ \alpha = \chi(d_\gamma)^\sigma g^\sigma \circ \alpha = \bar{\chi}(s)^\sigma g^\sigma \circ \alpha$. Now $\theta\big(-(Nz)^{-1}\big) = (-iNz)^{1/2}\theta(Nz)$, and so $(\theta^{2k})^X = \theta^{2k}(Nz)$. Since $\theta^\sigma = \theta$, we have $\big[(\theta^{2k})^X\big]^\sigma = \big[(\theta^{2k})^\sigma\big]^X$. We have $f^X = (\theta^{2k})^X g \circ \alpha = \theta^{2k}(Nz) g \circ \alpha$. We easily see that $g \circ \alpha \circ \delta = \bar{\chi}(d_\delta) g \circ \alpha$ for every $\delta \in \Gamma(1,\, MN)$ and $\theta^{2k}(Nz)$ belongs to $\mathscr{M}_k(2,\, N;\, \varphi\rho)$ with ρ as in our lemma. Thus $f^X \in \mathscr{M}_k(2,\, MN;\, \bar{\psi}\rho)$, and $(f^X)^\sigma = (\theta^{2k})^X (g \circ \alpha)^\sigma = \bar{\chi}(s)^\sigma \theta^{2k}(Nz) g^\sigma \circ \alpha = \bar{\chi}(s)^\sigma (f^\sigma)^X$. This proves (ii) and completes the proof.

7.7. We now prove Theorem 7.5(iia) for $\mathscr{M}_k$ and $\mathscr{S}_k$. Put $K = MN$. Given $\sigma \in \mathrm{Aut}(\mathbf{C})$, $f \in \mathscr{M}_k(M,\, N;\, \psi)$, and $\gamma \in \Gamma(M,\, N)$. take $\beta \in \Gamma(M,\, N)$ as in Lemma 7.6. Then $f^\sigma \|_k \gamma = (f\|_k\beta)^\sigma = \psi(d_\beta)^\sigma f^\sigma = \psi(d_\gamma)^\sigma f^\sigma$, and so $f^\sigma \in \mathscr{M}_k(M,\, N;\, \psi^\sigma)$, which proves Theorem 7.5(iia) for $\mathscr{M}_k$. Then the case of $\mathscr{S}_k$ follows immediately. Theorem 7.5(iib) will be proven in §7.9.

To prove Theorem 7.5 for $\mathscr{N}_k$, we first put

$$E_2(z) = \frac{1}{8\pi y} - \frac{1}{24} + \sum_{n=1}^{\infty}\Bigg(\sum_{0<d|n} d\Bigg)\mathbf{e}(nz), \tag{7.10}$$

$$D_k = (2\pi i)^{-1}\delta_k, \quad D_k^p = (2\pi i)^{-p}\delta_k^p, \tag{7.11}$$

where δ_k and δ_k^p are the differential operators defined in §6.7. Taking the logarithmic derivative of the expression $\Delta(z) = \mathbf{e}(z)\prod_{n=1}^{\infty}\big(1 - \mathbf{e}(nz)\big)^{24}$, we easily find that

$$\Delta^{-1} D_{12} \Delta = -24 E_2. \tag{7.12}$$

Therefore by (6.14b), $E_2\|_2\gamma = E_2$ for every $\gamma \in \Gamma(1)$, and so from (7.10) we see that E_2 satisfies (7.3a, b, c) with $\Gamma = \Gamma(1)$ and $k = 2$. Thus $E_2 \in \mathscr{N}_2^1\big(\Gamma(1),\, \mathbf{Q}\big)$. Also, E_2 is an Eisenstein series as will be shown in (8.14e).

Lemma 7.8. *For every subfield Φ of $\mathbf{C}$ and every congruence subgroup Γ of $\Gamma(1)$, which is assumed to be contained in Γ^θ if $k \notin 2\mathbf{Z}$, the following assertions hold:*

(i) *The operator D_k^p sends $\mathscr{N}_k^m(\Gamma,\, \Phi)$ into $\mathscr{N}_{k+2p}^{m+p}(\Gamma,\, \Phi)$; in particular, it sends $\mathscr{M}_k(\Gamma,\, \Phi)$ into $\mathscr{N}_{k+2p}^p(\Gamma,\, \Phi)$. Moreover, $(D_k^p f)^\sigma = D_k^p f^\sigma$ for every $\sigma \in$* $\mathrm{Aut}(\mathbf{C})$.

(ii) *Every element $f(z)$ of $\mathscr{N}_k(\Gamma,\, \Phi)$ can be written in the form*

$$f(z) = \sum_{0 \le p \le k/2} D_{k-2p}^p g_p + \begin{cases} cD_2^{(k/2)-1} E_2 & \text{if } k \in 2\mathbf{Z}, \\ 0 & \text{if } k \notin 2\mathbf{Z} \end{cases}$$

with $g_p \in \mathscr{M}_{k-2p}(\Gamma, \Phi)$ *and* $c \in \Phi$.

(iii) $\mathscr{N}_k = \mathscr{M}_k$ *if* $k < 2$.

PROOF. Denote by A_t the set of all functions f of the form

$$f(z) = \textstyle\sum_{a=0}^{t} (\pi y)^{-a} f_a(z) \tag{7.13}$$

with holomorphic functions f_a on $\mathfrak{H}$. We note that

$$D_k^p(\pi y)^{-a} = c(\pi y)^{-a-p} \text{ with } c \in \mathbf{Q}^\times \text{ if } k > a > 0, \tag{7.14}$$

which can easily be verified. From the definition of D_k and (6.14b), we easily see that (i) is true if $p = 1$. Then the general case can be proved by induction on p. To prove (iii), suppose f of (7.13) belongs to $\mathscr{N}_k(\Gamma, \Phi)$; suppose also $k \in \mathbf{Z}$ for the moment. For $\gamma \in SL_2(\mathbf{Q})$ we have $(y \circ \gamma)^{-a} = |j_\gamma|^{2a} y^{-a}$ and $\overline{j}_\gamma = j_\gamma - 2ic_\gamma y$, and so

$$f\|_k \gamma = \sum_{a=0}^{t} (\pi y)^{-a} j_\gamma^{a-k} (j_\gamma - 2ic_\gamma y)^a (f_a \circ \gamma), \tag{7.15}$$

which can be written $\sum_{a=0}^{t} (\pi y)^{-a} g_{a\gamma}$ with holomorphic functions $g_{a\gamma}$. Viewing this as a polynomial in y^{-1} and comparing the coefficients of y^{-t}, we obtain $g_{t\gamma} = j_\gamma^{2t-k}(f_t \circ \gamma)$, that is, $g_{t\gamma} = f_t\|_{k-2t}\gamma$, because the function y^{-1} is an algebraically independent variable over the field of meromorphic functions on $\mathfrak{H}$. In particular, $g_{t\gamma} = f_t$ if $\gamma \in \Gamma$, and so $f_t\|_{k-2t}\gamma = f_t$ for $\gamma \in \Gamma$. By condition (7.3c), $g_{t\gamma}$ has a Fourier expansion finite at ∞ for every $\gamma \in SL_2(\mathbf{Q})$. This shows that $f_t \in \mathscr{M}_{k-2t}(\Gamma)$. Therefore, if $t > 0$ and $f_t \neq 0$, then $k \geq 2t$. Consequently, $t = 0$ if $k < 2$, which proves (iii).

We prove (ii) for $f \in \mathscr{N}_k^t(\Gamma, \Phi)$ as above with $f_t \neq 0$ by induction on t. We have seen that $f_t \in \mathscr{M}_{k-2t}(\Gamma)$. This combined with the definition of $\mathscr{N}_k^t(\Gamma, \Phi)$ shows that $f_t \in \mathscr{M}_{k-2t}(\Gamma, \Phi)$. Suppose $k = 2t$; then $f_t \in \mathscr{M}_0(\Gamma, \Phi) = \Phi$. From (7.14) we see that $D_2^{t-1}E_2 = b(\pi y)^{-t} + q$ with $b \in \mathbf{Q}^\times$ and $q \in A_{t-1}$. Put $p = f - (f_t/b)D_2^{t-1}E_2$. Then $p \in \mathscr{N}_k^{t-1}(\Gamma, \Phi)$. In particular, if $k = 2$, then $p \in \mathscr{N}_2(\Gamma, \Phi) \cap A_0 = \mathscr{M}_2(\Gamma, \Phi)$, which settles the problem. If $k > 2$, we apply induction to p. Thus we obtain (ii) when $k = 2t$. It remains to consider the case $k > 2t > 0$. Then $f_t \in \mathscr{M}_{k-2t}(\Gamma, \Phi)$, and by (7.14) we see that $D_{k-2t}^t f_t = c(\pi y)^{-t} f_t + r$ with $c \in \mathbf{Q}^\times$ and $r \in A_{t-1}$. Put $g = f - c^{-1}D_{k-2t}^t f_t$. Then $g \in \mathscr{N}_k^{t-1}(\Gamma, \Phi)$, and applying induction to g, we can complete the proof of (ii).

So far we have assumed $k \in \mathbf{Z}$. The case $k \notin \mathbf{Z}$ can be proved in the same way; we have only to make the meaning of $f\|_k \alpha$ or j_γ^{-k} precise, which can easily be done. Our proof is now complete.

7.9. We now return to Theorem 7.5 and prove the assertions concerning $\mathscr{N}_k$. Given $\sigma \in \mathrm{Aut}(\mathbf{C})$ and f as in (7.3a), define f^σ formally by (7.5). Clearly $E_2^\sigma = E_2$. Also, we can easily verify that $(D_k f)^\sigma = D_k f^\sigma$ for every

$f \in \mathscr{N}_k$, and so $(D_k^p f)^\sigma = D_k^p f^\sigma$ for every p. Since Theorem 7.5(i) for $\mathscr{M}_k$ was proved, from Lemma 7.8(ii) we obtain Theorem 7.5(i) for $\mathscr{N}_k$. As for (iii), it is sufficient to prove it for $\alpha = \mathrm{diag}[e, 1]$ with $0 < e \in \mathbf{Q}$, $\alpha \in P$, and $\alpha \in \Gamma(1)$. The first two cases are clear. If $\alpha \in \Gamma(1)$, we have $E_2\|_2\,\alpha = E_2$, and so the desired result follows from Lemma 7.8(ii), (6.14c), and Theorem 7.5(iii) for $\mathscr{M}_k$.

To prove Theorem 7.5(iia) for $\mathscr{N}_k$, we consider the sum expression for f as in Lemma 7.8(ii), and take γ and β in $\Gamma(M, N)$ as in the first paragraph of §7.7. Since they are independent of the weight k, we have $(D_{k-2p}^p g_p)^\sigma\|_k\,\gamma = (D_{k-2p}^p g_p^\sigma)\|_k\,\gamma = D_{k-2p}^p(g_p^\sigma\|_{k-2p}\,\gamma) = D_{k-2p}^p(g_p\|_{k-2p}\,\beta)^\sigma$ by (6.14c); similarly $(D_2^{(k/2)-1}E_2)^\sigma\|_k\,\gamma = D_2^{(k/2)-1}(E_2\|_2\,\beta)^\sigma$, and so $f^\sigma\|_k\,\gamma = (f\|_k\,\beta)^\sigma$, from which we obtain Theorem 7.5(iia) for $\mathscr{N}_k$. Theorem 7.5(iv) for $\mathscr{N}_k$ was proved in the paragraph above Lemma 7.6.

It remains to prove Theorem 7.5(iib). From Lemma 7.8(ii) and (7.7) we obtain

$$\mathscr{N}_k\big(\Gamma(N)\big) = \mathscr{N}_k\big(\Gamma(N),\, \mathbf{Q}\big) \otimes_{\mathbf{Q}} \mathbf{C} \quad \text{if} \quad 24|N. \tag{7.16}$$

We apply Lemma 2.11 to the present setting with $\mathscr{X}_k\big(\Gamma(N_0),\, \mathbf{Q}_{\mathrm{ab}}\big), \mathbf{Q}_{\mathrm{ab}}$, and $\mathbf{C}$ as V, F, and K in that lemma, where $N_0 = 24MN$; we also take $\{\tilde{\alpha}\}$ there to be the set of the maps $f \mapsto f\|_k\gamma - \psi(d_\gamma)f$ for all $\gamma \in \Gamma(M, N)$. Then $W = V \otimes_F K = \mathscr{X}_k\big(\Gamma(N_0)\big)$ by (7.7), (7.7b), and (7.16). By Theorem 7.5(iii) that is already proved, V is stable under these maps. Then $\bigcap_{\alpha\in A}\mathrm{Ker}(\tilde{\alpha}) = \mathscr{X}_k(M,\, N;\, \psi)$ and $\bigcap_{\alpha\in A}\mathrm{Ker}(\alpha) = \mathscr{X}_k(M,\, N;\, \psi) \cap \mathscr{X}_k(\mathbf{Q}_{\mathrm{ab}})$. Therefore Lemma 2.11 gives Theorem 7.5(iib), and the proof of Theorem 7.5 is now complete.

We add here two remarks. First, in (5.14) we defined f_ρ for $f \in \mathscr{M}_k$. If we mean by ρ the complex conjugation in $\mathbf{C}$, then this coincides with f^ρ defined by (7.5). However, we employ f_ρ in addition to f^ρ for some notational reasons.

Next, we assumed $24|N$ in (7.7) and (7.7b) merely for expediency. In fact, we can prove better and more comprehensive results as follows.

Theorem 7.10. *Let $\mathscr{X}$ denote $\mathscr{M}$, $\mathscr{S}$, or $\mathscr{N}$ as in Theorem 7.5 and let $k \in 2^{-1}\mathbf{Z}, > 0$. Then the following assertions hold.*

(i) *For every congruence subgroup Γ of $\Gamma(1)$ the set $\mathscr{X}_k(\Gamma)$ is spanned by its $\mathbf{Q}_{\mathrm{ab}}$-rational elements, provided $\Gamma \subset \Gamma^\theta$ if $k \notin \mathbf{Z}$.*

(ii) *Let Γ denote any one of the groups $\Gamma_0(N)$, $\Gamma_1(N)$, $\Gamma(M, N)$, and $\Gamma(N)$ with positive integers M and N, where $\Gamma_0(N) = \{\gamma \in \Gamma(1) \,|\, c_\gamma \in N\mathbf{Z}\}$ and $\Gamma_1(N) = \{\gamma \in \Gamma_0(N) \,|\, d_\gamma - 1 \in N\mathbf{Z}\}$. Then $\mathscr{X}_k(\Gamma) = \mathscr{X}_k(\Gamma,\, \mathbf{Q}) \otimes_{\mathbf{Q}} \mathbf{C}$, provided $\Gamma \subset \Gamma^\theta$ if $k \notin \mathbf{Z}$.*

PROOF. To prove (i), take a multiple N_0 of 24 so that $\Gamma(N_0) \subset \Gamma$. We then apply Lemma 2.11 to the present setting by taking $\mathscr{X}_k(\Gamma(N_0), \mathbf{Q}_{ab})$, $\mathbf{Q}_{ab}$, and $\mathbf{C}$ to be V, F, and K in that lemma. We also take $\{\tilde{\alpha}\}$ there to be the set of the maps $f \mapsto f\|_k\gamma - f$ for all $\gamma \in \Gamma$. Then $W = V \otimes_F K = \mathscr{X}_k(\Gamma(N_0))$ by (7.7), (7.7b), and (7.16). By Theorem 7.5(iii), V is stable under these maps. Then $\bigcap_{\alpha\in A} \mathrm{Ker}(\tilde{\alpha}) = \mathscr{X}_k(\Gamma)$ and $\bigcap_{\alpha\in A} \mathrm{Ker}(\alpha) = \mathscr{X}_k(\Gamma, \mathbf{Q}_{ab})$, and so we obtain (i) from Lemma 2.11.

Given Γ as in (ii), we note

(7.17) $\mathscr{X}_k(\Gamma)^\sigma = \mathscr{X}_k(\Gamma)$ *for every* $\sigma \in \mathrm{Aut}(\mathbf{C})$.

This is included in Theorem 7.5(iia) if Γ is $\Gamma_0(N)$ or $\Gamma(M, N)$, because $\mathscr{X}_k(\Gamma_0(N)) = \mathscr{X}_k(1, N; \chi_0)$ and $\mathscr{X}_k(\Gamma(M, N)) = \mathscr{X}_k(M, N; \chi_0)$, where χ_0 is the trivial character. Now, as shown in the proof of Theorem 7.5(iv), $\mathscr{X}_k(\Gamma(N))$ is the sum of $\mathscr{X}_k(N, N; \psi)$ for all characters ψ of $(\mathbf{Z}/N\mathbf{Z})^\times$ such that $\psi(-1) = (-1)^{[k]}$, and so (7.17) for $\Gamma = \Gamma(N)$ follows from Theorem 7.5(iia). Also, (7.17) for $\mathscr{X} = \mathscr{N}$ follows from Lemma 7.8(ii) combined with (7.17) for $\mathscr{X} = \mathscr{M}$. Therefore we have only to prove (7.17) for $\mathscr{X} = \mathscr{M}$ or $\mathscr{S}$ when $\Gamma = \Gamma_1(N)$. Clearly

$$\mathscr{M}_k(\Gamma_1(N)) = \{f \in \mathscr{M}_k(\Gamma(N)) \mid f\|_k\gamma = f \text{ for every } \gamma \in \Gamma_1(N)\}.$$

Let $f \in \mathscr{M}_k(\Gamma_1(N))$, $\gamma \in \Gamma_1(N)$, and $\sigma \in \mathrm{Aut}(\mathbf{C})$. Taking $M = 1$ and $N = K$ in Lemma 7.6(i), we have $f^\sigma\|_k\gamma = (f\|_k\beta)^\sigma$ with some $\beta \in \Gamma_1(N)$. Then we find that $f^\sigma\|_k\gamma = f^\sigma$, and so $f^\sigma \in \mathscr{M}_k(\Gamma_1(N))$, which proves (7.17) when $\mathscr{X}_k(\Gamma) = \mathscr{M}_k(\Gamma_1(N))$. The case of $\mathscr{S}_k$ follows from this immediately, since $\mathscr{S}_k^\sigma = \mathscr{S}_k$.

To prove (ii), in view of (i), it is sufficient to show that $\mathscr{X}_k(\Gamma, \mathbf{Q}_{ab})$ is spanned by its $\mathbf{Q}$-rational elements. Take a $\mathbf{Q}_{ab}$-basis B of $\mathscr{X}_k(\Gamma, \mathbf{Q}_{ab})$. By Theorem 7.5(v) and Lemma 7.8(ii) we can find a finite extension K of $\mathbf{Q}$ contained in $\mathbf{Q}_{ab}$ such that every member of B is K-rational. By (7.17), $\mathscr{X}_k(\Gamma, K)^\sigma = \mathscr{X}_k(\Gamma, K)$ for every $\sigma \in \mathrm{Gal}(K/Q)$. Therefore $\mathscr{X}_k(\Gamma, K) = \mathscr{X}_k(\Gamma, \mathbf{Q}) \otimes_{\mathbf{Q}} K$ by Lemma 2.13. This proves (ii) and completes the proof.

Lemma 7.11. *Every element of* $\mathscr{N}_k$ *is slowly increasing at every cusp.*

PROOF. Given $f \in \mathscr{N}_k$, take the expression for f as in Lemma 7.8(ii). By Lemma 6.10, $D_{k-2p}^p g_p$ is slowly increasing at every cusp. Now take E_2 in place of g in the proof of Lemma 6.10. Then from the expression (7.10) we easily see that $\delta_2^p E_2$ is slowly increasing at every cusp, and so we obtain our lemma.

Lemma 7.12. *Given* $0 < t \in \mathbf{Q}$, *put* $\theta_t(z, \lambda) = \theta(0, tz; \lambda)$ *with* $\theta(u, z; \lambda)$ *as in (4.48b). Then for* $\alpha \in Sp(n, \mathbf{Q})$ *and* $r(z) = j_\alpha(z)^{1/2}$ *there exists an element* μ *of* $\mathscr{L}(\mathbf{Q}^n)$ *such that* $\theta_t(\alpha z, \lambda) = r(z)\theta_t(z, \mu)$. *Moreover, if* λ *is* $\mathbf{Q}_{ab}$*-valued, then so is* μ.

PROOF. Put $\beta = \mathrm{diag}[t1_n, 1_n]\alpha \cdot \mathrm{diag}[t1_n, 1_n]^{-1}$. Then $\beta \in Sp(n, \mathbf{Q})$, $t \cdot \alpha(z) = \beta(tz)$, $j_\beta(tz) = j_\alpha(z)$, and $\theta_t(\alpha z, \lambda) = \theta(0, \beta(tz); \lambda) = r(z)\theta(0, tz; \mu)$ with $\mu \in \mathscr{L}(\mathbf{Q}^n)$ determined for λ and β by (4.49g). This proves the first part of our lemma. Suppose λ is $\mathbf{Q}_{\mathrm{ab}}$-valued; then $\theta_t(z, \lambda) \in \mathscr{M}_{1/2}(\mathbf{Q}_{\mathrm{ab}})$, and so $\theta_t(z, \mu)$ belongs to $\mathscr{M}_{1/2}(\mathbf{Q}_{\mathrm{ab}})$ by Theorem 7.5(iii). This completes the proof when $n = 1$, since the quoted theorem concerns only that case. However, a similar fact is true in the case $n > 1$ too, as proved in [S00, Theorem 7.11]. We will later employ this lemma only in the case $n = 1$.

Lemma 7.13. (i) *Let* $f(z) = \sum_{n=0}^{\infty} a_n \mathbf{e}(nz/\nu) \in \mathscr{M}_k(\nu, N/\nu; \psi)$, *where* $0 < N \in \mathbf{Z}$, $\nu = 1$ *if* $k \in \mathbf{Z}$, *and* $\nu = 2$ *if* $k \notin \mathbf{Z}$; *we assume* $N \in 4\mathbf{Z}$ *if* $k \notin \mathbf{Z}$. *Let* s *be the conductor of* ψ. *Given a character* χ *modulo* rt, *where* $0 < t \in \mathbf{Z}$ *and* r *is the conductor of* χ, *put* $g(z) = \sum_{n=0}^{\infty} \chi(n) a_n \mathbf{e}(nz/\nu)$. *Let* t_0 *be the product of all the prime factors of* t. *Then* $g \in \mathscr{M}_k(\nu, M/\nu; \chi^2\psi)$, *where* M *is the least common multiple of* N, t_0^2, r^2, *and* rs.

(ii) *Given* f *as in* (i), *put* $h(z) = \sum_{(n,t)=1} a_n \mathbf{e}(nz/\nu)$ *with a positive integer* t. *Define* t_0 *as in* (i). *Then* $h \in \mathscr{M}_k(\nu, M/\nu; \psi)$, *where* M *is the least common multiple of* N *and* t_0^2.

PROOF. We first prove (i) when χ is primitive, that is, when $t = 1$. For $u \in \mathbf{Z}$ let $\xi(u) = \begin{bmatrix} 1 & \nu u/r \\ 0 & 1 \end{bmatrix}$. Using (2.3a), we easily see that $G(\bar{\chi})g = \sum_{u=1}^{r} \bar{\chi}(u) f\|_k \xi(u)$. Let $\gamma = \begin{bmatrix} a & b\nu \\ Mc/\nu & d \end{bmatrix} \in \Gamma(\nu, M/\nu)$ and $\gamma' = \begin{bmatrix} a' & b'\nu \\ Mc/\nu & d' \end{bmatrix}$ with $a' = a + Mcu/r$, $b' = b + du(1 - ad)/r - cd^2u^2M/r^2$, and $d' = d - cd^2uM/r$. Then $\gamma' \in \Gamma(\nu, M/\nu)$, $d' - d \in s\mathbf{Z}$, and $\xi(u)\gamma = \gamma'\xi(d^2u)$. Assuming that $k \in \mathbf{Z}$ for the moment, we have

$$(*) \qquad f\|_k \xi(u)\gamma = f\|_k \gamma' \xi(d^2u) = \psi(d) f\|_k \xi(d^2u),$$

and so

$$G(\bar{\chi})g\|_k\gamma = \sum_{u=1}^{r} \bar{\chi}(u) f\|_k \xi(u)\gamma = \psi(d) \sum_{u=1}^{r} \bar{\chi}(u) f\|_k \xi(d^2u) = \psi(d)\chi(d^2)G(\bar{\chi})g,$$

which proves (i) when $k \in \mathbf{Z}$ and $t = 1$.

If $k \notin \mathbf{Z}$, we have to justify $(*)$ by checking (4.40) for h_γ and $h_{\gamma'}$. First suppose r is odd; then $M = 4r^2q$ with $q \in \mathbf{Z}$, and so $d - d' \in 4qc\mathbf{Z}$. Thus $\varepsilon_d = \varepsilon_{d'}$ and $\left(\frac{Mc}{d}\right) = \left(\frac{4qc}{d}\right) = \left(\frac{4qc}{d'}\right) = \left(\frac{Mc}{d'}\right)$, from which we obtain $(*)$. Next suppose r is even; then $r = 2^\alpha r_1$ with $\alpha > 1$ and an odd r_1. Thus $M = 2^{2\alpha} r_1^2 q$ with $q \in \mathbf{Z}$ and $d - d' \in 4qc\mathbf{Z}$, and so we obtain $(*)$ in this case too. This completes the proof of (i) when $t = 1$.

We next prove (ii). Taking the prime decomposition of t, we can reduce the problem to the case where t is a prime number. Assuming this to be so, put $\ell(z) = \sum_{n=0}^{\infty} a_{tn} \mathbf{e}(tnz/\nu)$ and $\eta(u) = \begin{bmatrix} 1 & \nu u/t \\ 0 & 1 \end{bmatrix}$. Then $\sum_{u=1}^{t} f\|_k \eta(u) = t\ell$.

Define the matrices γ and γ' in the above proof of (i) with t in place of r and M defined as in (ii). Then $\gamma, \gamma' \in \Gamma(\nu, M/\nu)$, $\eta(u)\gamma = \gamma'\eta(d^2u)$, and we find that $f\|_k\eta(u)\gamma = f\|_k\gamma'\eta(d^2u) = \psi(d)f\|_k\eta(d^2u)$. Since d is prime to t, we easily see that $t\ell\|_k\gamma = \psi(d)t\ell$, and so $\ell \in \mathscr{M}_k(\nu, M/\nu; \psi)$. We have clearly $h = f - \ell$, from which we obtain (ii).

Finally to prove (i) in the general case, we have only to observe that it can be obtained by combining (ii) and the special case of (i) in which $t = 1$. This completes the proof.

In the proof of Lemma 5.5 we proved (5.8) only when ψ is primitive. That result combined with Lemma 7.3(ii) settles the case of imprimitive ψ.

Lemma 7.14. *For* $f(z) = \sum_{n=0}^{\infty} c(n)\mathbf{e}(nz/N) \in \mathscr{M}_k(\Gamma(N))$ *with* $N \in \nu\mathbf{Z}$, *put* $(Pf)(z) = (\nu/N)\sum_{u=1}^{N/\nu} f(z + \nu u)$, *where* ν *is as in Lemma 7.13. Then* $(Pf)(z) = \sum_{n=0}^{\infty} c(Nn/\nu)\mathbf{e}(nz/\nu)$, *and* $\langle Pf, h\rangle = \langle f, Ph\rangle$ *for every* $h \in \mathscr{S}_k(\Gamma(N))$. *Moreover, if* $f \in \mathscr{M}_k(N, N; \psi)$, *then* $Pf \in \mathscr{M}_k(\nu, N; \psi)$.

PROOF. The first equality for $(Pf)(z)$ is easy. Next, for $h \in \mathscr{S}_k(\Gamma(N))$ we have, by (6.12),

$$\langle Pf, h\rangle = (\nu/N)\Big\langle \sum_{u=1}^{N/\nu} f(z+\nu u),\, h\Big\rangle = (\nu/N)\Big\langle f,\, \sum_{u=1}^{N/\nu} h(z-\nu u)\Big\rangle = \langle f, Ph\rangle.$$

Suppose $f \in \mathscr{M}_k(N, N; \psi)$. Given $\gamma = \begin{bmatrix} a & b\nu \\ Nc & d \end{bmatrix} \in \Gamma(N, N)$ define $\gamma' = \begin{bmatrix} a' & b'\nu \\ Nc & d' \end{bmatrix}$ as in the proof of Lemma 7.13 with $r = 1$ and $M = \nu N$. Then we see that $\gamma' \in \Gamma(N, N)$, $d - d' \in \nu cN\mathbf{Z}$, and $(*)$ in the proof holds with $\xi(u) = \begin{bmatrix} 1 & \nu u \\ 0 & 1 \end{bmatrix}$, and so

$$(Pf)\|_k\gamma = (\nu/N)\sum_{u=1}^{N/\nu} f\|_k\xi(u)\gamma = \psi(d)(\nu/N)\sum_{u=1}^{N/\nu} f\|_k\xi(d^2u) = \psi(d)Pf.$$

Now $(Pf)(z + \nu) = Pf$, and $\Gamma(\nu, N)$ can be generated by $\Gamma(N, N)$ and $\xi(1)$, as will be proven in Lemma 8.18. Therefore $Pf \in \mathscr{M}_k(\nu, N; \psi)$, which completes the proof.

Lemma 7.15. *Let* K *be a multiple of* N, *and* A *a complete set of representatives for* $\Gamma(N)/\Gamma(K)$; *suppose* $N \in 2\mathbf{Z}$ *if* $k \notin \mathbf{Z}$. *For* $f \in \mathscr{M}_k(\Gamma(K))$ *put* $q(f) = \#(A)^{-1}\sum_{\gamma\in A} f\|_k\gamma$. *Then* $q(f) \in \mathscr{M}_k(\Gamma(N))$, $q(f)^\sigma = q(f^\sigma)$ *for every* $\sigma \in \mathrm{Aut}(\mathbf{C})$, *and* $\langle f, h\rangle = \langle q(f), h\rangle$ *for every* $h \in \mathscr{S}_k(\Gamma(N))$.

PROOF. That $q(f) \in \mathscr{M}_k(\Gamma(N))$ is easy. For $h \in \mathscr{S}_k(\Gamma(N))$ we have, by (6.12),

$$\langle q(f), h\rangle = \#(A)^{-1}\Big\langle \sum_{\gamma\in A} f\|_k\gamma,\, h\Big\rangle = \#(A)^{-1}\Big\langle f,\, \sum_{\gamma\in A} h\|_k\gamma^{-1}\Big\rangle = \langle f, h\rangle.$$

Let $\sigma \in \mathrm{Aut}(\mathbf{C})$. Then the proof of Lemma 7.6(i) shows that the element β there can be chosen so that $\gamma \mapsto \beta$ gives an automorphism of $\Gamma(N)/\Gamma(K)$. Writing β_γ for β, we obtain $q(f^\sigma) = \#(A)^{-1}\sum_{\gamma\in A} f^\sigma\|_k\gamma = \#(A)^{-1}\sum_{\gamma\in A} (f\|_k\beta_\gamma)^\sigma = q(f)^\sigma$. This completes the proof.

8. Dirichlet series and Eisenstein series

8.1. We are going to consider various types of Dirichlet series associated with modular forms. In order to study their analytic properties, the gamma function $\Gamma(s)$ is essential, and so we first recall its basic properties:

(8.1a) $\Gamma(s)$ *is a meromorphic function on the whole* $\mathbf{C}$.

(8.1b) $\Gamma(s)^{-1}$ *is an entire function.*

(8.1c) $\Gamma(s+1) = s\Gamma(s)$.

(8.1d) *The set of poles of Γ consists of 0 and all negative integers, and each pole is of order 1.*

(8.1e) $\Gamma(n) = (n-1)!$ *if* $0 < n \in \mathbf{Z}$.

These are well known. We will often be using

$$\Gamma(s)a^{-s} = \int_0^\infty e^{-at}t^{s-1}dt \quad \textit{if} \quad a \in \mathbf{C},\ \mathrm{Re}(a) > 0, \quad \textit{and} \quad \mathrm{Re}(s) > 0. \tag{8.2}$$

Here $a^{-s} = \exp(-s\log a)$ with the standard branch of $\log a$ for $\mathrm{Re}(a) > 0$. Indeed, the formula is well known for $0 < a \in \mathbf{R}$. Now the integral is meaningful for $\mathrm{Re}(a) > 0$, and defines a holomorphic function of a. Since it coincides with $\Gamma(s)a^{-s}$ for $0 < a \in \mathbf{R}$, we obtain (8.2) as stated.

Given $f(z) = \sum_{\xi\in\mathbf{Q}} a_\xi \mathbf{e}(\xi z) \in \mathscr{M}_k$ with an integral or a half-integral weight k, we put, ignoring a_0,

$$D(s, f) = \sum_{\xi>0} a_\xi \xi^{-s}, \tag{8.3}$$

$$R(s, f) = (2\pi)^{-s}\Gamma(s)D(s, f). \tag{8.4}$$

We can put $f(z) = \sum_{m=0}^\infty c_m \mathbf{e}(mz/N)$ as in Lemma 6.2. Then $D(s, f) = N^s\sum_{m=1}^\infty c_m m^{-s}$, and so from (ii) of the same lemma, we see that the right-hand side of (8.3) is convergent for $\mathrm{Re}(s) > k+1$, and so $D(s, f)$ is holomorphic for such s.

For $z = iy$ with $0 < y \in \mathbf{R}$ we have $f(iy) - a_0 = \sum_{\xi>0} a_\xi e^{-2\pi\xi y}$, and so in view of (8.2) we obtain

$$\int_0^\infty [f(iy) - a_0]y^{s-1}dy = (2\pi)^{-s}\Gamma(s)\sum_{\xi>0} a_\xi\xi^{-s} = R(s, f) \tag{8.5}$$

for $\mathrm{Re}(s) > k+1$. Termwise integration is justified, since the series for such s is absolutely convergent. Fixing a positive integer N, put

$$f^{\#}(z) = N^{-k/2}(-iz)^{-k}f\big(-(Nz)^{-1}\big), \tag{8.6}$$

where the branch of $(-iz)^{-k}$ is taken so that it is real and positive if $z \in \mathfrak{H} \cap i\mathbf{R}$. By Lemma 5.4, $f^{\#} \in \mathscr{M}_k$, and so we can put $f^{\#}(z) = \sum_{\xi \in \mathbf{Q}} b_\xi \mathbf{e}(\xi z)$. Thus $D(s,\, f^{\#})$ and $R(s,\, f^{\#})$ are meaningful.

Theorem 8.2. *In the above setting $R(s,\, f)$ and $R(s,\, f^{\#})$ can be continued as meromorphic functions to the whole s-plane with the following properties:*

$$N^{s/2}R(s,\, f) = -\frac{a_0}{s} + \frac{b_0}{s-k} + \text{an entire function}, \tag{8.7a}$$

$$R(k-s,\, f) = N^{s-k/2}R(s,\, f^{\#}). \tag{8.7b}$$

In particular, $R(s,\, f)$ is an entire function if $a_0 = b_0 = 0$.

This is a standard theorem first proved by Hecke in a somewhat different formulation. For the proof, see [S07, Theorem 3.2].

8.3. For a Dirichlet character ψ and a positive integer N we put, as we did in §2.9,

$$L_N(s,\, \psi) = \sum_{(n,\, N)=1} \psi(n)n^{-s}, \tag{8.8}$$

where the sum is extended over the positive integers n prime to N. If ψ is primitive, then $L_1(s,\, \psi)$ is the L-function of ψ, and is usually denoted by $L(s,\, \psi)$. Let χ be the primitive character associated with ψ. Then

$$L_N(s,\, \psi) = L(s,\, \chi)\prod_{p|N}\big(1-\chi(p)p^{-s}\big). \tag{8.9}$$

Thus the analytic properties of $L_N(s,\, \psi)$ can be reduced to those of $L(s,\, \chi)$, which can be summarized as follows. Let r be the conductor of χ and let $\chi(-1) = (-1)^\nu$ with $\nu = 0$ or 1. Put

$$R(s,\, \chi) = (r/\pi)^{(s+\nu)/2}\Gamma\big((s+\nu)/2\big)L(s,\, \chi). \tag{8.10}$$

Then $R(s,\, \chi)$ as a meromorphic function of s can be continued to the whole $\mathbf{C}$, and satisfies the functional equation

$$R(s,\, \chi) = W(\chi)R(1-s, \overline{\chi}) \quad \text{with} \quad W(\chi) = i^{-\nu}r^{-1/2}G(\chi), \tag{8.10a}$$

where $G(\chi)$ is defined by (2.3). Notice that $|W(\chi)| = 1$ because of (2.3c). Moreover, $R(s,\, \chi)$ is entire except when χ is the principal character, in which case it is holomorphic on $\mathbf{C}$ except for simple poles at $s = 0$ and 1, with residues 1 and -1 respectively.

These are well known, and in fact, can be obtained from Theorem 8.2 by taking f to be θ_ψ of (5.7). For simplicity we treat here the case of nontrivial χ. (If χ is the principal character, then $L(s,\, \chi) = \zeta(s)$, and its functional equation is well known.)

The notation being as in Lemma 5.5, put $f(z) = \theta_\psi(z/r)$, $f^*(z) = \theta_{\bar\psi}(z/r)$, $\alpha = \begin{bmatrix} 0 & -r^{-1} \\ r & 0 \end{bmatrix}$, and $\omega = \psi(-1)r^{-1/2}G(\psi)$. Since $j_\alpha^k(z) = r^k j_\iota^k(z)$, we have $f(-1/z)r^{-k}j_\iota^k(z/r)^{-1} = \theta_\psi(-1/rz)j_\alpha^k(z/r)^{-1} = (\theta_\psi\|_k\alpha)(z/r)$. By (5.9) this can be written

$$f(-1/z)z^{-\nu}(-iz)^{-1/2} = \omega f^*(z). \tag{8.10b}$$

Take $N = 1$ in (8.6). Then the left-hand side of (8.10b) is $i^{-\nu}f^\#$. Thus $f^\# = i^\nu\omega f^* = W(\psi)f^*$ with $W(\psi)$ of (8.10a). We see that $R(s,\, f) = (2\pi)^{-s}\Gamma(s) \cdot\sum_{n=1}^\infty \psi(n)n^\nu(n^2/2r)^{-s} = (r/\pi)^s\Gamma(s)\sum_{n=1}^\infty \psi(n)n^{\nu-2s}$, and so

$$R\bigl((s+\nu)/2,\, f\bigr) = R(s,\, \psi). \tag{8.10c}$$

By (8.7b) we have $R(k-s,\, f) = R(s,\, f^\#) = W(\psi)R(s,\, f^*)$. Substituting $(s+\nu)/2$ for $k-s$, we obtain (8.10a) with ψ in place of χ.

We add here a formula that will be needed in Section A3. In [S07, p. 20] it was shown that

$$R(k-s,\, f) = \int_1^\infty f(iy)y^{k-s-1}dy + \int_1^\infty f^\#(iy)y^{s-1}dy,$$

and so

$$R(s,\, \psi) = \int_1^\infty f(iy)y^{(s+\nu-2)/2}dy + W(\psi)\int_1^\infty f^*(iy)y^{(\nu-1-s)/2}dy. \tag{8.10d}$$

We have $f(iy) = \sum_{n=1}^\infty \psi(n)n^\nu \exp(-\pi n^2 y/r)$ and $f^*(iy)$ is the series of the same type with $\bar\psi$ in place of ψ.

8.4. There are two types of Eisenstein series with respect to $SL_2(\mathbf{Q})$. The first type can be defined for both integral and half-integral weights, but the second type can be defined only for integral weights. There is also the question whether the series involves a complex variable, which is usually denoted by s. In this book we first define the series with s, and specialize s to an element of $2^{-1}\mathbf{Z}$. We begin our discussion with two easy lemmas:

Lemma 8.5. + *The series* $\sum_{0\neq(m,n)\in\mathbf{Z}^2} |mz+n|^{-\sigma}$ *with a fixed* $z \in \mathbf{C},\ \notin \mathbf{R}$, *is convergent for* $2 < \sigma \in \mathbf{R}$.

PROOF. Let $L = \mathbf{Z}z + \mathbf{Z}$. For $0 < n \in \mathbf{Z}$ let Q_n be the parallelogram on the plane $\mathbf{C}$ whose vertices are $\pm(nz+n)$ and $\pm(nz-n)$. Then there are exactly $8n$ points of L lying on the sides of Q_n. Take $r \in \mathbf{R},\ > 0$, so that the circle $|z| = r$ is inside Q_1. Then $|\xi| \geq nr$ for any ξ of such $8n$ points, and so for $\sigma > 0$ we have $\sum_{0\neq\xi\in L} |\xi|^{-\sigma} \leq \sum_{n=1}^\infty 8n(nr)^{-\sigma} = 8r^{-\sigma}\sum_{n=1}^\infty n^{1-\sigma}$, which is convergent for $\sigma > 2$ as expected.

8.6. For two positive integers M and N we put (as we did in Theorem 7.5)

$$\Gamma(M,\, N) = \bigl\{\gamma \in \Gamma(1) \bigm| b_\gamma \in M\mathbf{Z},\ c_\gamma \in N\mathbf{Z}\bigr\}. \tag{8.11}$$

We also put

$$(8.11a)\qquad \Gamma_0(N) = \Gamma(1,\, N), \quad \Gamma_1(N) = \big\{\gamma \in \Gamma_0(N) \,\big|\, a_\gamma - 1 \in N\mathbf{Z}\big\}.$$

These are traditional. Notice that the condition $a_\gamma - 1 \in N\mathbf{Z}$ is equivalent to $d_\gamma - 1 \in N\mathbf{Z}$. For a character ψ modulo MN we put

$$(8.11b)\quad \mathscr{M}_k(M,\, N;\, \psi) = \big\{f \in \mathscr{M}_k \,\big|\, f\|_k\gamma = \psi(d_\gamma)f \text{ for every } \gamma \in \Gamma(M,\, N)\big\},$$

$$(8.11c)\qquad \mathscr{S}_k(M,\, N;\, \psi) = \mathscr{S}_k \cap \mathscr{M}_k(M,\, N;\, \psi),$$

$$(8.11d)\qquad \mathscr{M}_k(N,\, \psi) = \mathscr{M}_k(1,\, N;\, \psi), \quad \mathscr{S}_k(N,\, \psi) = \mathscr{S}_k(1,\, N;\, \psi),$$

where we assume that $M,\, N \in 2\mathbf{Z}$ if $k \notin \mathbf{Z}$. In fact, (8.11b) and (8.11c) are included in (7.6). Since $f\|_k(-1) = (-1)^{[k]}f$, we have $\mathscr{M}_k(M,\, N;\, \psi) = \{0\}$ if $\psi(-1) \neq (-1)^{[k]}$. Observing that $\gamma \mapsto a_\gamma \pmod{N}$ gives an isomorphism of $\Gamma_0(N)/\Gamma_1(N)$ onto $(\mathbf{Z}/N\mathbf{Z})^\times$ (if $N > 1$), we see that $\mathscr{M}_k\big(\Gamma_1(N)\big)$ resp. $\mathscr{S}_k\big(\Gamma_1(N)\big)$ is the direct sum of $\mathscr{M}_k(N,\, \psi)$ resp. $\mathscr{S}_k(N,\, \psi)$ for all characters ψ modulo N.

For two characters ψ and χ modulo N we have

$$(8.11e)\qquad \big\langle \mathscr{M}_k(N,\, \psi),\, \mathscr{S}_k(N,\, \chi)\big\rangle = 0 \quad \text{if} \quad \chi \neq \psi.$$

This can be seen by taking α of (6.12) to be an element γ of $\Gamma_0(N)$ such that $\psi(d_\gamma) \neq \chi(d_\gamma)$.

Lemma 8.7. *Put $\Gamma = \Gamma(M,\, N)$ and $\Gamma_\infty = \Gamma \cap P$ with fixed M and N; put also*

$$W_N^M = \begin{cases} \big\{(c,\, d) \in N\mathbf{Z} \times \mathbf{Z} \,\big|\, Mc\mathbf{Z} + d\mathbf{Z} = \mathbf{Z},\, d > 0\big\} & \text{if } MN > 1, \\ \{(1,\, 0)\} \cup \big\{(c,\, d) \in \mathbf{Z} \times \mathbf{Z} \,\big|\, c\mathbf{Z} + d\mathbf{Z} = \mathbf{Z},\, d > 0\big\} & \text{if } MN = 1. \end{cases}$$

Then each coset $\Gamma_\infty\alpha$ in $\Gamma_\infty\backslash\Gamma$ contains an element γ such that $(c_\gamma,\, d_\gamma) \in W_N^M$, and the map $\Gamma_\infty\alpha \mapsto (c_\gamma,\, d_\gamma)$ with such a γ gives a bijection of $\Gamma_\infty\backslash\Gamma$ onto W_N^M.

PROOF. Since $\Gamma_\infty = \big\{\gamma \in \Gamma \,\big|\, (0,\, 1)\gamma = \pm(0,\, 1)\big\}$ and Γ_∞ is generated by -1 and $\begin{bmatrix} 1 & M \\ 0 & 1 \end{bmatrix}$, we easily see that the map from $\Gamma_\infty\backslash\Gamma$ to W_N^M can be defined as described above and that it is injective. Conversely, let $(c,\, d) \in W_N^M$ with $MN > 1$. Then d is prime to Mc, and so we can find $a,\, b \in \mathbf{Z}$ such that $ad - Mbc = 1$. Put $\gamma = \begin{bmatrix} a & bM \\ c & d \end{bmatrix}$. Then $\gamma \in \Gamma$ and our map sends $\Gamma_\infty\gamma$ to $(c,\, d)$, which proves our lemma when $MN > 1$. The case $MN = 1$ requires only an additional observation that $(1,\, 0)$ corresponds to $\Gamma_\infty\iota$.

8.8. The first type of **Eisenstein series** is defined for any weight k and a congruence subgroup $\Gamma = \Gamma(M,\, N)$ with some M and N; we assume that $M \in 2\mathbf{Z}$ and $N \in 2\mathbf{Z}$ if $k \notin \mathbf{Z}$. We put $\Gamma_\infty = P \cap \Gamma$ and

$$(8.12)\qquad E_k(z,\, s) = E_k(z,\, s;\, \Gamma,\, \psi) = \sum_{\gamma \in \Gamma_\infty\backslash\Gamma} \psi(d_\gamma)y^s\|_k\,\gamma,$$

or more explicitly,

(8.12a) $$E_k(z, s; \Gamma, \psi) = y^s \sum_{\gamma \in \Gamma_\infty \backslash \Gamma} \psi(d_\gamma) j_\gamma^k(z)^{-1} |j_\gamma(z)|^{-2s}.$$

Here $z \in \mathfrak{H}$, $s \in \mathbf{C}$, $y = \mathrm{Im}(z)$, ψ is a character modulo MN such that $\psi(-1) = (-1)^{[k]}$, and j_γ^k is as in (5.1b). Clearly the sums of (8.12) and (8.12a) are formally well defined. By Lemma 8.7, for $\sigma = \mathrm{Re}(s)$ these sums are majorized by $y^\sigma \sum_{0 \neq (c,\, d) \in \mathbf{Z}^2} |cz+d|^{-2\sigma-k}$, which is convergent for $2\sigma+k > 2$ by Lemma 8.5. Thus $E_k(z, s)$ is defined as a holomorphic function of s at least for $\mathrm{Re}(s) > 1 - k/2$. Moreover, we easily see that

(8.13) $$E_k(\gamma z, s) = \psi(d_\gamma)^{-1} j_\gamma^k(z) E_k(z, s) \quad \text{for every} \quad \gamma \in \Gamma,$$

(8.13a) $$E_{-k}(z, s; \Gamma, \bar{\psi}) = y^k \overline{E_k(z, \bar{s} - k; \Gamma, \chi_0 \psi)},$$

where χ_0 is the principal character if $k \in \mathbf{Z}$ and $\chi_0(d) = \left(\frac{-1}{d}\right)$ if $k \notin \mathbf{Z}$; we use (5.3c) for the proof of (8.13a).

8.9. The second type of **Eisenstein series** is defined for an integral weight k. For $k \in \mathbf{Z}$, a positive integer N, and $(p, q) \in \mathbf{Z}^2$ we put

(8.14) $$\mathfrak{E}_k^N(z, s; p, q) = y^s \sum_{(m,\, n)} (mz+n)^{-k} |mz+n|^{-2s} \qquad (z \in \mathfrak{H},\ s \in \mathbf{C}),$$

where the sum is taken over all $(m, n) \in \mathbf{Z}^2$ such that $0 \neq (m, n) \equiv (p, q) \pmod{N\mathbf{Z}^2}$. From Lemma 8.5 we see that the series of (8.14) is convergent for $\mathrm{Re}(2s) + k > 2$. Also, we can easily verify that

(8.14a) $$j_\gamma(z)^{-k} \mathfrak{E}_k^N(\gamma z, s; p, q) = \mathfrak{E}_k^N\big(z, s; (p, q)\gamma\big) \quad \text{for every} \quad \gamma \in \Gamma(1),$$

(8.14b) $$\mathfrak{E}_{-k}^N(z, s; p, q) = y^k \overline{\mathfrak{E}_k^N(z, \bar{s} - k; p, q)},$$

(8.14c) $$\mathfrak{E}_k^N(rz, s; p, q) = \sum_{i=1}^{r} r^s \mathfrak{E}_k^{rN}(z, s; rp, q + iN) \qquad (0 < r \in \mathbf{Z}),$$

(8.14d) $$\mathfrak{E}_k^{hN}(z, s; hp, hq) = h^{-k-2s} \mathfrak{E}_k^N(z, s; p, q) \qquad (0 < h \in \mathbf{Z}).$$

Theorem 8.10. (i) *If $k \in \mathbf{Z}$, there is a real analytic function $F(z, s)$ of $(z, s) \in \mathfrak{H} \times \mathbf{C}$ which is holomorphic in s and coincides with*

$$s(s-1)\Gamma(s+k')\mathfrak{E}_k^N(z, s; p, q)$$

for $\mathrm{Re}(2s) + k > 2$, where $k' = \mathrm{Max}(k, 0)$. The factor $s(s-1)$ is unnecessary if $k \neq 0$. If $k = 0$, then $\Gamma(s)\mathfrak{E}_0^N(z, s; p, q)$ has residue πN^{-2} at $s = 1$, and $-\delta(p/N)\delta(q/N)$ at $s = 0$, where $\delta(x) = 1$ if $x \in \mathbf{Z}$, and $\delta(x) = 0$ otherwise.

(ii) *For every fixed $s \in \mathbf{C}$, $F(z, s)$ as a function of z belongs to $C_k\big(\Gamma(N)\big)$. Moreover, $F(z, s)$ is slowly increasing at every cusp locally uniformly in s.*

PROOF. In view of (8.14b) it is sufficient to prove the case $k \geq 0$. The first assertion can be proved in a rather elementary way as an application of Theorem 8.2. For details, the reader is referred to [S07, Theorem 9.7]. The first part of (ii) follows from (8.14a), since $\mathfrak{E}_k^N(z, s; p, q)$ depends only on

(p, q) modulo $N\mathbf{Z}^2$. The second part of (ii) will be proven in §A3.7 of the Appendix.

Notice that $\mathfrak{E}_k^N(z, 0; p, q)$ is meaningful for $k > 0$. In particular, put

(8.14e) $$E_k(z) = 2^{-1}(2\pi i)^{-k}\mathfrak{E}_k^1(z, 0; 0, 0) \qquad (2 \leq k \in 2\mathbf{Z}).$$

Then $E_k \in \mathscr{M}_k(\mathbf{Q})$ if $k > 2$; E_2 is exactly the function of (7.10); see [S07, §9.2].

8.11. For $k \in \mathbf{Z}$, $0 < N \in \mathbf{Z}$, and a primitive or an imprimitive character ψ modulo N such that $\psi(-1) = (-1)^k$, we put

(8.15) $$E_k^N(z, s; \psi) = y^s \sum_{0 \neq (m, n) \in \mathbf{Z}^2} \psi(n)(mNz + n)^{-k}|mNz + n|^{-2s},$$

where we put $\psi(n) = 0$ if n is not prime to N. Notice that the sum over (m, n) is 0 if $\psi(-1) \neq (-1)^k$. Again Lemma 8.5 guarantees the convergence of (8.15) for $\mathrm{Re}(s) > 1 - k/2$. We easily see that

(8.16) $$E_k^N(z, s; \psi) = \sum_{q=1}^{N} \psi(q)\mathfrak{E}_k^N(z, s; 0, q),$$

and so from (8.14a) we obtain

(8.16a) $$j_\gamma(z)^{-k}E_k^N(\gamma z, s; \psi) = \psi(a_\gamma)E_k^N(z, s; \psi) \text{ for every } \gamma \in \Gamma_0(N).$$

From (8.16) and Theorem 8.10 we immediately obtain

Theorem 8.12. (i) *There is a real analytic function $F_\psi(z, s)$ of $(z, s) \in \mathfrak{H} \times \mathbf{C}$ such that $F_\psi(z, s) = s(s-1)\Gamma(s + k')E_k^N(z, s; \psi)$ for $\mathrm{Re}(2s) + k > 2$, where $k' = \mathrm{Max}(k, 0)$; moreover F_ψ is holomorphic in s. The factor $s(s-1)$ is unnecessary if $k \neq 0$ or ψ is nontrivial. If $k = 0$ and ψ is trivial, then $\Gamma(s)E_0^N(z, s; \psi)$ has residue $\pi N^{-2}\varphi(N)$ at $s = 1$, and $-\delta(1/N)$ at $s = 0$, where φ is Euler's function.*

(ii) *For every fixed $s \in \mathbf{C}$, $F_\psi(z, s)$ as a function of z belongs to $C_k(\Gamma(N))$, and is slowly increasing at every cusp locally uniformly in s.*

8.13. Returning to (8.12) with $k \in \mathbf{Z}$, by Lemma 8.7 we have

(8.17) $$E_k(z, s; \Gamma, \psi) = y^s \sum_{(c,d) \in W_N^M} \psi(d)(cz + d)^{-k}|cz + d|^{-2s}$$

for $\Gamma = \Gamma(M, N)$. Let us now assume that $M|N$ and ψ is a character modulo N. Given $mNz + n$ as in (8.15), put $(m, n) = \pm r(m', n')$ with $0 < r \in \mathbf{Z}$ and relatively prime m' and n' such that $n' > 0$. If $\psi(n) \neq 0$, then $(m'N, n') \in W_N^M$. Therefore we find that

(8.18) $$E_k^N(z, s; \psi) = 2L_N(2s + k, \psi)E_k(z, s; \Gamma, \psi),$$

and so the analytic properties of $E_k(z, s; \Gamma, \psi)$ can be obtained from those of $L_N(s, \psi)$ and $E_k^N(z, s; \psi)$. We note here only its residue at $s = 1$.

(8.19) *For $\Gamma = \Gamma(M, N)$ and ψ as above and $k \geq 0$, $E_k(z, s; \Gamma, \psi)$ has nonzero residue at $s = 1$ only if $k = 0$ and ψ is trivial, in which case the residue is $(3/\pi)N^{-1}\prod_{p|N}(1+p^{-1})^{-1}$.*

This follows from Theorem 8.12, since if χ_0 is the principal character, then

(8.19a) $$L_N(2, \chi_0) = \zeta(2)\prod_{p|N}(1-p^{-2}) \quad \text{and} \quad \zeta(2) = \pi^2/6.$$

The case $k \notin \mathbf{Z}$ is more complex and difficult. Indeed, we have the following theorem which will be proven in §A3.6 of the Appendix.

Theorem 8.14. *Suppose $k \notin \mathbf{Z}$ and $\Gamma = \Gamma(M, N)$ with $N \in M\mathbf{Z} \subset 2\mathbf{Z}$; put $\kappa = 2k$, $\lambda = (1-\kappa)/2$, and $\lambda_0 = 0$ or 1 according as λ is even or odd. For $z \in \mathfrak{H}$ and $s \in \mathbf{C}$ put*

$$F^*(z, s) = (2s - \lambda - 1)L_N(4s - 2\lambda, \psi^2)E_k(z, s; \Gamma, \psi) \cdot \begin{cases} \Gamma(s)\Gamma\big(s + (1-\lambda-\lambda_0)/2\big) & (\kappa \leq 1), \\ \Gamma(s+k)\Gamma\big(s + (\lambda_0 - \lambda)/2\big) & (\kappa \geq -1). \end{cases}$$

Then $F^*(z, s)$ can be continued as a holomorphic function of s to the whole s-plane. Moreover, for any fixed $s \in \mathbf{C}$, $F^*(z, s)$ as a function of z belongs to $C_k\big(\Gamma(N)\big)$. Moreover, $F^*(z, s)$ is slowly increasing at every cusp, locally uniformly in s. The factor $2s-\lambda-1$ is necessary only if $(|\kappa|+1)/2$ is odd and ψ^2 is trivial, in which case $F^*(z, s)$ at $s = (\lambda+1)/2$ is a nonzero function on $\mathfrak{H}$, whose nature is described in Theorem 8.16 below when $k > 0$.

Theorem 8.15. (i) *If $0 < k \in \mathbf{Z}$, then $E_k(z, s; \Gamma, \psi)$ with $\Gamma = \Gamma(M, N)$ and ψ as in (8.12) is finite at $s = 0$ and $E_k(z, 0; \Gamma, \psi)$ belongs to $\mathscr{M}_k(\mathbf{Q}_{ab})$ except when $k = 2$ and ψ is trivial, in which case it is a nonholomorphic element of $\mathscr{N}_2^1(\mathbf{Q}_{ab})$. Moreover, $E_k(z, 0; \Gamma, \psi)^\sigma = E_k(z, 0; \Gamma, \psi^\sigma)$ for every $\sigma \in \mathrm{Gal}(\mathbf{Q}_{ab}/\mathbf{Q})$.*

(ii) *If $0 < k \in \mathbf{Z}$, then $E_k^N(z, s; \psi)$ is finite at $s = 1-k$ and $E_k^N(z, 1-k; \psi)$ belongs to $\pi\mathscr{M}_k(\mathbf{Q}_{ab})$ except when $k = 2$ and $N = 1$, in which case it is a nonholomorphic element of $\pi\mathscr{N}_2^1(\mathbf{Q}_{ab})$. Moreover,*

$$\big\{i^k G(\psi)^{-1}\pi^{-1}E_k^N(z, 1-k; \psi)\big\}^\sigma = i^k G(\psi^\sigma)^{-1}\pi^{-1}E_k^N(z, 1-k; \psi^\sigma)$$

for every $\sigma \in \mathrm{Gal}(\mathbf{Q}_{ab}/\mathbf{Q})$, where $G(\psi)$ is the Gauss sum defined in §2.7.

(iii) *If $3/2 \leq k \notin \mathbf{Z}$, then $E_k(z, s; \Gamma, \psi)$ of (8.12) is finite at $s = 0$, and $E_k(z, 0; \Gamma, \psi)$ belongs to $\mathscr{M}_k(\mathbf{Q}_{ab})$ except when $k = 3/2$ and ψ^2 is trivial. Moreover, when it belongs to $\mathscr{M}_k(\mathbf{Q}_{ab})$ we have $E_k(z, 0; \Gamma, \psi)^\sigma = E_k(z, 0; \Gamma, \psi^\sigma)$ for every $\sigma \in \mathrm{Gal}(\mathbf{Q}_{ab}/\mathbf{Q})$.*

(iv) *If $3/2 \leq k \notin \mathbf{Z}$, then $L_N(4s + 2k - 1, \psi^2)E_k(z, s; \Gamma, \psi)$ is finite at $s = 1-k$ and its value at $s = 1-k$ is π times a nonzero element of $\mathscr{M}_k(\mathbf{Q}_{ab})$ except when $k = 3/2$ and ψ^2 is trivial. Denote this $\mathbf{Q}_{ab}$-rational element times $2^k i^{[k]}G(\psi)^{-1}$ by $C_k^*(\Gamma, \psi)$, that is,*

$$C_k^*(\Gamma,\,\psi) = 2^k i^{[k]} G(\psi)^{-1}\pi^{-1}\big[L_N(4s+2k-1,\,\psi^2)E_k(z,\,s;\,\Gamma,\,\psi)\big]_{s=1-k}.$$

Then $C_k^*(\Gamma,\,\psi)^\sigma = C_k^*(\Gamma,\,\psi^\sigma)$ *for every* $\sigma \in \mathrm{Gal}(\mathbf{Q}_{\mathrm{ab}}/\mathbf{Q})$.

PROOF. Suppose $k \in \mathbf{Z}$. From (8.16) and (8.18) we obtain

$$2(\pi i)^{-k}L_N(k,\,\psi)E_k(z,\,0;\,\Gamma,\,\psi) = \sum_{q=1}^{N}\psi(q)(\pi i)^{-k}\mathfrak{E}_k^N(z,\,0;\,0,\,q).$$

Let $A(\psi)$ denote the right-hand side times $G(\psi)^{-1}$, and let $\sigma \in \mathrm{Aut}(\mathbf{C})$. In [S07, (9.4)] we gave a Fourier expansion of $(\pi i)^{-k}\mathfrak{E}_k^N(z,\,0;\,p,\,q)$, which shows that $A(\psi)$ is $\mathbf{Q}_{\mathrm{ab}}$-rational. Applying σ to the expansion and employing (2.17b), we find that $A(\psi)^\sigma = A(\psi^\sigma)$. Combining this with Lemma 2.10, we obtain (i). Assertion (ii) can be derived similarly from (8.16) and [S07, (9.14)]. In the case $k \notin \mathbf{Z}$ assertion (iii) will be proven in §A3.8 and (iv) in §A3.9 of the Appendix.

Notice that $4s+2k-1$ becomes $3-2k$ at $s = 1-k$, and $0 \geq 3-2k \in 2\mathbf{Z}$ if $k \geq 3/2$. Since $\Gamma(s/2)L_N(s,\,\psi^2)$ is finite for $\mathrm{Re}(s) \leq 0$, we see that $L_N(m,\,\psi^2) = 0$ for every $m \in 2\mathbf{Z},\, \leq 0$. Thus $L_N(4s+2k-1,\,\psi^2) = 0$ at $s = 1-k$. For this reason, Theorem 8.15(iv) cannot be given as a statement on the value of $E_k(z,\,s;\,\Gamma,\,\psi)$ at $s = 1-k$. Similarly, Theorem 8.15(ii) cannot be stated in terms of $E_k(z,\,s;\,\Gamma,\,\psi)$ at $s = 1-k$.

Theorem 8.16. *Let the notation be as in Theorem 8.14 with* $k \notin \mathbf{Z}$; *suppose that* $k > 0$, λ *is even, and* ψ^2 *is trivial. Then* $F^*(z,\,s)$ *is nonzero at* $s = (\lambda+1)/2$, *and the following assertions hold:*

(i) *If* $k = 1/2$, *then* $\lambda = 0$, *and* $F^*(z,\,1/2)$ *belongs to* $\pi^{3/2}\mathscr{M}_{1/2}(\mathbf{Q}_{\mathrm{ab}})$. *More precisely,* $F^*(z,\,1/2) = \pi^{3/2}\sum_{\xi\in\mathbf{Q}}\mu(\xi)\mathbf{e}(t\xi^2 z/2)$ *with* $0 < t \in \mathbf{Q}$ *and a* $\mathbf{Q}_{\mathrm{ab}}$*-valued element* μ *of* $\mathscr{L}(\mathbf{Q})$.

(ii) *If* $k > 1/2$, *then* $k = 2p + 1/2$ *with* $0 < p \in \mathbf{Z}$, *and* $F^*\big(z,\,(\lambda+1)/2\big)$ *belongs to* $\pi^{3/2+p}\mathscr{N}_k^p(\mathbf{Q}_{\mathrm{ab}})$.

PROOF. Assertion (i) will be proven in §A3.10 of the Appendix. To prove (ii), we take the operator δ_k^p of (6.13d) and D_k^p of (7.11). Then we have

(8.20) $\quad D_k^p E_k(z,\,s;\,\Gamma,\,\psi) = (-4\pi)^{-p}\varepsilon_k(s)E_{k+2p}(z,\,s-p;\,\Gamma,\,\psi)$ with

$$\varepsilon_k(s) = \prod_{a=0}^{p-1}(s+k+a) = \Gamma(s+k+p)/\Gamma(s+k).$$

(This is true even for integral k.) Indeed, we can easily verify (by induction on p, for example) that $\delta_k^p y^s = (2i)^{-p}\varepsilon_k(s)y^{s-p}$. By (6.14c), $\delta_k^p(y^s\|_k\gamma) = (\delta_k^p y^s)\|_{k+2p}\gamma = (2i)^{-p}\varepsilon_k(s)y^{s-p}\|_{k+2p}\gamma$ for every $\gamma \in \Gamma$, and so we obtain (8.20) at least formally. To justify termwise differentiation of the infinite series of (8.12), we note a well-known principle on the validity of $(d/dx)\sum_{\gamma\in A}f_\gamma(x) = \sum_{\gamma\in A}df_\gamma/dx$. If it is applied to (8.12) p times, then we see that (8.20) holds at least for sufficiently large $\mathrm{Re}(s)$. Since δ_k^p commutes with $\partial/\partial\bar{s}$, we easily see that the left-hand side of (8.20) is meromorphic in s when $E_k(z, s)$ is defined, and the same is true with the right-hand side. This proves (8.20) in the domain of s in which both sides of (8.20) are meaningful as meromorphic functions. (There are alternative methods that are applicable to the case of many complex variables instead of a single z. We refer the reader to Lemma A in [S02, vol. III, p. 922].)

Now in the setting of (ii) put $-\lambda = 2p$. Then $0 < p \in \mathbf{Z}$ and $k = 2p + 1/2$, and so from (8.20) we obtain $(2i)^p \delta_{1/2}^p E_{1/2}(z, s+p) = \varepsilon_0 E_k(z, s)$ with $\varepsilon_0 = \Gamma(s+k)/\Gamma(s+p+1/2)$. Let F_k^* denote F^* of Theorem 8.14. Then, after verifying the cancellation of all gamma factors, we see that $F_k^*(z, s) = (2i)^p \delta_{1/2}^p F_{1/2}^*(z, s+p)$, and therefore $F_k^*\bigl(z, (\lambda+1)/2\bigr) = (2i)^p \delta_{1/2}^p F_{1/2}^*(z, 1/2)$ $= (-4\pi)^p D_{1/2}^p F_{1/2}^*(z, 1/2)$. By Lemma 7.8(i) we see that $F_k^*\bigl(z, (\lambda+1)/2\bigr)$ belongs to $\pi^{3/2+p}\mathscr{N}_k^p(\mathbf{Q}_{\mathrm{ab}})$. This proves (ii) and completes the proof.

Lemma 8.17. *Let $0 < m \in \mathbf{Z}$ and $f \in \mathscr{M}_k(M, N; \psi)$ in the notation of §8.6; suppose $2|M$ and $2|N$ if $k \notin \mathbf{Z}$. Then the following assertions hold.*

(i) *Put $g_1(z) = f(mz)$ and $g_2(z) = f(z/m)$. Then $g_1 \in \mathscr{M}_k(M, mN; \chi)$ and $g_2 \in \mathscr{M}_k(mM, N; \chi)$, where $\chi = \psi$ if $k \in \mathbf{Z}$ and $\chi(a) = \left(\dfrac{m}{a}\right)\psi(a)$ if $k \notin \mathbf{Z}$. Moreover, $g_2 \in \mathscr{M}_k(mM, N/m; \chi)$ if $2m|N$.*

(ii) *The map $f \mapsto f\|_k\iota$ is a bijection of $\mathscr{M}_k(M, N; \psi)$ onto $\mathscr{M}_k(N, M; \psi^{-1})$ and $(f\|_k\iota)\|_k\iota = (-1)^{[k]}f$.*

(iii) *For $f \in \mathscr{M}_k(2, K/2; \psi)$ with $k \notin \mathbf{Z}$ and $0 < K \in 4\mathbf{Z}$ define f^τ by $f^\tau(z) = f(-4/Kz)j_\iota^k(z)^{-1}$. Then $f \mapsto f^\tau$ is a bijection of $\mathscr{M}_k(2, K/2; \psi)$ onto $\mathscr{M}_k(2, K/2; \zeta)$, where $\zeta(a) = \left(\dfrac{K}{a}\right)\psi(a)^{-1}$ and $(f^\tau)^\tau = (-1)^{[k]}(K/4)^k f$.*

(iv) *The forms g_1, g_2, $f\|_k\iota$ and f^τ belong to $\mathscr{S}_k$ if $f \in \mathscr{S}_k$.*

Proof. For $\alpha = \begin{bmatrix} a & b \\ c & d \end{bmatrix} \in \Gamma(M, mN)$ with $d > 0$ put $\beta = \begin{bmatrix} a & mb \\ c/m & d \end{bmatrix}$. Then $\beta \in \Gamma(M, N)$ and $g_1(\alpha z) = f\bigl(\beta(mz)\bigr) = \psi(d)j_\beta^k(mz)f(mz)$. We have clearly $j_\beta^{[k]}(mz) = j_\alpha^{[k]}(z)$ and $h_\beta(mz) = \left(\dfrac{m}{d}\right)h_\alpha(z)$ by (4.40). Thus $g_1 \in \mathscr{M}_k(M, mN; \chi)$ in view of Lemma 5.4. As for g_2, taking $\begin{bmatrix} a & b/m \\ cm & d \end{bmatrix}$ in place of β, we obtain the desired results in the same way. To prove (ii), put $\delta = \iota^{-1}\gamma\iota$ for $\gamma \in \Gamma(M, N)$. Then $\delta \in \Gamma(N, M)$ and $a_\delta = d_\gamma$, and so (ii) can easily be verified. In the proof of (i) suppose $2m|N$; then we see that $g_2 \in \mathscr{M}_k(mM, N/m; \chi)$. Now, to prove (iii), take $M = 2$, $N = K/2$, and $m = K/4$. Then $g_2(z) = f(4z/K)$ and $f^\tau = g_2\|_k\iota$. Thus $g_2 \in \mathscr{M}_k(K/2, 2; \chi)$ with $\chi(a) = \left(\dfrac{K}{a}\right)\psi(a)$, and $f^\tau \in \mathscr{M}_k(2, K/2; \chi^{-1})$ by (ii). Now $(f^\tau)^\tau(z) =$

$f(z)j_\iota^k(-4/Kz)^{-1}j_\iota^k(z)^{-1}$. We easily see that $j_\iota^{[k]}(-4/Kz)j_\iota^{[k]}(z) = (-4/K)^{[k]}$ and $h_\iota(-4/Kz)h_\iota(z)$ is $(4/K)^{1/2}$ times an element t of $\mathbf{T}$. Taking $z = iy$ with $y > 0$, we find that $t = 1$, and so $j_\iota^k(-4/Kz)j_\iota^k(z) = (-1)^{[k]}(4/K)^k$, which completes the proof of (iii). Assertion (iv) follows immediately from the last statement of §5.8.

Lemma 8.18. *Let M, N, K be positive integers such that $K \in MN\mathbf{Z}$. Then $\Gamma(M, N)$ can be generated by* $\begin{bmatrix} 1 & M \\ 0 & 1 \end{bmatrix}$, $\begin{bmatrix} 1 & 0 \\ N & 1 \end{bmatrix}$, *and $\Gamma(K, K)$.*

PROOF. This is not so easy as it looks. We put $U = \prod_p SL_2(\mathbf{Z}_p)$, $E(M) = \prod_p E_p(M)$, $E'(N) = \prod_p E'_p(N)$, and $D(M, N) = \prod_p D_p(M, N)$, where p runs over all prime numbers and

$$E_p(M) = \left\{ \begin{bmatrix} 1 & b \\ 0 & 1 \end{bmatrix} \,\middle|\, b \in M\mathbf{Z}_p \right\}, \quad E'_p(N) = \left\{ \begin{bmatrix} 1 & 0 \\ c & 1 \end{bmatrix} \,\middle|\, c \in N\mathbf{Z}_p \right\},$$
$$D_p(M, N) = \left\{ \gamma \in SL_2(\mathbf{Z}_p) \,\middle|\, b_\gamma \in M\mathbf{Z}_p,\ c_\gamma \in N\mathbf{Z}_p \right\}.$$

Then $D(M, N)$ can be generated by $E(M)$, $E'(N)$, and $D(K, K)$. To prove this, we first note that $D_p(M, N) = D_p(K, K)$ if $p \nmid K$. Thus it is sufficient to show that $D_p(M, N)$ is generated by $E_p(M)$, $E'_p(N)$, and $D_p(K, K)$ if $p|K$. If $p \nmid MN$, then $D_p(M, N) = SL_2(\mathbf{Z}_p)$, and the fact is easy to verify, noting that

$$\begin{bmatrix} 0 & -1 \\ 1 & 0 \end{bmatrix} = \begin{bmatrix} 1 & -1 \\ 0 & 1 \end{bmatrix} \begin{bmatrix} 1 & 0 \\ 1 & 1 \end{bmatrix} \begin{bmatrix} 1 & -1 \\ 0 & 1 \end{bmatrix}.$$

Suppose $p|MN$ and let $\alpha = \begin{bmatrix} a & b \\ c & d \end{bmatrix} \in D_p(M, N)$. Then $a \in \mathbf{Z}_p^\times$ and

$$\begin{bmatrix} a & b \\ c & d \end{bmatrix} = \begin{bmatrix} 1 & 0 \\ a^{-1}c & 1 \end{bmatrix} \begin{bmatrix} a & 0 \\ 0 & a^{-1} \end{bmatrix} \begin{bmatrix} 1 & ba^{-1} \\ 0 & 1 \end{bmatrix},$$

and so we obtain the desired fact. Now put

$$H = \textstyle\prod_p H_p, \quad H_p = \left\{ w \in D_p(K, K) \,\middle|\, w - 1 \prec K\mathbf{Z}_p \right\}.$$

Then H is a normal subgroup of U. Let $\beta \in \Gamma(M, N)$. Since $\beta \in D(M, N)$, we have $\beta = u_1 \cdots u_m$ with $u_i \in E(M) \cup E'(N) \cup D(K, K)$. By strong approximation in $SL_2(\mathbf{Q})$ (see [S71, Lemmas 1.38 and 6.15] or [S10, Theorem 10.21], for example) we can put $u_i = \gamma_i v_i$ with $\gamma_i \in SL_2(\mathbf{Q})$ and $v_i \in H$. If $u_i \in E(M)$, then clearly we can take γ_i of the form $\gamma_i = \begin{bmatrix} 1 & Mb \\ 0 & 1 \end{bmatrix}$ with $b \in \mathbf{Z}$; similarly, if $u_i \in E'(N)$, then we can take $\gamma_i = \begin{bmatrix} 1 & 0 \\ Nc & 1 \end{bmatrix}$ with $c \in \mathbf{Z}$. If $u_i \in D(K, K)$, then $\gamma_i - u_i \prec K\mathbf{Z}_p$ for every $p|K$, and so $\gamma_i \in \Gamma(K, K)$. Put $\gamma = \gamma_1 \cdots \gamma_m$. Then we see that $\beta = \gamma z$ with $z \in H$, that is, $\gamma^{-1}\beta \in \Gamma(K, K)$. This proves our lemma.

This lemma can be generalized to the case of $Sp(n, F)$ with an arbitrary algebraic number field F; see [S93, Lemma 3b.4].

8.19. Let us add some technical and historical remarks on the Eisenstein series discussed in this section. Let $E_k(z, s)$ denote any function belonging to types (8.12), (8.14), and (8.15). If the series expressing $E_k(z, s)$ is absolutely convergent at $s = 0$, then obviously $E_k(z, 0)$ is a holomorphic function in z, which one can define without the parameter s. For example, if it is of type (8.12) with $k \in \mathbf{Z}$, the function is $\sum_{\gamma \in \Gamma_\infty \backslash \Gamma} \psi(d_\gamma) j_\gamma^k(z)^{-1}$, which is meaningful only for $k \geq 3$. In order to include the cases $k = 1$ and 2, Hecke introduced in [H27] the parameter s and proved that $E_k(z, s)$ can be continued analytically to a neighborhood of 0, and obtained the explicit forms of the Fourier expansions of the value at $s = 0$. Similar results in the case of half-integral weight were obtained by Maass. However, neither Hecke nor Maass investigated analytic continuation of $E_k(z, s)$ as a meromorphic function in s on the whole complex plane.

Analytic continuation of $E_k(z, s)$, especially in the case $k = 0$, was investigated by several researchers, Rankin [Ra39] for example. The proof of the fact alone is not difficult. However, it is important to show that $E_k(z, s)$ is slowly increasing at every cusp locally uniformly in s, which guarantees the convergence of the integral of the type (8.27) below, a highly nontrivial fact that almost all authors took for granted without proof. I found that the Fourier expansion of E_k involving both z and s was also important, and confluent hypergeometric functions were quite effective in obtaining such basic pieces of information. In fact, I investigated such Fourier expansions in [S75], and eventually similar expansions, as well as analytic continuation, of Eisenstein series on $\mathfrak{H}_n$. In §A2 of the Appendix we will give an exposition of confluent hypergeometric functions, and employing them, we will discuss in the next section various properties of $E_k(z, s)$ in the context of eigenforms of the operator L_k of (6.13c).

8.20. For $f \in \mathcal{M}_k$ with $k \in 2^{-1}\mathbf{Z}$ we put

$$f(z) = \sum_{0 \leq \xi \in \mathbf{Q}} \mu_f(\xi)\mathbf{e}(\xi z). \tag{8.21}$$

Let $f \in \mathcal{M}_k(K, M; \psi)$ and $g \in \mathcal{M}_\ell(K', M'; \varphi)$ with characters ψ and φ. We naturally assume (see §8.6)

$$\psi(-1) = (-1)^{[k]}, \qquad \varphi(-1) = (-1)^{[\ell]}. \tag{8.22}$$

We now define a Dirichlet series $D(s; f, g)$ by

$$D(s; f, g) = \sum_{0 < \xi \in \mathbf{Q}} \mu_f(\xi)\mu_g(\xi)\xi^{-s-(k+\ell)/2}. \tag{8.23}$$

By Lemma 6.2 the right-hand side is convergent for $\mathrm{Re}(s) > (k + \ell)/2 + 1$. Thus $D(s; f, g)$ is holomorphic for such s. We assume:

(8.24) *$k \geq \ell$ and f is a cusp form.*

To find an integral expression for (8.23), we first take g_ρ defined by (5.14) with $n = 1$ and g in place of f there, and observe that

$$f(z)\overline{g_\rho(z)} = \sum_{0<\xi}\sum_{0\le\eta} \mu_f(\xi)\mu_g(\eta)e^{2\pi i(\xi-\eta)x}e^{-2\pi(\xi+\eta)y}$$

for $z = x + iy$. We can find a positive integer N such that φ and ψ are characters modulo N, $f \in \mathscr{M}_k(N, N; \psi)$, and $g \in \mathscr{M}_\ell(N, N; \varphi)$. Then $f(z)$ and $g(z)$ depend only on z (mod $N\mathbf{Z}$), and both $\mu_f(\xi)$ and $\mu_g(\xi)$ are nonzero only when $\xi \in N^{-1}\mathbf{Z}$. Thus

$$\int_0^N f(z)\overline{g_\rho(z)}dx = N \sum_{0<\xi\in N^{-1}\mathbf{Z}} \mu_f(\xi)\mu_g(\xi)e^{-4\pi\xi y}.$$

Multiplying this by $y^{s+(k+\ell)/2-1}$ and employing (8.2), we find that

$$(8.25)\qquad \int_0^\infty\int_0^N f(z)\overline{g_\rho(z)}dx \cdot y^{s+(k+\ell)/2-1}dy = N(4\pi)^{-s-(k+\ell)/2}\Gamma\big(s+(k+\ell)/2\big)D(s;\, f,\, g)$$

for $\mathrm{Re}(s) > (k+\ell)/2 + 1$, since termwise integration is justified in view of the convergence of (8.23). Put $\Gamma = \Gamma(N, N)$, $\Gamma_\infty = P \cap \Gamma$, $\Psi = \Gamma_\infty\backslash\mathfrak{H}$, and $\Phi = \Gamma\backslash\mathfrak{H}$. Since $\{\, x + iy \,|\, 0 \le x < N\}$ represents Ψ, the left-hand side of (8.25) can be written $\int_\Psi f\overline{g_\rho}y^{s+(k+\ell)/2+1}\mathbf{d}z$. Let A be a complete set of representatives for $\Gamma_\infty\backslash\Gamma$. Then $\bigsqcup_{\alpha\in A}\alpha\Phi$ represents Ψ, and so (replacing $s + (k+\ell)/2 + 1$ by s)

$$\int_\Psi f(z)\overline{g_\rho(z)}y^s\mathbf{d}z = \sum_{\alpha\in A}\int_{\alpha\Phi} f(z)\overline{g_\rho(z)}y^s\mathbf{d}z = \int_\Phi\Big\{\sum_{\alpha\in A}(f\overline{g_\rho}y^s)\circ\alpha\Big\}\mathbf{d}z,$$

provided $\int_\Psi |f\overline{g_\rho}y^s|\mathbf{d}z < \infty$.

Taking g as f of (5.16), we see that $g_\rho \in \mathscr{M}_\ell(N, N; \bar\varphi)$. For $\alpha \in \Gamma$ we easily verify that

$$(f\overline{g_\rho}y^s)\circ\alpha = \omega(d_\alpha)\overline{j_\alpha^{k-\ell}(z)}^{\,-1}|j_\alpha(z)|^{2k-2s}f\overline{g_\rho}y^s \quad \text{with}$$

$$(8.26)\qquad \omega(d) = \begin{cases} (\psi\varphi)(d)\left(\dfrac{-1}{d}\right) & \text{if } k \in \mathbf{Z} \text{ and } \ell \notin \mathbf{Z}, \\ (\psi\varphi)(d) & \text{otherwise.}\end{cases}$$

Thus the last integral over Φ can be written

$$\int_\Phi f\overline{g_\rho}\cdot y^{s-k}\sum_{\alpha\in A}\omega(d_\alpha)\overline{j_\alpha^{k-\ell}(z)}^{\,-1}|j_\alpha(z)|^{2k-2s}y^k\mathbf{d}z = \int_\Phi f\overline{g_\rho}\overline{E_{k-\ell}(z,\,\bar s - k;\,\Gamma,\,\bar\omega)}y^k\mathbf{d}z$$

with $E_{k-\ell}$ of (8.12). Notice that $\omega(-1) = (-1)^{[k-\ell]}$ by (8.22). Substituting $s + (k+\ell)/2 + 1$ for s, we obtain

$$(8.27)\qquad N(4\pi)^{-s-(k+\ell)/2}\Gamma\big(s+(k+\ell)/2\big)D(s;\, f,\, g) = \int_\Phi f\overline{g_\rho}\overline{E_{k-\ell}(z,\,\bar s + (\ell-k)/2 + 1;\,\Gamma,\,\bar\omega)}y^k\mathbf{d}z$$

for $\mathrm{Re}(s) > (k+\ell)/2+1$.

The expression of a series of type (8.23) by an integral of type (8.27) was first given in [Ra39] by Rankin when $g = f_\rho$ and $k \in \mathbf{Z}$. Thus this technique may be called **Rankin's transformation.** We will later use this transformation at various places.

Let us now put

$$(8.28)\qquad \mathscr{D}_N(s;\, f,\, g) = D(s;\, f,\, g)\cdot \begin{cases} L_N(2s+2,\,\omega) & \text{if } k-\ell \in \mathbf{Z},\\ L_N(4s+3,\,\omega^2) & \text{if } k-\ell \notin \mathbf{Z},\end{cases}$$

$$(8.29)\quad \tilde{\Gamma}(s) = \Gamma\left(s+\frac{k+\ell}{2}\right)\Gamma\left(s+1+\frac{k-\ell}{2}\right)\cdot\begin{cases} 1 & \text{if } k-\ell\in\mathbf{Z},\\ \Gamma\left(s+\dfrac{3}{4}+\dfrac{\lambda_0}{2}\right) & \text{if } k-\ell\notin\mathbf{Z},\end{cases}$$

where λ_0 is 0 or 1 according as $[k-\ell]$ is even or odd.

Theorem 8.21. *Under (8.24) the product $\tilde{\Gamma}(s)\mathscr{D}_N(s;\, f,\, g)$ can be continued to the whole s-plane as a meromorphic function, which is holomorphic except for possible simple poles at the following points: $s=0$ only if $k=\ell$ and $\varphi\psi$ is trivial; $s=-1$ only if $k=\ell\in\mathbf{Z}$ and $N=1$; $s=-1/4$ only if $\psi^2\varphi^2$ is trivial and $k-\ell-1/2\in 2\mathbf{Z}$. The residue of $\mathscr{D}_N(s;\, f,\, g)$ at $s=0$ is $\pi^{2+k}\Gamma(k)^{-1}\langle g_\rho,\, f\rangle r_0$, where r_0 is a positive rational number that depends on the choice of the levels of f and g.*

PROOF. Suppose $k-\ell\in\mathbf{Z}$; then by Theorem 8.12, $s(s-1)\Gamma(s+k-\ell)E_{k-\ell}^N(z,\, s;\, \omega)$ is entire in s; moreover, for each fixed s the product as a function of z is slowly increasing at every cusp locally uniformly in s. Since f is a cusp form, we see that

$$s(s+1)L_N(2s+2,\,\omega)\Gamma\bigl(s+1+(k-\ell)/2\bigr)$$

times the right-hand side of (8.27) is meaningful for the reason explained in §6.5; $s(s+1)$ is unnecessary if $k\neq\ell$ or ω is nontrivial. From the reasoning there we also see that the integral is convergent locally uniformly in s. Suppose $k=\ell$ and ω is trivial; then a pole may occur at $s=0$ and $s=-1$. By Theorem 8.12 we find that the residue of $\mathscr{D}_N(s;\, f,\, g)$ at $s=0$ is

$$N^{-1}\Gamma(k)^{-1}(4\pi)^k\mu(\Phi)\langle g_\rho,\, f\rangle\cdot 2^{-1}\pi N^{-2}\varphi(N).$$

By (6.6), $\mu(\Phi)\in\pi\mathbf{Q}$, and so we obtain the residue as stated in our theorem.

Next suppose $k-\ell\notin\mathbf{Z}$. We apply Theorem 8.14 to $E_{k-\ell}(\cdots)$ in (8.27). Define κ, λ, and $F^*(z,\, s)$ as in that theorem. Then we find that $\kappa=2k-2\ell$, $\lambda=-[k-\ell]$, and $F^*\bigl(z,\, s+(\ell-k)/2+1\bigr)$ equals

$$(2s+1/2)L_N(4s+3,\omega^2)\Gamma\left(s+\tfrac{k-\ell}{2}\right)\Gamma\left(s+\tfrac{3}{4}+\tfrac{\lambda_0}{2}\right)E_{k-\ell}\left(z,s+\tfrac{\ell-k}{2}+1;\Gamma,\bar{\omega}\right).$$

Then we obtain the results in this case as stated in our theorem. This completes the proof.

If f and g of §8.20 are eigenfunctions of Hecke operators of integral weight, then $D(s;\, f,\, g)$ has an Euler product that can be given as follows.

Lemma 8.22. *Let* $f(z) = \sum_{n=1}^{\infty} a(n)\mathbf{e}(nz) \in \mathscr{S}_k(N,\, \psi)$ *and* $g(z) = \sum_{n=0}^{\infty} b(n)\mathbf{e}(nz) \in \mathscr{M}_\ell(M,\, \varphi)$ *with* $k,\, \ell \in \mathbf{Z}$. *Suppose that these are normalized eigenfunctions of Hecke operators in the sense that we can put*

$$\sum_{n=1}^{\infty} a(n)n^{-s} = \prod_p \left[1 - a(p)p^{-s} + \psi(p)p^{k-1-s}\right]^{-1},$$

$$\sum_{n=1}^{\infty} b(n)n^{-s} = \prod_p \left[1 - b(p)p^{-s} + \varphi(p)p^{\ell-1-s}\right]^{-1},$$

where $\prod_p$ *means the product over all the prime numbers* p *(see [S71,Theorem 3.43]). Taking an indeterminate* X, *put, for each prime number* p,

$$X^2 - a(p)X + \psi(p)p^{k-1} = (X - \alpha_p)(X - \beta_p),$$
$$X^2 - b(p)X + \varphi(p)p^{\ell-1} = (X - \gamma_p)(X - \delta_p)$$

with complex numbers $\alpha_p,\, \beta_p,\, \gamma_p,\, \delta_p$. *Then*

$$L_{NM}(2s + 2 - k - \ell,\, \psi\varphi) \sum_{n=1}^{\infty} a(n)b(n)n^{-s}$$
$$= \prod_p \left[(1 - \alpha_p\gamma_p p^{-s})(1 - \alpha_p\delta_p p^{-s})(1 - \beta_p\gamma_p p^{-s})(1 - \beta_p\delta_p p^{-s})\right]^{-1}.$$

Proof. We have $\sum_{n=0}^{\infty} a(p^n)X^n = \left(1 - a(p)X + \psi(p)p^{k-1}X^2\right)^{-1}$, and so

$$\sum_{n=0}^{\infty} (\alpha_p^n - \beta_p^n)X^n = (1 - \alpha_p X)^{-1} - (1 - \beta_p X)^{-1}$$
$$= (\alpha_p - \beta_p)X\left[(1 - \alpha_p X)(1 - \beta_p X)\right]^{-1} = (\alpha_p - \beta_p)\sum_{n=0}^{\infty} a(p^n)X^{n+1}.$$

Thus we obtain $a(p^n) = (\alpha_p^{n+1} - \beta_p^{n+1})/(\alpha_p - \beta_p)$ if $\alpha_p \neq \beta_p$. Similarly, $b(p^n) = (\gamma_p^{n+1} - \delta_p^{n+1})/(\gamma_p - \delta_p)$ if $\gamma_p \neq \delta_p$. Now $\sum_{m=1}^{\infty} a(m)b(m)m^{-s} = \prod_p \left\{\sum_{n=0}^{\infty} a(p^n)b(p^n)p^{-ns}\right\}$ and

$$(\alpha_p - \beta_p)(\gamma_p - \delta_p)\sum_{n=0}^{\infty} a(p^n)b(p^n)X^n = \sum_{n=0}^{\infty} (\alpha_p^{n+1} - \beta_p^{n+1})(\gamma_p^{n+1} - \delta_p^{n+1})X^n$$
$$= \frac{\alpha_p\gamma_p}{1 - \alpha_p\gamma_p X} - \frac{\alpha_p\delta_p}{1 - \alpha_p\delta_p X} - \frac{\beta_p\gamma_p}{1 - \beta_p\gamma_p X} + \frac{\beta_p\delta_p}{1 - \beta_p\delta_p X}$$
$$= \frac{(\alpha_p - \beta_p)(\gamma_p - \delta_p)(1 - \alpha_p\beta_p\gamma_p\delta_p X^2)}{(1 - \alpha_p\gamma_p X)(1 - \alpha_p\delta_p X)(1 - \beta_p\gamma_p X)(1 - \beta_p\delta_p X)}.$$

This gives the desired result when $\alpha_p \neq \beta_p$ and $\gamma_p \neq \delta_p$. However, we have $a(p^n) = \sum_{i=0}^n \alpha_p^{n-i}\beta_p^i$ and $b(p^n) = \sum_{i=0}^n \gamma_p^{n-i}\delta_p^i$ unconditionally, and so our result is a formula for $\sum_{n=0}^\infty \bigl(\sum_{i=0}^n \alpha_p^{n-i}\beta_p^i\bigr)\bigl(\sum_{i=0}^n \gamma_p^{n-i}\delta_p^i\bigr)X^n$, which is valid even if $\alpha_p \neq \beta_p$ or $\gamma_p \neq \delta_p$. This completes the proof.

Theorem 8.23. *Given* $f(z) = \sum_{n=1}^\infty a(n)\mathbf{e}(nz) \in \mathscr{S}_k(N,\,\psi)$ *and a primitive or an imprimitive Dirichlet character* χ, *put*

$$\mathscr{D}(s;\,f,\,\chi) = L_N(2s-2k+2,\,\chi^2\psi^2)\sum_{n=1}^\infty \chi(n)a(n^2)n^{-s} \tag{8.30}$$

and $\chi(-1) = (-1)^\mu$ *with* $\mu = 0$ *or* 1. *Put also* $\tilde{\Gamma}(s) = \Gamma(s/2)\Gamma\bigl((s+1)/2\bigr)\Gamma\bigl((s-k-1+\lambda_0)/2\bigr)$, *where* $\lambda_0 = 0$ *or* 1 *according as* $k-\mu-1$ *is even or odd. Then* $\tilde{\Gamma}(s)\mathscr{D}(s;\,f,\chi)$ *can be continued to a meromorphic function on the whole complex plane, which is holomorphic except for a possible simple pole at* $s = k$, *that occurs only if* $\psi^2\chi^2$ *is trivial and* $k-\mu-1 \in 2\mathbf{Z}$.

PROOF. Put $\theta_1(z) = 2^{-1}\sum_{m\in\mathbf{Z}}\chi(m)m^\mu\mathbf{e}(m^2z)$. Then $\theta_1(z) = \theta_\chi(2z)$ with θ_χ of (5.7). Suppose χ is defined modulo r; then by Lemmas 5.5 and 8.17, $\theta_1 \in \mathscr{M}_\ell(2,\,4r^2;\chi_1)$, where $\ell = \mu + 1/2$ and $\chi_1(a) = \left(\dfrac{8}{a}\right)\chi(a)$. We then see that

$$D(s;\,f,\,\theta_1) = \sum_{m=1}^\infty \chi(m)a(m^2)m^{-2s-k-1/2},$$

and so

$$\mathscr{D}(s;\,f,\chi) = L(2s-2k+2,\,\chi^2\psi^2)D(s/2-k/2-1/4;\,f,\,\theta_1).$$

Comparing this with (8.28), we find that

$$\mathscr{D}(s;\,f,\chi) = \mathscr{D}_M(s/2-k/2-1/4;\,f,\,\theta_1),$$

where M is the least common multiple of N and $4r^2$. Thus we obtain the desired result from Theorem 8.21.

The above theorem was essentially given in [S75, Theorem 1], which stated that another pole at $s = k-1$ might occur, but that is not the case as shown here. The pole at $s = k$ can indeed happen; see Theorem 2 and the discussion on page 97 of that paper. Notice that if $k-\mu-1 \in 2\mathbf{Z}$, then $(\psi\chi)(-1) = -1$, and so $\psi\chi$ is nontrivial.

Lemma 8.24. *Suppose that* f *of Theorem 8.23 is a normalized Hecke eigenform as in Lemma 8.22; define* α_p *and* β_p *as in that lemma. Then*

$$\mathscr{D}(s;\,f,\,\chi) = \prod_p \bigl[(1-\chi(p)\alpha_p^2p^{-s})(1-\chi(p)\alpha_p\beta_pp^{-s})(1-\chi(p)\beta_p^2p^{-s})\bigr]^{-1}.$$

PROOF. We have $\sum_{n=1}^{\infty} \chi(n)a(n^2)n^{-s} = \prod_p \{ \sum_{m=0}^{\infty} \chi(p)^m a(p^{2m})p^{-2ms} \}$, and

$$(\alpha_p - \beta_p) \sum_{m=0}^{\infty} a(p^{2m})X^m = \sum_{m=0}^{\infty} (\alpha_p^{2m+1} - \beta_p^{2m+1})X^m$$
$$= \frac{\alpha_p}{1 - \alpha_p^2 X} - \frac{\beta_p}{1 - \beta_p^2 X} = \frac{(\alpha_p - \beta_p)(1 + \alpha_p\beta_p X)}{(1 - \alpha_p^2 X)(1 - \beta_p^2 X)}$$
$$= (\alpha_p - \beta_p)(1 - \alpha_p^2\beta_p^2 X^2)\{(1 - \alpha_p^2 X)(1 - \alpha_p\beta_p X)(1 - \beta_p^2 X)\}^{-1}.$$

Thus we obtain the desired equality. The result is valid even if $\alpha_p = \beta_p$ for the reason explained at the end of the proof of Lemma 8.22.

Comparing this with the case $f = g$ in Lemma 8.22, we obtain

$$\text{(8.31)} \quad L(s - k + 1, \chi\psi)\mathscr{D}(s; f, \chi) = L(2s - 2k + 2, \chi^2\psi^2) \sum_{n=1}^{\infty} \chi(n)a(n)^2 n^{-s}$$
$$= L(2s - 2k + 2, \chi^2\psi^2)D(s - k; f, h),$$

$$\text{(8.32)} \qquad L(s - k + 1, \chi\psi) \sum_{n=1}^{\infty} \chi(n)a(n^2)n^{-s} = \sum_{n=1}^{\infty} \chi(n)a(n)^2 n^{-s},$$

where $h(z) = \sum_{n=1}^{\infty} \chi(n)a(n)\mathbf{e}(nz)$. By Lemma 7.13, $h \in \mathscr{S}_k(M, \psi\chi^2)$ with a multiple M of N, and so the analytic continuation of $D(s, f, h)$ follows from Theorem 8.21. Therefore we can derive the analytic continuation of $\mathscr{D}(s; f, \chi)$ also by combining this with (8.31), but this gives a weaker result than Theorem 8.23, because of the factor $L(s - k + 1, \chi\psi)$.

9. Eisenstein series as automorphic eigenforms

9.1. Given a congruence subgroup Γ of $\Gamma(1)$ and a weight k (not necessarily ≥ 0) we consider a C^∞ function f on $\mathfrak{H}$ satisfying the following three conditions:

(9.1a) $\qquad f\|_k \gamma = f$ *for every* $\gamma \in \Gamma$;

(9.1b) $\qquad L_k f = \lambda f$ *with* $\lambda \in \mathbf{C}$, *where* L_k *is as in (6.13c);*

(9.1c) $\qquad f$ *is slowly increasing at every cusp, that is,* f *satisfies (6.9a).*

Such an f is called an **automorphic eigenform,** or simply, an **eigenform** of L_k belonging to the **eigenvalue** λ. It is also called a **Maass form**, as Maass introduced this type of function and made some fundamental contributions in [Ma49] and [Ma53]. We denote by $\mathfrak{A}_k(\Gamma, \lambda)$ the set of all such functions f. We naturally assume that $\Gamma \subset \Gamma^\theta$ if $k \notin \mathbf{Z}$. We put

$$\text{(9.2)} \qquad \mathfrak{A}_k(\lambda) = \bigcup_{N=1}^{\infty} \mathfrak{A}_k(\Gamma(2N), \lambda).$$

By (6.14d), for every $f \in \mathfrak{A}_k(\lambda)$ and $\alpha \in SL_2(\mathbf{Q})$, we see that $j_\alpha(z)^{-k} f(\alpha z) \in \mathfrak{A}_k(\lambda)$. This is clear if $k \in \mathbf{Z}$, but if $k \notin \mathbf{Z}$, then we have to invoke Theorem 4.7(4); cf. Lemma 5.4. If f is holomorphic, then $L_k f = 4\delta_{k-2}\varepsilon f = 0$, since $\varepsilon = -y^2\partial/\partial\bar{z}$, and so from Lemma 6.4(i) we see that

(9.3) *$\mathscr{M}_k$ consists of the holomorphic elements of $\mathfrak{A}_k(0)$.*

As the title of this section indicates, we are mainly interested in the significance of Eisenstein series among eigenforms in general; we will not investigate much about the so-called cusp eigenforms, though we will define them and prove a few of their elementary properties.

If $f \in \mathfrak{A}_k(\Gamma, \lambda)$, we have $f(z + b) = f(z)$ for $b \in N\mathbf{Z}$ for some positive integer N, for the same reason as in (3.6a). Therefore, putting $r = N^{-1}$, we see that $f(x + iy)$ as a function of x has a Fourier expansion

$$f(x+iy) = \sum_{h \in r\mathbf{Z}} c_h(y)\mathbf{e}(hx) \tag{9.4}$$

with C^∞ functions $c_h(y)$ of y. It should be noted that h may be negative. Moreover, termwise partial differentiation is valid (see [S07, §A2]), and so

$$(\partial/\partial x)^a(\partial/\partial y)^b f(x+iy) = \sum_{h \in r\mathbf{Z}} (2\pi i h)^a (\partial/\partial y)^b c_h(y)\mathbf{e}(hx)$$

for every a and b. Therefore, applying L_k to (9.4), we find that c_h is a solution of the differential equation

$$\big(y^2(d/dy)^2 + ky \cdot d/dy - 4\pi^2 h^2 y^2 + 2\pi hky + \lambda\big)c(y) = 0. \tag{9.5}$$

If $h \neq 0$, the solutions of this equation are given by Whittaker functions, which we use in the form

$$V(y;\, \alpha,\, \beta) = e^{-y/2}\,\Gamma(\beta)^{-1}\, y^\beta \int_0^\infty e^{-yt}(1+t)^{\alpha-1}t^{\beta-1}dt, \tag{9.6}$$

where $0 < y \in \mathbf{R}$ and $(\alpha, \beta) \in \mathbf{C}^2$. The last integral is convergent for $\mathrm{Re}(\beta) > 0$. In fact, $V(y;\, \alpha,\, \beta)$ can be defined as a holomorphic function of (α, β) on the whole $\mathbf{C}^2$. In §A2 of the Appendix we give an exposition of some basic facts on this function.

Given k and $\lambda \in \mathbf{C}$, we take $(\alpha, \beta) \in \mathbf{C}^2$ so that

$$k = \alpha - \beta, \qquad \lambda = \beta(1-\alpha), \tag{9.7}$$

and define a function $W_k(t, \lambda)$ for $t \in \mathbf{R}^\times$ by

$$W_k(t,\, \lambda) = \begin{cases} V(4\pi t;\, \alpha,\, \beta) & \text{if } \ t > 0, \\ |4\pi t|^{-k} V(|4\pi t|;\, \beta,\, \alpha) & \text{if } \ t < 0. \end{cases} \tag{9.8}$$

If (α, β) is a solution of (9.7), then the other solution is $(1-\beta, 1-\alpha)$ (which may be equal to (α, β)), but $W_k(t, \lambda)$ is determined by k and λ, since $V(y;\, 1-\beta,\, 1-\alpha) = V(y;\, \alpha,\, \beta)$, as will be shown in Lemma A2.2 of the Appendix.

Returning to $c_h(y)$ of (9.4), we have

$$c_h(y) = r\int_0^{1/r} f(x+iy)\mathbf{e}(-hx)dx, \tag{9.9}$$

and so (9.1c) implies that $c_h(y) = O(y^B)$ as $y \to \infty$ with $B \in \mathbf{R}$. By Lemma A2.4 of the Appendix every solution c of (9.5) such that $c(y) = O(y^B)$ as $y \to \infty$ is a constant times $W_k(hy,\,\lambda)$, and vice versa. Thus $c_h(y) = b_h W_k(hy,\,\lambda)$ with $b_h \in \mathbf{C}$, and so

$$f(x+iy) = b_0(y) + \sum_{0\neq h\in r\mathbf{Z}} b_h W_k(hy,\,\lambda)\mathbf{e}(hx) \tag{9.10}$$

with a C^∞ function b_0. We call this the **Fourier expansion** of f, and b_0 the **constant term** of f. (The word "constant" is used with respect to the variable x, and $b_0(y)$ may involve y nontrivially.) Let $\alpha \in SL_2(\mathbf{Q})$. Then $j_\alpha(z)^{-k}f(\alpha z)$ belongs to $\mathfrak{A}_k(\lambda)$, and so has a Fourier expansion of the type (9.10). We call f a **cusp form** if the constant term of $j_\alpha(z)^{-k}f(\alpha z)$ is 0 for every $\alpha \in SL_2(\mathbf{Q})$, and we denote by $\mathfrak{S}_k(\lambda)$ the set of cusp forms of $\mathfrak{A}_k(\lambda)$, and put $\mathfrak{S}_k(\Gamma,\,\lambda) = \mathfrak{A}_k(\Gamma,\,\lambda) \cap \mathfrak{S}_k(\lambda)$.

Taking $h = 0$ in (9.5), we obtain a differential equation

$$y^2b_0'' + kyb_0' + \lambda b_0 = 0, \tag{9.11}$$

which is easy to solve. Indeed, let $X^2 + (k-1)X + \lambda = (X-p)(X-q)$ with $p,\,q \in \mathbf{C}$, and $p \neq q$ if and only if $4\lambda \neq (k-1)^2$; if $4\lambda = (k-1)^2$, then $X^2 + (k-1)X + \lambda = (X-p)^2$ with $p = (1-k)/2$. With these p and q equation (9.11) has linearly independent solutions as follows:

(9.12a) y^p and y^q with $p+q = 1-k$ and $pq = \lambda$ if $4\lambda \neq (k-1)^2$,

(9.12b) y^p and $y^p \log y$ with $p = (1-k)/2$ if $4\lambda = (k-1)^2$.

Thus b_0 is a **C**-linear combination of these two functions in each case. We call the eigenvalue λ **critical** in the latter case.

Lemma 9.2. *For f, b_0, and b_h as in (9.10), the following assertions hold:*

(i) *There exist two positive constants M and m such that $|b_h| \leq M|h|^{k/2+m}$ for every $h \in r\mathbf{Z}$, $\neq 0$. Moreover, we can take $m = 0$ if f is a cusp form.*

(ii) *There exist positive constants A, B, and C such that*

$$y^{k/2}\sum_{0\neq h\in r\mathbf{Z}} \bigl|b_h W_k(hy,\,\lambda)\bigr| \leq Ae^{-By} \quad \textit{if} \quad y \geq C.$$

(iii) *$\mathfrak{S}_k(\lambda)$ consists of all the elements of $\mathfrak{A}_k(\lambda)$ that are rapidly decreasing at every cusp.*

(iv) *$\mathfrak{S}_k(0) = \mathscr{S}_k$.*

PROOF. We have

$$b_h W_k(hy,\,\lambda) = r\int_0^{1/r} f(x+iy)\mathbf{e}(-hx)dx,$$

and so by Lemma 6.4(iv) we have

$$\left|y^{k/2}b_h W_k(hy,\,\lambda)\right| \le A(y^c + y^{-c}) \tag{9.13}$$

with positive constants A and c independent of h. In Lemma A2.5 of the Appendix we will show that $\lim_{y\to\infty} e^{y/2}V(y;\,\alpha,\,\beta) = 1$. Thus we can find a constant $d > 1$ such that $|V(y;\,\alpha,\,\beta)| \ge 2^{-1}e^{-y/2}$ and $|V(y;\,\beta,\,\alpha)| \ge 2^{-1}e^{-y/2}$ if $y > d$, and so by (9.8) we obtain

$$|W_k(hy,\,\lambda)| \ge \begin{cases} 2^{-1}e^{-2\pi hy} & \text{if } h > 0 \text{ and } 4\pi hy > d, \\ 2^{-1}e^{-2\pi|h|y}|4\pi hy|^{-k} & \text{if } h < 0 \text{ and } 4\pi|h|y > d. \end{cases} \tag{9.14}$$

Dividing (9.13) by this, we can find a positive constant M independent of h such that

$$|b_h| \le Me^{2\pi|h|y}y^{-k/2}(y^c + y^{-c}) \cdot \begin{cases} 1 & \text{if } h > 0 \text{ and } 4\pi hy > d, \\ |hy|^k & \text{if } h < 0 \text{ and } 4\pi|h|y > d. \end{cases}$$

Taking $y = d/|2\pi h|$, we obtain $|b_h| \le M'e^d|h|^{k/2+c}$ with a constant M' independent of h. This proves the first part of (i).

By Lemma A2.2(i) of the Appendix we have, for any positive number y_0,

$$|V(y;\,\alpha,\,\beta)| + |V(y;\,\beta,\,\alpha)| \le A_1e^{-y/2} \quad \text{for } y > y_0 \tag{9.15}$$

with a positive constant A_1. Combining this with (i), we obtain

$$y^{k/2}\sum_{0\ne h\in r\mathbf{Z}} \left|b_h W_k(hy,\,\lambda)\right| \le A_2y^{k/2}\sum_{0\ne h\in r\mathbf{Z}} |h|^{k/2+m}e^{-2\pi|h|y}.$$

for $y > 1/2$ with a constant $A_2 > 0$. Put $|h| = rn$ with $0 < n \in \mathbf{Z}$. Then the last sum is majorized by $2r^ae^{-\pi ry}\sum_{n=1}^{\infty} n^ae^{-\pi rny}$ with an integer $a \ge k/2+m$. We have $\sum_{n=1}^{\infty} n^ax^n = xP_a(x)/(1-x)^{a+1}$ with a polynomial P_a of degree $a-1$, and so we obtain the estimate of (ii).

If $f \in \mathfrak{A}_k(\lambda)$, then (ii) is applicable to $f\|_k\alpha$ for every $\alpha \in SL_2(\mathbf{Q})$. If $f \in \mathfrak{S}_k(\lambda)$ in particular, (ii) implies that f is rapidly decreasing at every cusp. Conversely, if an element f of $\mathfrak{A}_k(\lambda)$ is rapidly decreasing at every cusp, then from (9.9) with $h = 0$ we see that $\lim_{y\to\infty} y^cb_0(y) = 0$ for every $c \in \mathbf{R}$, and so $b_0 = 0$. This is so for $f\|_k\alpha$ in place of f for every $\alpha \in SL_2(\mathbf{Q})$. Thus $f \in \mathfrak{S}_k(\lambda)$. This proves (iii).

Returning to (i), suppose f is a cusp form; then Lemma 6.4(v) combined with (iii) shows that $|y^{k/2}f|$ is bounded on $\mathfrak{H}$, and so we can take $c = 0$ in (9.13). Thus $|b_h| \le M|h|^{k/2}$, which proves the last part of (i).

As for (iv), we have $\mathscr{S}_k \subset \mathfrak{S}_k(0)$ by (9.3) combined with (iii). Let $f \in \mathfrak{S}_k(0)$. As will be shown in Lemma 9.3 below, εf is rapidly decreasing at every cusp. Therefore, by Corollary 6.9(ii), $f \in \mathscr{S}_k$. This proves (iv) and completes the proof.

Remark. For $h > 0$ and $\lambda = 0$ we have $W_k(hy,\,0) = V(4\pi hy;\,k,\,0) = e^{-2\pi hy}$ by (A2.3) of the Appendix, and so $W_k(hy,\,0)\mathbf{e}(hx) = \mathbf{e}(hz)$. Thus the Fourier expansion of an element of $\mathscr{M}_k$ is a special case of (9.10), and (i) of the above lemma includes Lemma 6.2(ii, iii) as special cases.

Lemma 9.3. *Let* ε *and* δ_k *be as in (6.13a, b). Then the following assertions hold:*

(i) $$\varepsilon\mathfrak{A}_k(\Gamma, \lambda) \subset \mathfrak{A}_{k-2}(\Gamma_*, \lambda - k + 2).$$

(ii) $$\delta_k\mathfrak{A}_k(\Gamma, \lambda) \subset \mathfrak{A}_{k+2}(\Gamma^*, \lambda + k).$$

(iii) *These inclusions are true with* $\mathfrak{S}$ *in place of* $\mathfrak{A}$.

(iv) $\mathfrak{S}_k(\Gamma, \lambda)$ *contains a nonholomorphic function only if* $0 < \lambda \in \mathbf{R}$.

For the moment we have $\Gamma_* = \Gamma^* = \Gamma$. In §9.5 we will generalize the notion of A congruence subgroup, and explain the meaning of Γ_* and Γ^*.

Proof. Let $f \in \mathfrak{A}_k(\Gamma, \lambda)$. Then, from (6.14a, b, e) we easily see that εf and $\delta_k f$ satisfy (9.1a, b) with k and λ modified as in (i) and (ii). Thus our task is to show that they are slowly increasing at every cusp. We see that $W_k(hy, \lambda)\mathbf{e}(hx)$ equals $\varphi_A(x + iy; k, \lambda)$ of (A2.9) of the Appendix with $A = 2\pi h$, and so from (A2.10) and (A2.11) of the Appendix we obtain

$$\varepsilon\{W_k(hy, \lambda)\mathbf{e}(hx)\} = W_{k-2}(hy, \lambda + 2 - k)\mathbf{e}(hx) \cdot \begin{cases} (8\pi ih)^{-1}\lambda & \text{if } h > 0, \\ (8\pi ih)^{-1} & \text{if } h < 0, \end{cases}$$

$$\delta_k\{W_k(hy, \lambda)\mathbf{e}(hx)\} = W_{k+2}(hy, \lambda + k)\mathbf{e}(hx) \cdot \begin{cases} 2\pi ih & \text{if } h > 0, \\ 2\pi ih(\lambda + k) & \text{if } h < 0. \end{cases}$$

Therefore, if f is as in (9.10), then

$$\varepsilon f = \varepsilon b_0 + \sum_{h \neq 0} c_h W_{k-2}(hy, \lambda + 2 - k)\mathbf{e}(hx)$$

with $c_h = (8\pi ih)^{-1}\lambda b_h$ if $h > 0$ and $c_h = (8\pi ih)^{-1}b_h$ if $h < 0$. By Lemma 9.2(i), $|c_h| = M'|h|^{(k-2)/2+m}$ with positive constants M' and m. Since the technique of the proof of Lemma 9.2(ii) is applicable to $\varepsilon f - \varepsilon b_0$, we have $y^{(k-2)/2}|\varepsilon f - \varepsilon b_0| = O(e^{-By})$ as $y \to \infty$ with some $B > 0$. Observe that εb_0 is a function of the same type as b_0 with $(k-2, \lambda+2-k)$ in place of (k, λ). These are applicable to $\varepsilon(f\|_k\alpha)$ with any $\alpha \in SL_2(\mathbf{Q})$. Since $\varepsilon(f\|_k\alpha) = (\varepsilon f)\|_{k-2}\,\alpha$ by (6.14a), we see that εf is slowly increasing at every cusp. Thus $\varepsilon f \in \mathfrak{A}_{k-2}(\Gamma, \lambda - k + 2)$, which is (i). The proof of (ii) can be given in a similar way. Suppose $f \in \mathfrak{S}_k(\Gamma, \lambda)$. Then $b_0 = 0$, and so $|y^{(k-2)/2}(\varepsilon f)\|_{k-2}\,\alpha| = O(e^{-By})$ as $y \to \infty$. Thus εf is rapidly decreasing at every cusp, and the same is true with $\delta_k f$. This proves (iii). To prove (iv), let $0 \neq f \in \mathfrak{S}_k(\Gamma, \lambda)$. Then, by (6.19), $\lambda\langle f, f\rangle = \langle f, L_k f\rangle = \langle L_k f, f\rangle = \bar{\lambda}\langle f, f\rangle$; also, $\langle f, L_k f\rangle \geq 0$ by (6.20). Therefore, in view of Lemma 9.2(iv) we obtain (iv). This completes the proof.

Returning to Corollary 6.9(ii), we see that the assumption on εf is unnecessary in view of (iii) of the above lemma.

Theorem 9.4. *The vector space* $\mathfrak{A}_k(\Gamma, \lambda)$ *is finite-dimensional.*

PROOF. We first note that given $0 < r < 1$ and two positive integers a and p, we have

$$\sum_{m=p}^{\infty} m^a x^m \le C_{a,r} p^a x^p \quad \text{for } 0 \le x \le r \tag{9.16}$$

with a constant $C_{a,r}$ independent of p and x. Indeed,

$$\sum_{m=p}^{\infty} m^a x^{m-p} = \sum_{n=0}^{\infty} (n+p)^a x^n \le \sum_{i=0}^{a} \binom{a}{i} p^{a-i} \sum_{n=0}^{\infty} n^i r^n$$

for $0 \le x \le r$, which proves (9.16). Next, take a finite subset X of $\Gamma(1)$ such that $\mathfrak{H} = \bigcup \Gamma X T$ with $T = \{x + iy \in \mathbf{C} \mid |x| \le 1/2,\ y > 1/2\}$, as we did in the proof of Lemma 6.4. By Lemma 9.2(iii) and Lemma 6.4(v), $|y^{k/2} f|$ is bounded on $\mathfrak{H}$. Given $f \in \mathfrak{S}_k(\Gamma, \lambda)$ and $\xi \in X$, put $M_f = \mathrm{Max}_{z \in \mathfrak{H}} |y^{k/2} f(z)|$ and $f\|_k \xi = \sum_{0 \ne h \in r\mathbf{Z}} b_{h,\xi} W_k(hy, \lambda)\mathbf{e}(hx)$ with the same r for all $\xi \in X$ and all $f \in \mathfrak{S}_k(\Gamma, \lambda)$. Since $|y^{k/2}(f\|_k\xi)| = |(y^{k/2} f) \circ \xi| \le M_f$, from (9.9) we obtain $y^{k/2}|b_{h,\xi} W_k(hy, \lambda)| \le M_f$. Dividing this by (9.14) and employing the technique of a few lines below (9.14), we find that $|b_{h,\xi}| \le BM_f |h|^{k/2}$ with a constant B independent of h, ξ, and f. Fix an integer $p > 1$ and suppose $b_{h,\xi} = 0$ for all $\xi \in X$ and all h such that $|h| < rp$. Then by (9.15),

$$|y^{k/2}(f\|_k\xi)| \le y^{k/2} \sum_{|h| \ge p} |b_{h,\xi} W_k(hy, \lambda)| \le DM_f \sum_{|h| \ge rp} |h|^{k/2} e^{-2\pi|h|y}$$

for $y > 1/2$ with a constant D independent of f and ξ. The last sum is majorized by $2r^a \sum_{m=p}^{\infty} m^a e^{-2\pi rmy}$ with a positive integer $a \ge k/2$. By (9.16) this is $\le 2r^a C p^a e^{-2\pi rpy}$ for $y > 1/2$ with some C independent of p. Thus

$$|y^{k/2}(f\|_k\xi)| \le 2r^a CDM_f p^a e^{-2\pi rpy} \quad \text{for } y > 1/2. \tag{9.17}$$

Now, given $z \in \mathfrak{H}$, take $\gamma \in \Gamma$, $\xi \in X$, and $w \in T$ so that $z = \gamma\xi w$. Then $|y^{k/2} f(z)| = |y^{k/2} f(\gamma\xi w)| = |v^{k/2}(f\|_k\xi)(w)|$, where $v = \mathrm{Im}(w)$. Therefore, by (9.17) we have

$$|y^{k/2} f(z)| \le 2r^a CDM_f p^a e^{-2\pi rpv} \le 2r^a C_{a,1} DM_f p^a e^{-\pi rp}, \tag{9.17a}$$

since $v > 1/2$. The last quantity of (9.17a) tends to 0 as $p \to \infty$. Thus if p is sufficiently large, we obtain $M_f = 0$. This means that for $f \in \mathfrak{S}_k(\Gamma, \lambda)$ if $b_{h,\xi} = 0$ for all $\xi \in X$ and all h such that $|h| < rp$ with a large enough p, then $f = 0$. This shows that $\dim\left[\mathfrak{S}_k(\Gamma, \lambda)\right]$ is finite. Since an element of $\mathfrak{A}_k(\Gamma, \lambda)/\mathfrak{S}_k(\Gamma, \lambda)$ is determined by the constant terms of $f\|_k\xi$ for all $\xi \in X$, we see that $\dim\left[\mathfrak{A}_k(\Gamma, \lambda)/\mathfrak{S}_k(\Gamma, \lambda)\right] \le 2^{\#(X)}$. This proves our theorem.

9.5. So far it was unnecessary to specify a branch of j_α^k for an arbitrary α in $SL_2(\mathbf{Q})$ when $k \notin \mathbf{Z}$, but in the following treatment, that is not satisfactory. To make it more specific, for a weight k we define a group G_k as follows: $G_k = SL_2(\mathbf{Q})$ if $k \in \mathbf{Z}$. If $k \notin \mathbf{Z}$, G_k consists of all couples (α, q), where

$\alpha \in SL_2(\mathbf{Q})$ and q is a holomorphic function on $\mathfrak{H}$ such that $q(z)^2 = tj_\alpha(z)^{2k}$ with a root of unity t. We make G_k a group by the law of multiplication

$$(\alpha,\, q)(\alpha',\, q') = \bigl(\alpha\alpha',\, q(\alpha' z)q'(z)\bigr). \tag{9.18}$$

We define a projection map $\mathrm{pr} : G_k \to SL_2(\mathbf{Q})$ by $\mathrm{pr}(\alpha,\, q) = \alpha$ if $k \notin \mathbf{Z}$, and take the identity map to be pr if $k \in \mathbf{Z}$. We put $P_k = \bigl\{\alpha \in G_k \,\big|\, \mathrm{pr}(\alpha) \in P\bigr\}$. (This is different from what was defined in Lemma 2.2(iii), since we are in the one-dimensional case.) Notice that P_k is isomorphic to $P \times \mathbf{T}_0$, where $\mathbf{T}_0$ is the group of all roots of unity. Thus P_k is commutative.

If Γ is a congruence subgroup of Γ^θ, then the map $\gamma \mapsto (\gamma,\, j_\gamma^k)$ for $k \notin \mathbf{Z}$ with j_γ^k as in (5.1b) is an injection of Γ into G_k. We identify Γ with its image under this map, and view Γ as a subgroup of G_k.

For $\xi = (\gamma,\, q) \in G_k$, $z \in \mathfrak{H}$, and a function f on $\mathfrak{H}$ we put $a_\xi = a_\gamma$, $b_\xi = b_\gamma$, $c_\xi = c_\gamma$, $d_\xi = d_\gamma$, $j_\xi^k(z) = q(z)$, $\xi z = \gamma z$, and

$$(f\|_k\xi)(z) = q(z)^{-1} f(\gamma z). \tag{9.19}$$

From (9.18) we easily obtain

$$f\|_k(\xi\eta) = (f\|_k\xi)\|_k\eta. \tag{9.19a}$$

We also define elements ξ_* of G_{k-2} and ξ^* of G_{k+2} by

$$\xi_* = (\gamma,\, qj_\gamma^{-2}), \qquad \xi^* = (\gamma,\, qj_\gamma^{2}). \tag{9.20}$$

Then from (6.14a, b) we obtain

$$\varepsilon(f\|_k\xi) = (\varepsilon f)\|_{k-2}\xi_*, \qquad \delta_k(f\|_k\xi) = (\delta_k f)\|_{k+2}\xi^*. \tag{9.21}$$

We easily see that (6.12) is valid for $\alpha \in G_k$. Also, we have

(9.21a) *Let the notation be as in Theorem 7.5, and let Φ be a subfield of* $\mathbf{C}$ *containing* $\mathbf{Q}_{\mathrm{ab}}$. *Then* $\mathscr{X}_k(\Phi)$ *is stable under the map* $f \mapsto f\|_k\xi$ *for every* $\xi \in G_k$.

This follows immediately from Theorem 7.5(iii).

By a **congruence subgroup** of G_k we mean a subgroup Γ of G_k that contains $\Gamma(N)$ (viewed as a subgroup of G_k) for some even N as a subgroup of finite index, and such that pr restricted to Γ is injective. For such a Γ and $\xi \in G_k$ we see that $\xi\Gamma\xi^{-1}$ is a congruence subgroup of G_k. This follows from Theorem 4.7(4).

Given a congruence subgroup Γ of G_k, we can define $\mathfrak{A}_k(\Gamma,\, \lambda)$ by (9.1a, b, c). If $\Gamma(N) \subset \Gamma$ as above, then $\mathfrak{A}_k(\Gamma,\, \lambda) \subset \mathfrak{A}_k\bigl(\Gamma(N),\, \lambda\bigr)$, and so what we have done in §9.1 is applicable to the elements of $\mathfrak{A}_k(\Gamma,\, \lambda)$; also, $\mathfrak{S}_k(\Gamma,\, \lambda)$ can be defined in an obvious fashion. Lemma 9.2 and Theorem 9.4 are valid in this generalized sense. In Lemma 9.3 we take $\Gamma_* = \bigl\{\xi_* \,\big|\, \xi \in \Gamma\bigr\}$ and $\Gamma^* = \bigl\{\xi^* \,\big|\, \xi \in \Gamma\bigr\}$. These are congruence subgroups of G_{k-2} and G_{k+2}.

We note here an alternative way of treating factors of automorphy of non-integral weight. In [S74] we developed an axiomatic (and algebraic) theory of automorphic forms of an arbitrary weight. The advantage of this method is that we can prove a certain trace formula for Hecke operators, which is quite practicable. Indeed, in [N77] Niwa computed the traces of some Hecke operators of half-integral weight, and investigated the structure of Hecke algebras effectively. Though we do not discuss this theory here, those researchers interested in the computation of the trace of a Hecke operator of half-integral weight may be encouraged to look at [S74].

9.6. Given a congruence subgroup Γ of G_k, we put $\Gamma_\infty = P_k \cap \Gamma$ and

$$E_k(z,\, s;\, \Gamma) = \sum_{\alpha \in \Gamma_\infty \backslash \Gamma} y^s \|_k \, \alpha, \tag{9.22}$$

where $z \in \mathfrak{H}$, $s \in \mathbf{C}$, and $y = \mathrm{Im}(z)$; we assume that $j_\gamma^k = 1$ for $\gamma \in \Gamma_\infty$. Then the sum of (9.22) is formally well defined, and convergent for $\mathrm{Re}(s) > 1 - k/2$ by Lemma 8.7. This series is called the **Eisenstein series** of Γ.

Let Γ' be a congruence subgroup of G_k contained in Γ, and let $\Gamma'_\infty = \Gamma' \cap P_k$. Then we easily see that

$$[\Gamma_\infty : \Gamma'_\infty] E_k(z,\, s;\, \Gamma) = \sum_{\alpha \in \Gamma' \backslash \Gamma} E_k(z,\, s;\, \Gamma') \|_k \, \alpha. \tag{9.23}$$

Take a multiple N of 4 so that $\Gamma(N) \subset \Gamma$. Let Ψ_k be the set of all characters ψ modulo N such that $\psi(-1) = (-1)^{[k]}$. Then we easily see that

$$2\#(\Psi_k) E_k\bigl(z,\, s;\, \Gamma(N)\bigr) = \sum_{\psi \in \Psi_k} E_k\bigl(z,\, s;\, \Gamma(N,\, N),\, \psi\bigr). \tag{9.24}$$

This combined with (9.23) reduces the problems on analytic properties of $E_k(z,\, s;\, \Gamma)$ to those of $E_k\bigl(z,\, s;\, \Gamma(N,\, N),\, \psi\bigr)$. Therefore from Theorems 8.12 and 8.14 we see that $E_k(z,\, s;\, \Gamma)$ can be continued as a meromorphic function of s to the whole complex plane. We will give more precise statements in Theorem 9.9.

In view of (9.21) we can easily verify, by termwise differentiation, that

$$\varepsilon E_k(z,\, s;\, \Gamma) = (-si/2) E_{k-2}(z,\, s+1;\, \Gamma_*), \tag{9.25a}$$

$$\delta_k E_k(z,\, s;\, \Gamma) = \bigl(-(s+k)i/2\bigr) E_{k+2}(z,\, s-1;\, \Gamma^*), \tag{9.25b}$$

$$L_k E_k(z,\, s;\, \Gamma) = s(1-k-s) E_k(z,\, s;\, \Gamma). \tag{9.25c}$$

Strictly speaking, termwise differentiation is first justified for sufficiently large $\mathrm{Re}(s)$, but meromorphic continuation of both sides guarantees the equalities on the whole s-plane. For more details, see the paragraph below (8.20).

Given a congruence subgroup Γ of G_k, the projection map pr gives a bijection of $P_k \backslash G_k / \Gamma$ onto $P \backslash SL_2(\mathbf{Q})/\mathrm{pr}(\Gamma)$, which corresponds to the $\mathrm{pr}(\Gamma)$-equivalence classes of cusps as observed in §6.3. We call each coset $P_k \xi \Gamma$

with $\xi \in G_k$ a **cusp-class** of Γ, and call it **regular** if $j_\eta^k = 1$ for every $\eta \in P_k \cap \xi\Gamma\xi^{-1}$. This is the condition on $P_k\xi\Gamma$, and is independent of the choice of ξ. Then $E_k(z, s; \xi\Gamma\xi^{-1})$ is well defined.

Lemma 9.7. *Let Γ be a congruence subgroup of G_k, $\tilde{Y}$ a complete set of representatives for $P_k \backslash G_k/\Gamma$, and Y the set of all $\xi \in \tilde{Y}$ such that $P_k\xi\Gamma$ is regular. Then the following assertions hold:*

(i) *Let $f \in \mathfrak{A}_k(\Gamma, \lambda)$ and $\xi \in \tilde{Y}$. Then the Fourier expansion of $f\|_k\xi^{-1}$ has a nontrivial constant term only if $\xi \in Y$.*

(ii) *For $\xi, \eta \in Y$ we have*

$$E_k(z, s; \xi\Gamma\xi^{-1})\|_k\xi\eta^{-1} = \delta_{\xi\eta}y^s + f_{\xi\eta}(s)y^{1-k-s} + \sum_{0\neq h\in p\mathbf{Z}} g_{\xi\eta}(h, s, y)\mathbf{e}(hx),$$

where $\delta_{\xi\eta}$ is Kronecker's delta, $f_{\xi\eta}$ and $g_{\xi\eta}$ are meromorphic functions in s, and $0 < p \in \mathbf{Q}$.

PROOF. Put $\Gamma_\xi = \xi\Gamma\xi^{-1}$. Let $\alpha \in P_k \cap \Gamma_\xi$ and $\beta = \mathrm{pr}(\alpha)$. Then $\beta \in P \cap \mathrm{pr}(\Gamma_\xi)$, and so $\beta = \pm\begin{bmatrix} 1 & b \\ 0 & 1 \end{bmatrix}$ with $b \in \mathbf{Q}$. Thus $y^s\|_k\,\alpha = (j_\alpha^k)^{-1}y^s$. Since $f\|_k\,\xi^{-1}\alpha = f\|_k\,\xi^{-1}$ in the setting of (i), $f\|_k\,\xi^{-1}$ has a nontrivial constant term only if $j_\alpha^k = 1$. This proves (i). To prove (ii), for $a \in \mathbf{Q}$ define an element $r(a)$ of P_k by

$$r(a) = \left(\begin{bmatrix} 1 & a \\ 0 & 1 \end{bmatrix}, 1\right). \tag{9.26}$$

Then $r(\mathbf{Q}) \cap \Gamma_\xi = r(q\mathbf{Z})$ with $0 < q \in \mathbf{Q}$. (This is because $\Gamma(N) \subset \Gamma$ for some even N.) Take a subset Φ of Γ so that $1 \notin \Phi$ and $\Phi \cup \{1\}$ is a complete set of representatives for $(P_k \cap \Gamma_\xi)\backslash\Gamma_\xi/r(q\mathbf{Z})$. Then 1 and the elements $\varphi r(a)$ with $\varphi \in \Phi$ and $a \in q\mathbf{Z}$ represent $(P_k \cap \Gamma_\xi)\backslash\Gamma_\xi$ without overlap. Therefore

$$E_k(z, s; \Gamma_\xi) = y^s + \sum_{\varphi\in\Phi}\sum_{m\in\mathbf{Z}} y^s\|_k\,\varphi r(qm).$$

For a fixed $\varphi \in \Phi$ put $c = c_\varphi$ and $d = d_\varphi$. Then $c \neq 0$ and $j_\varphi^k(z)/(cz+d)^k \in \mathbf{T}$, and so

$$\sum_{m\in\mathbf{Z}} y^s\|_k\,\varphi r(qm) = ty^sc^{-2s-k}\sum_{m\in\mathbf{Z}}(z + c^{-1}d + qm)^{-s-k}(\bar{z} + c^{-1}d + qm)^{-s}$$

with an element $t \in \mathbf{T}$ determined by φ. By Lemma A2.3 of the Appendix the last sum over $\mathbf{Z}$ has an expansion of the form

$$i^{-k}(2\pi/q)^{2s+k}\sum_{n\in\mathbf{Z}}\mathbf{e}(q^{-1}n(x + c^{-1}d) + q^{-1}|n|iy)g_n(q^{-1}y;\, s+k,\, s)$$

with g_n as in (A2.4). For $n = 0$, from (A2.4) we obtain $y^{1-2s-k}h(s)$ with a meromorphic function $h(s)$. Multiplying by ty^sc^{-2s-k} and taking the sum over all $\varphi \in \Phi$, we obtain the Fourier expansion of $E_k(z, s; \Gamma_\xi)$ as stated in our lemma in the case $\xi = \eta$.

Next, let $\xi \neq \eta \in Y$. Let Z be a complete set of representatives for $(P_k \cap \Gamma_\xi)\backslash \xi\Gamma\eta^{-1}/(r(\mathbf{Q}) \cap \Gamma_\eta)$. Then the elements $\zeta r(a)$ with $\zeta \in Z$ and $r(a) \in r(\mathbf{Q}) \cap \Gamma_\eta$ represent $(P_k \cap \Gamma_\xi)\backslash \xi\Gamma\eta^{-1}$ without overlap, because of the following simple fact:

(9.27) *If* $\alpha \in SL_2(\mathbf{Q})$ *and* $\alpha \begin{bmatrix} 1 & x \\ 0 & 1 \end{bmatrix} \alpha^{-1} \in P$ *with* $x \in \mathbf{Q}^\times$, *then* $\alpha \in P$.

Indeed, suppose $x \neq 0$ and $\alpha \begin{bmatrix} 1 & x \\ 0 & 1 \end{bmatrix} \alpha^{-1} = \begin{bmatrix} v^{-1} & u \\ 0 & v \end{bmatrix}$. Then $c_\alpha = vc_\alpha$ and $c_\alpha x + d_\alpha = vd_\alpha$. If $c_\alpha \neq 0$, then $v = 1$ and $x = 0$, a contradiction. This proves (9.27). Therefore the same argument as above establishes the Fourier expansion of $E_k(z, s; \Gamma_\xi)\|_k \xi\eta^{-1}$; the only new feature is that y^s does not appear. This proves (ii) and completes the proof.

Lemma 9.8. *Let* Q *be a finite set of functions* $q(z, s)$ *of the form* $q(z, s) = E(z, s)\|_k \alpha$ *with* E *of type (8.12a) or (9.22) and* $\alpha \in G_k$, *and let* $g(z, s) = \sum_{q \in Q} f_q(s) q(z, s)$ *with meromorphic functions* f_q *on* $\mathbf{C}$. *Then, for every* $s_0 \in \mathbf{C}$ *there exists an integer* m *and a neighborhood* V *of* s_0 *such that* $(s - s_0)^m g(z, s)$ *is a real analytic function on* $\mathfrak{H} \times V$ *that is holomorphic in* s, *and, as a function of* z, *is slowly increasing at every cusp, locally uniformly in* $s \in V$. *In particular, if* g *is finite at* $s = s_0$, *then* $g(z, s_0)$ *is an element of* $\mathfrak{A}_k(\lambda)$ *with* $\lambda = s_0(1 - k - s_0)$.

PROOF. In view of (9.23) and (9.24), it is sufficient to prove our lemma when $q = E\|_k\alpha$ with a function E of type (8.12). Then our first assertion follows immediately from Theorems 8.12 and 8.14. From (9.25c) and (6.14d) we obtain $L_k g(z, s) = s(1 - k - s)g(z, s)$, and so $L_k g(z, s_0) = \lambda g(z, s_0)$ with $\lambda = s_0(1 - k - s_0)$ if g is finite at $s = s_0$. Moreover, those theorems show that m and V can be taken in such a way that $(s - s_0)^m g(z, s)$ is slowly increasing at every cusp, locally uniformly in $s \in V$. This shows that $g(z, s_0)$, if finite, belongs to $\mathfrak{A}_k(\lambda)$. This proves our lemma.

Theorem 9.9. *Given a congruence subgroup* Γ *of* G_k, *there exist a non-zero entire function* $A(s)$ *and a real analytic function* $B(z, s)$ *on* $\mathfrak{H} \times \mathbf{C}$ *holomorphic in* s *such that* $A(s)E_k(z, s; \Gamma) = B(z, s)$. *Moreover,* $E_k(z, s; \Gamma)$ *is holomorphic in* s *except at the points given in* (1), (2), (1′), (2′), *and* (3′) *below, and* $E_0(z, s; \Gamma)$ *has a pole as described in* (4).

(1) $k \in \mathbf{Z}$ *and* $-k/2 \leq \mathrm{Re}(s) < (1 - k)/2$.

(2) $k \in \mathbf{Z}$ *and* s *is an integer such that* $s \leq -(k + \nu)/2$, *where* ν *is* 0 *or* 1 *determined by* $k + \nu \in 2\mathbf{Z}$.

(1′) $k \notin \mathbf{Z}$ *and* $(1 - 2k)/4 \leq \mathrm{Re}(s) < (1 - k)/2$.

(2′) $k \notin \mathbf{Z}$ *and* s *is an element of* $4^{-1}\mathbf{Z}$ *such that* $0 \geq 2s + k - 1/2 \in \mathbf{Z}$.

(3′) $k \notin \mathbf{Z}$ *and* $|k| - 1/2 \in 2\mathbf{Z}$; *then* $E_k(z, s; \Gamma)$ *has at most a simple pole at* $s = (3 - 2k)/4$.

(4) *If* $k = 0$, *then* $E_0(z, s; \Gamma)$ *has a simple pole at* $s = 1$, *and the residue is* π^{-1} *times a positive rational number.*

PROOF. By (9.23) and (9.24) our problem can be reduced to the functions of type (8.12). Therefore the existence of $A(s)$ and $B(z, s)$ follows from Theorems 8.12 and 8.14 combined with those formulas.

Suppose $k \in \mathbf{Z}$. We employ Theorem 8.12 and (8.18), which involves a character ψ such that $\psi(-1) = (-1)^k$. Thus $\psi(-1) = (-1)^\nu$ with ν as in (2). Let ψ' be the primitive character associated with ψ. It is well known that $L(s, \psi')$ can be 0 only if $0 < \mathrm{Re}(s) < 1$ or $0 \geq s + \nu \in 2\mathbf{Z}$. We have $L_N(2s + k, \psi)$ in (8.18), and so we have to consider $1 - \psi'(p)p^{-2s-k}$, which becomes 0 only if $\mathrm{Re}(2s + k) = 0$. Therefore $L_N(2s + k, \psi) = 0$ only if $-k/2 \leq \mathrm{Re}(s) < (1 - k)/2$ or $0 \geq 2s + k + \nu \in 2\mathbf{Z}$. Thus, from (8.18), (9.23), and (9.24) we see that $E_k(z, s; \Gamma)$ is finite except at the points of (1) and (2).

Suppose $k = 0$; then $E_0^N(z, s; \chi_0) = 2\zeta(2s)E_0\big(z, s; \Gamma(1)\big)$ by (8.18), where χ_0 is the principal character. From Theorem 8.12 we see that $E_0\big(z, s; \Gamma(1)\big)$ has a simple pole at $s = 1$ with residue $\pi/[2\zeta(2)]$, which equals $3/\pi$. This combined with (9.23) proves (4).

Next suppose $k \notin \mathbf{Z}$. We employ Theorem 8.14, in which $L_N(4s-1+2k, \psi^2)$ appears. In this case $N > 1$. For the same reason as in the case $k \in \mathbf{Z}$ we see that it becomes 0 only if $(1-2k)/4 \leq \mathrm{Re}(s) < (1-k)/2$ or $0 \geq 4s - 1 + 2k \in 2\mathbf{Z}$. Also, a factor $2s - \lambda - 1$ appears in Theorem 8.14, where $\lambda = 1/2 - k$. This is necessary only if $|k| + 1/2$ is odd. Thus $E_k(z, s; \Gamma)$ may have a simple pole at $s = (3 - 2k)/4$ if $|k| + 1/2$ is odd. Therefore we have conditions (1′), (2′), and (3′) when $k \notin \mathbf{Z}$. This completes the proof.

9.10. Let Γ be a congruence subgroup of G_k and X a finite subset of G_k such that $G_k = \bigsqcup_{\xi \in X} \Gamma\xi P_k$. (This is consistent with (6.10). Then we can take $\tilde{Y} = \{\xi^{-1} \mid \xi \in X\}$ in Lemma 9.7.) Let $f \in \mathfrak{A}_k(\Gamma, \lambda)$ and $g \in \mathfrak{A}_k(\Gamma, \mu)$. Assuming that both λ and μ are noncritical, denote by $\{p\}_\lambda$ the set $\{p, q\}$ with p, q such that $p + q = 1 - k$ and $pq = \lambda$. Then for each $\xi \in X$ we put

$$f\|_k\xi = \sum_{p \in \{p\}_\lambda} a_{p,\xi}y^p + \text{nonconstant terms}, \tag{9.28a}$$

$$g\|_k\xi = \sum_{p \in \{p\}_\mu} b_{p,\xi}y^p + \text{nonconstant terms} \tag{9.28b}$$

with $a_{p,\xi}$, $b_{p,\xi} \in \mathbf{C}$. If both λ and μ are critical, we put

$$f\|_k\xi = a_\xi y^p + a'_\xi y^p \log y + \text{nonconstant terms}, \tag{9.28c}$$

$$g\|_k\xi = b_\xi y^p + b'_\xi y^p \log y + \text{nonconstant terms} \tag{9.28d}$$

with a_ξ, b_ξ, a'_ξ, $b'_\xi \in \mathbf{C}$, where $p = (1-k)/2$. We also put $Q_\xi = P \cap \mathrm{pr}(\xi^{-1}\Gamma\xi)$ and $R = \bigcap_{\xi \in X} Q_\xi$.

Theorem 9.11. *With f and g as above, suppose $\mu = \bar{\lambda}$ and λ is not critical. Fix one $p \in \{p\}_\lambda$ and put $q = 1 - k - p$. Then*

$$\sum_{\xi \in X} \nu_\xi (\bar{a}_{p,\xi} b_{\bar{q},\xi} - \bar{a}_{q,\xi} b_{\bar{p},\xi}) = 0, \tag{9.29a}$$

where $\nu_\xi = \left[\{\pm 1\}Q_\xi : \{\pm 1\}R\right]^{-1}$. If $\lambda = \mu$ and λ is critical, then

$$\sum_{\xi \in X} \nu_\xi (\bar{a}_\xi b'_\xi - \bar{a}'_\xi b_\xi) = 0. \tag{9.29b}$$

Proof. The idea of the proof is the same as in the proof of Theorem 6.8. Define T_r and M_r as in (6.17), and take a sufficiently large r so that the sets $\xi(Q_\xi \backslash T_r)$ for $\xi \in X$ can be embedded into $\Gamma\backslash\mathfrak{H}$ without overlap. Also, take a union J of small neighborhoods of elliptic fixed points on $\Gamma\backslash\mathfrak{H}$. Let K be the complement of $\bigcup_{\xi \in X} \xi(Q_\xi \backslash T_r) \cup J$ in $\Gamma\backslash\mathfrak{H}$. Then K is a compact manifold with boundary, and

$$\partial K = \sum_{\xi \in X} \xi(Q_\xi \backslash M_r) - \partial J.$$

Let φ be a 1-form on $\mathfrak{H}$ that is C^∞ and Γ-invariant. Then

$$\int_K d\varphi = \int_{\partial K} \varphi = \sum_{\xi \in X} \nu_\xi \int_{B_r} \varphi \circ \xi - \int_{\partial J} \varphi, \tag{9.30}$$

where $B_r = R\backslash M_r$ with a natural orientation. Take $\varphi = \bar{f} \cdot \varepsilon g \cdot y^{k-2} d\bar{z}$. Then by (6.18) with εg as h, we have

$$d\varphi = (2i)^{-1} \bar{f} \cdot L_k g \cdot y^k \mathbf{d}z + 2i\overline{\varepsilon f} \cdot \varepsilon g \cdot y^{k-2} \mathbf{d}z$$

with $\mathbf{d}z$ viewed as a 2-form. Putting similarly $\psi = \bar{g} \cdot \varepsilon f \cdot y^{k-2} d\bar{z}$, we find that

$$d\varphi + \overline{d\psi} = (2i)^{-1}(\bar{f} \cdot L_k g - \overline{L_k f} \cdot g) y^k \mathbf{d}z = 0,$$

since $L_k f = \lambda f$ and $L_k g = \bar{\lambda} g$. Applying (9.30) to this form, we obtain

$$\sum_{\xi \in X} \nu_\xi \int_{B_r} (\varphi + \bar{\psi}) \circ \xi - \int_{\partial J} (\varphi + \bar{\psi}) = 0.$$

We now take the expansions (9.28a, b) into consideration. We have $\varphi \circ \xi = \overline{f\|_k \xi} \cdot \varepsilon(g\|_k \xi) y^{k-2} d\bar{z}$, and a similar formula holds for $\bar{\psi} \circ \xi$. Fix one $p \in \{p\}_\lambda$ and put $q = 1 - k - p$. Then for $\mu = \bar{\lambda}$ we have $\{p\}_\mu = \{\bar{p}, \bar{q}\}$, and so

$$\begin{aligned} 2i\varphi \circ \xi &= (\bar{a}_{p,\xi} y^{\bar{p}} + \bar{a}_{q,\xi} y^{\bar{q}})(\bar{p} b_{\bar{p},\xi} y^{\bar{p}+1} + \bar{q} b_{\bar{q},\xi} y^{\bar{q}+1}) y^{k-2} d\bar{z} \\ &\quad + \text{nonconstant terms}, \\ -2i\bar{\psi} \circ \xi &= (b_{\bar{p},\xi} y^{\bar{p}} + b_{\bar{q},\xi} y^{\bar{q}})(\bar{p}\bar{a}_{p,\xi} y^{\bar{p}+1} + \bar{q}\bar{a}_{q,\xi} y^{\bar{q}+1}) y^{k-2} dz \\ &\quad + \text{nonconstant terms}. \end{aligned}$$

Since $\int_{B_r} y^s(dx \pm dy) = -\int_0^h r^s dx$ with a constant $h > 0$ independent of ξ, we have

$$\int_{B_r} 2i(\varphi+\bar{\psi})\circ\xi = (\bar{p}-\bar{q})(\bar{a}_{p,\xi}b_{\bar{q},\xi} - \bar{a}_{q,\xi}b_{\bar{p},\xi})\int_0^h dx + \text{nonconstant terms}.$$

By Lemma 9.2(ii) the sum of the nonconstant terms of (9.28a, b, c, d) are $O(e^{-cy})$ as $y\to\infty$ with some $c>0$, and so the same is true for the nonconstant terms of $\varphi\circ\xi$ and $\bar{\psi}\circ\xi$, and even for their integrals over B_r. We have $q-p=1-k-2p\neq 0$, since λ is noncritical. Therefore, taking the limit as $r\to\infty$ and making J shrink to the elliptic points, we obtain (9.29a). When λ is critical, the constant terms involving $\log y$ cancel each other, and so equality (9.29b) can be proved in a similar way. Our proof is now complete.

9.12. We put

(9.31a) $\quad \mathfrak{N}_k(\lambda) = \{g\in\mathfrak{A}_k(\lambda)\mid \langle f,\,g\rangle = 0 \text{ for every } f\in\mathfrak{S}_k(\lambda)\},$

(9.31b) $\quad \mathfrak{N}_k(\Gamma,\,\lambda) = \{g\in\mathfrak{A}_k(\Gamma,\,\lambda)\mid \langle f,\,g\rangle = 0 \text{ for every } f\in\mathfrak{S}_k(\Gamma,\,\lambda)\},$

where Γ is a congruence subgroup of G_k. The inner product $\langle f,\,g\rangle$ is meaningful in view of (9.1c), Lemma 9.2(iii), and what we said in §6.5. From (6.12) we see that $\mathfrak{N}_k(\lambda)\|_k\,\alpha = \mathfrak{N}_k(\lambda)$ for every $\alpha\in G_k$. We have also

(9.32a) $$\mathfrak{A}_k(\lambda) = \mathfrak{S}_k(\lambda)\oplus\mathfrak{N}_k(\lambda),$$

(9.32b) $$\mathfrak{A}_k(\Gamma,\,\lambda) = \mathfrak{S}_k(\Gamma,\,\lambda)\oplus\mathfrak{N}_k(\Gamma,\,\lambda),$$

(9.33) $$\mathfrak{N}_k(\Gamma,\,\lambda) = \mathfrak{N}_k(\lambda)\cap\mathfrak{A}_k(\Gamma,\,\lambda).$$

Indeed, (9.32b) is easy, since $\mathfrak{A}_k(\Gamma,\,\lambda)$ is of finite dimension, as proved in Theorem 9.4. Clearly the left-hand side of (9.33) contains the right-hand side. To prove the opposite inclusion, let $g\in\mathfrak{N}_k(\Gamma,\,\lambda)$ and $f\in\mathfrak{S}_k(\lambda)$. Take a normal congruence subgroup Δ of Γ so that $f\in\mathfrak{S}_k(\Delta,\,\lambda)$. By (9.32b), $g=p+q$ with $p\in\mathfrak{S}_k(\Delta,\,\lambda)$ and $q\in\mathfrak{N}_k(\Delta,\,\lambda)$. For $\gamma\in\Gamma$ we have $g=g\|_k\gamma = p\|_k\gamma + q\|_k\gamma$. We easily see that $p\|_k\gamma\in\mathfrak{S}_k(\Delta,\,\lambda)$ and $q\|_k\gamma\in\mathfrak{N}_k(\Delta,\,\lambda)$ by virtue of (6.12), and so (9.32b) with Δ in place of Γ shows that $p\|_k\gamma = p$ and $q\|_k\gamma = q$, and so $p\in\mathfrak{S}_k(\Gamma,\,\lambda)$ and $q\in\mathfrak{N}_k(\Gamma,\,\lambda)$. By (9.32b), $g=q\in\mathfrak{N}_k(\Delta,\,\lambda)$, and so $\langle g,\,f\rangle = 0$, which shows that $g\in\mathfrak{N}_k(\lambda)$. This proves (9.33). Since $\mathfrak{A}_k(\lambda)$ is the union of $\mathfrak{A}_k(\Gamma,\,\lambda)$ for all Γ, we obtain (9.32a) from (9.32b) and (9.33).

9.13. Lemma 9.8 shows that a function of type $E_k(z,\,s;\,\Gamma)\|_k\alpha$, if finite, belongs to $\mathfrak{A}_k(\lambda)$ with $\lambda = s(1-k-s)$. We are going to show that the function actually belongs to $\mathfrak{N}_k(\lambda)$, and moreover, $\mathfrak{N}_k(\lambda)$ is generated by such functions for almost all values of λ.

To prove the first statement, given $f\in\mathfrak{S}_k(\lambda_1)$ with any $\lambda_1\in\mathbf{C}$, take an even integer $N>2$ so that $f\in\mathfrak{S}_k\bigl(\Gamma(N),\,\lambda_1\bigr)$. We have expansion (9.10) with $r=1/N$. Since the constant term of f is 0, we have $\int_0^N\overline{f(z)}dx = 0$, and so $\int_0^\infty\int_0^N\overline{f(z)}dx\,y^{s+k+2}dy = 0$. Let $\Gamma(N)_\infty = P_k\cap\Gamma(N)$, $\Psi = \Gamma(N)_\infty\backslash\mathfrak{H}$, and $\Phi = \Gamma(N)\backslash\mathfrak{H}$. Then we see that $\int_0^\infty\int_0^N \cdots y^{-2}dxdy = \int_\Psi \cdots \mathbf{d}z$. Since Ψ can be given by $\bigsqcup_{\gamma\in R}\gamma\Phi$ with $R = \Gamma(N)_\infty\backslash\Gamma(N)$, we have by Rankin's transformation, at least formally,

$$0 = \int_\Psi \overline{f(z)}y^{s+k}\mathbf{d}z = \sum_{\gamma\in R}\int_\Phi (\bar{f}y^{s+k})\circ\gamma\mathbf{d}z$$

$$= \int_\Phi \overline{f(z)}\sum_{\gamma\in R}\overline{j_\gamma^k(z)}|j_\gamma(z)|^{-2s-2k}y^{s+k}\mathbf{d}z = \int_\Phi \overline{f(z)}E_k\bigl(z,\, s;\, \Gamma(N)\bigr)y^k\mathbf{d}z.$$

This can be justified for sufficiently large $\mathrm{Re}(s)$. Indeed, $E_k\bigl(z,\, s;\, \Gamma(N)\bigr)$ is majorized by

$$y^\sigma\sum_{\gamma\in R}|j_\gamma(z)|^{-2\sigma} = E_0\bigl(z,\, \sigma;\, \Gamma(N)\bigr)$$

for $\sigma\in\mathbf{R}$, which, if finite, is slowly increasing at every cusp. Since f, being an element of $\mathfrak{S}_k(\lambda_1)$, is rapidly decreasing at every cusp, our formal calculation is justified for sufficiently large σ. Combining this with (9.23), for every $f\in\mathfrak{S}_k(\lambda_1)$ and every congruence subgroup Γ of G_k we have

$$\int_{\Gamma\backslash\mathfrak{H}} \overline{f(z)}E_k(z,\, s;\, \Gamma)y^k\mathbf{d}z = 0 \qquad (9.34)$$

at least for sufficiently large $\mathrm{Re}(s)$. By Lemma 9.8, for every $s_0\in\mathbf{C}$ there is an integer m and a neighborhood V of s_0 such that $(s-s_0)^m E_k(z,\, s;\, \Gamma)$ is slowly increasing at every cusp, locally uniformly in $s\in V$. Therefore the left-hand side of (9.34) is meaningful as a meromorphic function of s on the whole $\mathbf{C}$, and also is valid whenever $E_k(z,\, s;\, \Gamma)$ is finite at s. Thus $E_k(z,\, s;\, \Gamma)$, if finite at s, belongs to $\mathfrak{N}_k(\lambda_1)$ with any $\lambda_1\in\mathbf{C}$. For the moment λ_1 is unrelated to s, but we will eventually take $\lambda_1 = s(1-k-s)$.

9.14. Let Γ be a congruence subgroup of G_k. Taking Y as in Lemma 9.7, we denote by $\mathfrak{E}_k(\Gamma)$ the $\mathbf{C}$-linear span of $E_k(z,\, s;\, \xi\Gamma\xi^{-1})\|_k\xi$ for all $\xi\in Y$. Given $s_0\in\mathbf{C}$, we denote by $\mathfrak{E}_k[s_0,\, \Gamma]$ the subset of $\mathfrak{E}_k(\Gamma)$ consisting of all $g(z,\, s)$ in $\mathfrak{E}_k(\Gamma)$ that are finite at $s = s_0$, and put $\mathfrak{E}_k(s_0,\, \Gamma) = \bigl\{g(z,\, s_0)\bigm| g\in\mathfrak{E}_k[s_0,\, \Gamma]\bigr\}$. We also denote by $\mathfrak{E}_k^*[s_0,\, \Gamma]$ the set of all $g\in\mathfrak{E}_k(\Gamma)$ that have at most a simple pole at s_0, and by $\mathfrak{E}_k^*(s_0,\, \Gamma)$ the set of the residues at s_0 of the elements of $\mathfrak{E}_k^*[s_0,\, \Gamma]$. Let us now prove

$$\mathfrak{E}_k(s_0,\, \Gamma) + \mathfrak{E}_k^*(s_0,\, \Gamma)\subset\mathfrak{N}_k(\Gamma,\, \lambda)\quad\text{with } \lambda = s_0(1-k-s_0). \qquad (9.35)$$

Indeed, from Lemma 9.8 it follows that both $\mathfrak{E}_k(s_0,\, \Gamma)$ and $\mathfrak{E}_k^*(s_0,\, \Gamma)$ are contained in $\mathfrak{A}_k(\Gamma,\, \lambda)$. Now the elements of $\mathfrak{E}_k(\Gamma)$ are functions of the type $g(z,\, s)$ of Lemma 9.8, and so formula (9.34) can be generalized to

$$\int_{\Gamma\backslash\mathfrak{H}} \overline{f(z)}g(z,\, s)y^k\mathbf{d}z = 0 \qquad (9.36)$$

for every $f\in\mathfrak{S}_k(\lambda_1)$ with any $\lambda_1\in\mathbf{C}$ in the sense that the left-hand side is meromorphic in s on the whole $\mathbf{C}$, and is valid whenever $g(z,\, s)$ is finite at

s. This shows that $\mathfrak{E}_k(s_0, \Gamma) \subset \mathfrak{N}_k(\Gamma, \lambda)$. Considering $(s - s_0)g$ instead of g, we see similarly that $\mathfrak{E}_k^*(s_0, \Gamma) \subset \mathfrak{N}_k(\Gamma, \lambda)$, and so we obtain (9.35).

Lemma 9.15. (i) *The symbols being as in* §9.14, *we have* $\dim \mathfrak{E}_k(\Gamma) = \#Y$.

(ii) *The map* $g(z, s) \mapsto g(z, s_0)$ *is a bijection of* $\mathfrak{E}_k[s_0, \Gamma]$ *onto* $\mathfrak{E}_k(s_0, \Gamma)$, *provided* $s_0 \neq (1-k)/2$.

(iii) $\mathfrak{E}_k(\Gamma) = \mathfrak{E}_k[s_0, \Gamma]$ *if* $\mathrm{Re}(s_0) \geq (1-k)/2$ *except in the following two cases:* (a) $s_0 = 1$ *and* $k = 0$; (b) $s_0 = (3-2k)/4$ *and* $|k| - 1/2 \in 2\mathbf{Z}$.

PROOF. Let $g(z, s) = \sum_{\xi \in Y} c_\xi E_k(z, s; \xi\Gamma\xi^{-1})\|_k \xi$ with $c_\xi \in \mathbf{C}$. Suppose $g \in \mathfrak{E}_k[s_0, \Gamma]$. Then from Lemma 9.7 we obtain, for every $\eta \in Y$,

$$g(z, s_0)\|_k \eta^{-1} = c_\eta y^{s_0} + \Bigl(\sum_{\xi \in Y} c_\xi f_{\xi\eta}\Bigr)(s_0) y^{1-k-s_0} + \cdots .$$

If $s_0 \neq (1-k)/2$, then $s_0 \neq 1 - k - s_0$, and so if $g(z, s_0) = 0$, we have $c_\eta = 0$ for every $\eta \in Y$. This proves (ii). In particular, if $g = 0$, then $c_\xi = 0$ for every $\xi \in Y$, which proves (i). Assertion (iii) follows immediately from Theorem 9.9.

Theorem 9.16. (i) *The notation being as in* §9.14, *suppose that* $\mathfrak{E}_k(\Gamma) = \mathfrak{E}_k[s_0, \Gamma] = \mathfrak{E}_k[\bar{s}_0, \Gamma]$ *and* $s_0 \neq (1-k)/2$; *let* $\lambda = s_0(1-k-s_0)$. *Then* $\mathfrak{N}_k(\Gamma, \lambda) = \mathfrak{E}_k(s_0, \Gamma)$ *and* $\dim \mathfrak{N}_k(\Gamma, \lambda) = \#Y$.

(ii) *In the setting of* (i) *let* $f \in \mathfrak{A}_k(\Gamma, \lambda)$ *and* $f\|_k \xi^{-1} = a_\xi y^{s_0} + a'_\xi y^{1-k-s_0} +$ nonconstant terms. *If* $a_\xi = 0$ *for every* $\xi \in Y$, *then* f *is a cusp form.*

PROOF. Given $f \in \mathfrak{A}_k(\Gamma, \lambda)$ and $g \in \mathfrak{A}_k(\Gamma, \bar{\lambda})$, for each $\xi \in Y$ put

$$f\|_k \xi^{-1} = a_\xi y^{s_0} + a'_\xi y^{1-k-s_0} + \text{nonconstant terms},$$
$$g\|_k \xi^{-1} = b_\xi y^{\bar{s}_0} + b'_\xi y^{1-k-\bar{s}_0} + \text{nonconstant terms}.$$

(We can consider such expansions even for $\xi \in \tilde{Y}$, but the constant term is 0 by Lemma 9.7(i) if $\xi \notin Y$.) By Theorem 9.11 (with $X = \{\xi^{-1} \mid \xi \in \tilde{Y}\}$) we have

$$\sum_{\xi \in Y} \nu_{\xi^{-1}} \bigl(a_\xi \bar{b}'_\xi - a'_\xi \bar{b}_\xi\bigr) = 0, \tag{9.37}$$

where ν_* is as in that theorem. Moreover, the map

$$f \mapsto (a_\xi, a'_\xi)_{\xi \in Y} \tag{9.38}$$

gives an injection of $\mathfrak{A}_k(\Gamma, \lambda)/\mathfrak{S}_k(\Gamma, \lambda)$ into $\mathbf{C}^{2\kappa}$, where $\kappa = \#Y$. A similar statement holds with $\bar{\lambda}$ in place of λ. By Lemma 9.15 and our assumption, $\dim \mathfrak{E}_k(s_0, \Gamma) = \dim \mathfrak{E}_k(\bar{s}_0, \Gamma) = \kappa$. Each nonzero element g of $\mathfrak{E}_k(\bar{s}_0, \Gamma)$ defines a nontrivial linear relation on (a_ξ, a'_ξ) by (9.37), since $\mathfrak{E}_k(\bar{s}_0, \Gamma) \cap \mathfrak{S}_k(\bar{\lambda}) = \{0\}$ by (9.32b) and (9.35). Therefore the elements of $\mathfrak{E}_k(\bar{s}_0, \Gamma)$ produce κ linearly independent relations on (a_ξ, a'_ξ), which means that $\dim \mathfrak{N}_k(\Gamma, \lambda) = \dim \left[\mathfrak{A}_k(\Gamma, \lambda)/\mathfrak{S}_k(\Gamma, \lambda)\right] \le \kappa$. Since $\mathfrak{E}_k(s_0, \Gamma) \subset \mathfrak{N}_k(\Gamma, \lambda)$, this proves (i).

Let f be as in (ii). By (i) we can put $f(z) = g(z, s_0) + h(z)$ with $g \in \mathfrak{E}_k(s_0, \Gamma)$ and $h \in \mathfrak{S}_k(\Gamma, \lambda)$. Take this g as g in the proof of Lemma 9.15. Then our assumption $a_\xi = 0$ means $c_\xi = 0$ in that lemma, and so $g = 0$. This proves (ii).

Theorem 9.17. *The notation being as in* §9.14, *define a* $\mathbf{C}^Y$*-valued function* $\mathbf{E}_k$ *on* $\mathfrak{H} \times \mathbf{C}$ *by* $\mathbf{E}_k(z, s; \Gamma) = \big(E_k(z, s; \xi\Gamma\xi^{-1})\|_k\, \xi\big)_{\xi \in Y}$. *Then there exists an* $\mathrm{End}(\mathbf{C}^Y)$*-valued meromorphic function* $\Phi_k(s, \Gamma)$ *such that*

$$\mathbf{E}_k(z, s; \Gamma) = \Phi_k(s, \Gamma)\mathbf{E}_k(z, 1-k-s; \Gamma), \tag{9.39a}$$

$$\Phi_k(1-k-s, \Gamma)\Phi_k(s, \Gamma) = 1. \tag{9.39b}$$

Moreover, there is a diagonal element A of $\mathrm{End}(\mathbf{C}^Y)$, *depending only on Γ and Y, whose diagonal entries are positive integers such that*

$$\overline{\Phi_k(s, \Gamma)}A \cdot {}^t\Phi_k(1-k-\bar{s}, \Gamma) = A. \tag{9.39c}$$

PROOF. Put $\Gamma_\xi = \xi\Gamma\xi^{-1}$ and $E_\xi(s) = E_k(z, s; \Gamma_\xi)\|_k\, \xi$. By Lemma 9.7, for $\xi, \eta \in Y$ we have

$$E_\xi(s)\|_k\, \eta^{-1} = \delta_{\xi\eta}y^s + f_{\xi\eta}(s)y^{1-k-s} + \cdots \tag{9.40a}$$

with meromorphic functions $f_{\xi\eta}$ on $\mathbf{C}$, and so

$$E_\xi(1-k-s)\|_k\, \eta^{-1} = \delta_{\xi\eta}y^{1-k-s} + f_{\xi\eta}(1-k-s)y^s + \cdots. \tag{9.40b}$$

Therefore

$$\Big\{E_\xi(1-k-s) - \sum_{\zeta\in Y} f_{\xi\zeta}(1-k-s)E_\zeta(s)\Big\}\|_k\, \eta^{-1} \tag{9.41}$$
$$= 0 \cdot y^s + \Big\{\delta_{\xi\eta} - \sum_{\zeta\in Y} f_{\xi\zeta}(1-k-s)f_{\zeta\eta}(s)\Big\}y^{1-k-s} + \cdots.$$

We can easily find a nonempty open subset W of $\mathbf{C}$ such that $\mathfrak{E}_k(\Gamma) = \mathfrak{E}_k[s, \Gamma] = \mathfrak{E}_k[\bar{s}, \Gamma] = \mathfrak{E}_k[1-k-s, \Gamma] = \mathfrak{E}_k[1-k-\bar{s}, \Gamma]$, $f_{\xi\eta}(1-k-s)$ is finite, and $s \ne (1-k)/2$ for every $s \in W$. Then by Theorem 9.16(i) the left-hand side of (9.41) without $\|_k\, \eta^{-1}$ for such an s belongs to $\mathfrak{N}_k(\Gamma, \lambda)$ with $\lambda = s(1-k-s)$. By Theorem 9.16(ii) it must be a cusp form, and so

$$\sum_{\zeta\in Y} f_{\xi\zeta}(1-k-s)f_{\zeta\eta}(s) = \delta_{\xi\eta} \quad \text{and} \quad E_\xi(1-k-s) = \sum_{\zeta\in Y} f_{\xi\zeta}(1-k-s)E_\zeta(s)$$

for every $s \in W$. Writing $\Phi_k(s, \Gamma)$ for the matrix $[f_{\xi\eta}(s)]$, we obtain (9.39a, b). Next, $E_\xi(1-k-\bar{s})$ belongs to $\mathfrak{A}_k(\Gamma, \bar{\lambda})$, and so from (9.37), (9.40a), and (9.40b) with ζ in place of ξ we obtain

$$\sum_{\eta\in Y} \nu_{\eta^{-1}}\big(\delta_{\xi\eta}\delta_{\zeta\eta} - \overline{f_{\xi\eta}(s)}f_{\zeta\eta}(1-k-\bar{s})\big) = 0.$$

Let $A = \mathrm{diag}[\nu_{\eta^{-1}}]_{\eta \in Y}$. Viewing this as an element of $\mathrm{End}(\mathbf{C}^Y)$, we obtain (9.39c). This completes the proof.

Theorem 9.18. *The notation being as in Theorem 9.17, suppose* $\lambda = \mu^2$ *with* $\mu = (1-k)/2$. *Given a congruence subgroup* Γ *of* G_k, *let* $\mathfrak{E}'_k(\Gamma)$ *denote the space spanned by* $(\partial g/\partial s)(z, \mu)$ *for* $g \in \mathfrak{E}_k[\mu, \Gamma]$ *and* $\mathfrak{E}^0_k(\Gamma)$ *the space consisting of* $(\partial g/\partial s)(z, \mu)$ *for all* $g \in \mathfrak{E}_k[\mu, \Gamma]$ *such that* $g(z, \mu) = 0$. *Further let* κ_+ *resp.* κ_- *be the multiplicity of* 1 *resp.* -1 *in the eigenvalues of* $\Phi_k(\mu, \Gamma)$. *Then* $\#Y = \kappa_+ + \kappa_-$, $\dim \mathfrak{E}_k(\mu, \Gamma) = \kappa_+$, $\dim \mathfrak{E}^0_k(\Gamma) = \kappa_-$, *and*

$$\mathfrak{E}'_k(\Gamma) \subset \mathfrak{N}_k(\Gamma, \lambda) = \mathfrak{E}_k(\mu, \Gamma) \oplus \mathfrak{E}^0_k(\Gamma).$$

Moreover, $\mathfrak{E}_k(\mu, \Gamma)$ *consists of the elements of* $\mathfrak{N}_k(\Gamma, \lambda)$ *that do not involve* $y^\mu \log y$.

PROOF. Put $\kappa = \#Y$. By Lemma 9.15(iii), $\mathfrak{E}_k(\Gamma) = \mathfrak{E}_k[\mu, \Gamma]$ and so every function appearing in this proof is finite at $s = \mu$; also, $\dim \mathfrak{E}_k[\mu, \Gamma] = \kappa$ by Lemma 9.15(i). We easily see that the elements of $\mathfrak{E}'_k(\Gamma)$ satisfy (9.1a, b), and also (9.1c), in view of Lemma 9.8. Also, equality (9.34) holds with an element $g(z, s)$ of $\mathfrak{E}_k[\mu, \Gamma]$ in place of $E_k(z, s; \Gamma)$, and the integral is uniformly convergent in a neighborhood of $s = \mu$. Therefore we see that $(\partial g/\partial s)(z, \mu) \in \mathfrak{N}_k(\lambda)$, and so $\mathfrak{E}'_k(\Gamma) \subset \mathfrak{N}_k(\Gamma, \lambda)$. From (9.39b) we obtain $\Phi_k(\mu, \Gamma)^2 = 1$. Thus the eigenvalues of $\Phi_k(\mu, \Gamma)$ are ± 1. Let $E_\xi(s)$ be as in the proof of Theorem 9.17. Then from (9.40a) we obtain

$$E_\xi(\mu)\|_k\, \eta^{-1} = \bigl(\delta_{\xi\eta} + f_{\xi\eta}(\mu)\bigr) y^\mu + \cdots,$$
$$(\partial E_\xi/\partial s)(\mu)\|_k\, \eta^{-1} = \bigl(\delta_{\xi\eta} - f_{\xi\eta}(\mu)\bigr) y^\mu \log y + (df_{\xi\eta}/ds)(\mu) y^\mu + \cdots$$

for every $\xi, \eta \in Y$. Let $g(z, s) = \sum_{\xi \in Y} c_\xi E_\xi(s) \in \mathfrak{E}_k[\mu, \Gamma]$ with $c_\xi \in \mathbf{C}$. Put $c = (c_\xi)_{\xi \in Y}$. Then $g(z, \mu) = 0$ if and only if ${}^t\Phi_k(\mu, \Gamma)c = -c$, which means that $\dim \mathfrak{E}_k(\mu, \Gamma) = \kappa - \kappa_- = \kappa_+$. If ${}^t\Phi_k(\mu, \Gamma)c = -c$, then $g(z, \mu) = 0$ and $(\partial g/\partial s)(z, \mu)\|_k \eta^{-1} = 2c_\eta y^\mu \log y + b_\eta \cdot y^\mu + \cdots$ with some b_η, which shows that $\dim \mathfrak{E}^0_k(\Gamma) = \kappa_-$. Since no element of $\mathfrak{E}_k(\mu, \Gamma)$ involves $y^\mu \log y$, we see that $\mathfrak{E}_k(\mu, \Gamma)$ and $\mathfrak{E}^0_k(\Gamma)$ form a direct sum.

Next, the notation being as in (9.28c, d), consider the map $f \mapsto (a_\xi, a'_\xi)_{\xi \in Y}$ defined for $f \in \mathfrak{A}_k(\Gamma, \lambda)$. This sends $\mathfrak{A}_k(\Gamma, \lambda)$ into $\mathbf{C}^{2\kappa}$ with kernel $\mathfrak{S}_k(\Gamma, \lambda)$. Let d be the dimension of the image space. Then $d = \dim \mathfrak{N}_k(\Gamma, \lambda)$ and (9.29b) shows that $d \le 2\kappa - d$, and so $d \le \kappa$. Since $\mathfrak{E}_k(\mu, \Gamma) \oplus \mathfrak{E}^0_k(\Gamma)$ is a subspace of $\mathfrak{N}_k(\Gamma, \lambda)$ of dimension $\kappa_+ + \kappa_- = \kappa$, we can establish all the statements of our theorem.

Theorem 9.19. *Let* $\lambda = s_0(1 - k - s_0)$ *with* $s_0 \in \mathbf{C}$. *The notation being as in* §9.14, *suppose that* $s_0 \ne (1-k)/2$, $\lambda \in \mathbf{R}$, *and* $\mathfrak{E}_k(\Gamma) = \mathfrak{E}^*_k[s_0, \Gamma]$. *Then* $\dim \mathfrak{N}_k(\Gamma, \lambda) = \#Y$ *and* $\mathfrak{N}_k(\Gamma, \lambda) = \mathfrak{E}_k(s_0, \Gamma) \oplus \mathfrak{E}^*_k(s_0, \Gamma)$.

PROOF. Put $\kappa = \#Y$. For $p \in \mathfrak{E}_k(\Gamma)$ denote by $\rho(p)$ the residue of p at $s = s_0$. Then ρ is a $\mathbf{C}$-linear map of $\mathfrak{E}_k(\Gamma)$ onto $\mathfrak{E}_k^*(s_0, \Gamma)$ with kernel $\mathfrak{E}_k[s_0, \Gamma]$, and so by Lemma 9.15(i,ii), $\dim \mathfrak{E}_k(s_0, \Gamma) + \dim \mathfrak{E}_k^*(s_0, \Gamma) = \kappa$. Let $h \in \mathfrak{E}_k(s_0, \Gamma) \cap \mathfrak{E}_k^*(s_0, \Gamma)$. Then $h(z) = g(z, s_0) = \rho(p)$ with $g \in \mathfrak{E}_k[s_0, \Gamma]$ and $p \in \mathfrak{E}_k(\Gamma)$. Put $g = \sum_{\xi \in Y} a_\xi E_\xi(s)$ and $p = \sum_{\xi \in Y} b_\xi E_\xi(s)$ with $a_\xi, b_\xi \in \mathbf{C}$ and $E_\xi(s) = E_k(z, s; \xi\Gamma\xi^{-1})\|_k \xi$. By (9.40a) we have, for $\eta \in Y$,

$$\begin{aligned} h\|_k \eta^{-1} &= a_\eta y^{s_0} + \sum_{\xi \in Y} a_\xi f_{\xi\eta}(s_0) y^{1-k-s_0} + \cdots \\ &= 0 \cdot y^{s_0} + \sum_{\xi \in Y} b_\xi \big[\mathrm{Res}_{s=s_0} f_{\xi\eta}(s)\big] y^{1-k-s_0} + \cdots, \end{aligned}$$

and so $a_\eta = 0$ for every $\eta \in Y$. Thus $h = 0$, and consequently, $\mathfrak{E}_k(s_0, \Gamma)$ and $\mathfrak{E}_k^*(s_0, \Gamma)$ form a direct sum of dimension κ. Take again the map of $\mathfrak{A}_k(\Gamma, \lambda)$ into $\mathbf{C}^{2\kappa}$ with kernel $\mathfrak{S}_k(\Gamma, \lambda)$ given by (9.38). Let $m = \dim \mathfrak{N}_k(\Gamma, \lambda)$, which is the dimension of the image space of (9.38). Since $\bar{\lambda} = \lambda$, relation (9.37) shows that $m \le 2\kappa - m$, and so $m \le \kappa$. We have seen that the left-hand side of (9.35) has dimension κ, and so we obtain our theorem.

Corollary 9.20. *Let Γ and Y be as in Lemma 9.7. Then* $\dim \mathfrak{N}_k(\Gamma, \lambda) = \#Y$ *for every* λ.

PROOF. If λ is critical, this is included in Theorem 9.18. Suppose λ is not critical. Let s_0 be a solution of $X^2 + (k-1)X + \lambda = 0$. Then the other solution is $1 - k - s_0$. Replacing s_0 by $1 - k - s_0$ if necessary, we may assume that $\mathrm{Re}(s_0) \ge (1-k)/2$. Excluding cases (a) and (b) of Lemma 9.15(iii), we obtain $\dim \mathfrak{N}_k(\Gamma, \lambda) = \#Y$ from Theorem 9.16. In cases (a) and (b) we obtain the desired result from Theorem 9.19.

9.21. We now return to holomorphic modular forms on $\mathfrak{H}$. Given a congruence subgroup Γ of G_k, we denote by $\mathscr{M}_k(\Gamma)$ resp. $\mathscr{S}_k(\Gamma)$ the set of elements f of $\mathscr{M}_k$ resp. $\mathscr{S}_k$ such that $f\|_k \gamma = f$ for every $\gamma \in \Gamma$. This is consistent with what we already have if $k \in \mathbf{Z}$ or if $k \notin \mathbf{Z}$ and $\Gamma \subset \Gamma^\theta$. We have then

(9.42a) $\mathscr{M}_k(\Gamma)$ *consists of the holomorphic elements of* $\mathfrak{A}_k(\Gamma, 0)$,

(9.42b) $\mathscr{S}_k(\Gamma) = \mathfrak{S}_k(\Gamma, 0)$.

These follow immediately from (9.3) and Lemma 9.2(iv). We put

(9.43a) $$\mathscr{E}_k(\Gamma) = \big\{ f \in \mathscr{M}_k(\Gamma) \,\big|\, \langle f, g \rangle = 0 \text{ for every } g \in \mathscr{S}_k(\Gamma) \big\},$$

(9.43b) $$\mathscr{E}_k = \bigcup_{N=1}^{\infty} \mathscr{E}_k\big(\Gamma(2N)\big).$$

Theorem 9.22. *For every congruence subgroup Γ of G_k with $k > 0$ we have*

(9.44) $$\mathscr{E}_k(\Gamma) = \mathscr{M}_k(\Gamma) \cap \mathfrak{N}_k(0),$$

(9.45a) $$\mathscr{M}_k = \mathscr{S}_k \oplus \mathscr{E}_k,$$

(9.45b) $$\mathscr{M}_k(\Gamma) = \mathscr{S}_k(\Gamma) \oplus \mathscr{E}_k(\Gamma),$$

(9.46) $$\mathscr{E}_k(\Gamma) = \mathscr{M}_k(\Gamma) \cap \mathfrak{E}_k(0, \Gamma) \text{ if } k \geq 1,$$

(9.47) $$\mathscr{E}_k(\Gamma) = \mathfrak{E}_k(0, \Gamma) \text{ if } k > 2 \text{ or } k = 1,$$

(9.48) $$\mathscr{E}_{1/2}(\Gamma) = \mathfrak{E}^*_{1/2}(1/2, \Gamma).$$

PROOF. By (9.42b) the right-hand side of (9.44) is contained in $\mathscr{E}_k(\Gamma)$. Conversely, let $f \in \mathscr{E}_k(\Gamma)$ and $g \in \mathfrak{S}_k(\Gamma', 0)$ with any $\Gamma' \subset \Gamma$. Let R be a complete set of representatives for $\Gamma' \backslash \Gamma$. Then by (6.12) we have

$$[\Gamma : \Gamma']\langle f, g\rangle = \sum_{\alpha \in R} \langle f, g\|_k \alpha\rangle = \left\langle f, \sum_{\alpha \in R} g\|_k \alpha \right\rangle = 0,$$

since $\sum_{\alpha \in R} g\|_k \alpha \in \mathfrak{S}_k(\Gamma, 0) = \mathscr{S}_k(\Gamma)$ by (9.42b). Thus $f \in \mathfrak{N}_k(0)$, which proves (9.44). Formula (9.45b) follows immediately from the definition of $\mathscr{E}_k(\Gamma)$. Then clearly (9.45a) holds. Take $s_0 = 0$ in Lemma 9.15 and Theorem 9.16(i); then we find that $\mathfrak{E}_k(0, \Gamma) = \mathfrak{N}_k(\Gamma, 0)$ for $k \geq 3/2$, which combined with (9.44) proves (9.46) for such k. Next, take $k = 1$ in Theorem 9.18. Then $\mathscr{M}_1 \cap \mathfrak{N}_1(\Gamma, 0) = \mathscr{M}_1 \cap \mathfrak{E}_1(0, \Gamma)$, which gives (9.46) for $k = 1$. Formulas (9.47) and (9.48) will be poven in the proof of the following theorem.

Theorem 9.23. *Let $E_k(z, s)$ denote the analytic continuation of any series of type (8.12) or (9.22) with $k > 0$. Then the following assertions hold:*

(i) *$E_k(z, s)$ is finite at $s = 0$.*

(ii) *$E_k(z, 0)$ belongs to $\mathscr{M}_k(\mathbf{Q}_{ab})$ if $k > 2$ or $k = 1$.*

(iii) *If $k = 2$, then $E_2\big(z, 0; \Gamma(N, N), \psi\big)$ belongs to $\mathscr{M}_2(\mathbf{Q}_{ab})$ except when ψ is trivial, in which case it belongs to $\mathscr{N}_2^1(\mathbf{Q}_{ab})$.*

(iv) *If $k = 3/2$, then $E_{3/2}\big(z, 0; \Gamma(N, N), \psi\big)$ belongs to $\mathscr{M}_{3/2}(\mathbf{Q}_{ab})$ except when ψ^2 is trivial.*

(v) *If $k = 1/2$, then $E_{1/2}(z, s; \Gamma)$ has at most a simple pole at $s = 1/2$, and the residue belongs to $\pi^{-1}\mathscr{M}_{1/2}(\mathbf{Q}_{ab})$. More explicitly, the residue is of the form $\pi^{-1}\sum_{\xi \in \mathbf{Q}} \lambda(\xi)\mathbf{e}(t\xi^2 z/2)$ with $0 < t \in \mathbf{Q}$ and a $\mathbf{Q}_{ab}$-valued element λ of $\mathscr{L}(\mathbf{Q})$.*

PROOF. By (9.23), (9.24), and Theorem 7.5(iii) the problems can be reduced to the case of $E_k(z, s; \Gamma, \psi)$. Assertions (i), (ii), and (iv) follow from Theorem 8.15(iii) if $3/2 \leq k \notin \mathbf{Z}$. Suppose $k \in \mathbf{Z}$; then (8.18) reduces the problem to $E_k^N(z, s; \psi)$. Indeed, since $\psi(-1) = (-1)^k$, we have $L_N(k, \psi) \in \pi^k \mathbf{Q}_{ab}^\times$ by Lemma 2.9, and so we obtain (i) and (ii) from Theorem 8.15(i). Combining these results with (9.46), we obtain (9.47). Assertion (iii) follows from Theorem 8.15(i).

Suppose $k = 1/2$; let F^* be as in Theorem 8.14. Then

$$F^*(z, s) = (2s-1)\Gamma(s)\Gamma(s+1/2)L_N(4s, \psi^2)E_{1/2}(z, s; \Gamma, \psi).$$

By Lemma 2.10, $L_N(2, \psi^2) \in \pi^2\mathbf{Q}_{\rm ab}^\times$. Therefore from Theorem 8.16(i) we see that $E_k(z, s; \Gamma, \psi)$ has at most a simple pole at $s = 1/2$, and the residue is π^{-1} times a $\mathbf{Q}_{\rm ab}$-rational theta series as given in that theorem. Thus we obtain (v), and see that $\mathfrak{E}_k(\Gamma) = \mathfrak{E}_k^*[1/2, \Gamma]$ and $\mathfrak{E}_k^*(1/2, \Gamma) \subset \mathscr{M}_k(\Gamma)$. Therefore, by Theorem 9.19, $\mathfrak{N}_k(\Gamma, 0) = \mathfrak{E}_k(1/2, \Gamma) \oplus \mathfrak{E}_k^*(1/2, \Gamma)$. Thus, to prove (9.48), it is sufficient to show that the only holomorphic element of $\mathfrak{E}_k(1/2, \Gamma)$ is 0. For that purpose, take $h(z) = g(z, 1/2)$ with $g \in \mathfrak{E}_k[1/2, \Gamma]$. Put $g = \sum_{\xi\in Y} a_\xi E_\xi(s)$ with $a_\xi \in \mathbf{C}$ as in the proof of Theorem 9.19. Then $h\|_k\eta^{-1} = a_\eta y^{1/2} + c_\eta + \cdots$ with $c_\eta \in \mathbf{C}$ for every $\eta \in Y$. If h is holomorphic, then $a_\eta = 0$ for every $\eta \in Y$, and so $h = 0$ as expected. This proves (9.48) and completes the proof of our theorem.

From (v) above and (9.48) we obtain

(9.49) *$\mathscr{E}_{1/2}(\Gamma)$ is spanned by some functions of the form $\sum_{\xi\in\mathbf{Q}} \lambda(\xi)\mathbf{e}(t\xi^2 z/2)$ with $0 < t \in \mathbf{Q}$ and an element λ of $\mathscr{L}(\mathbf{Q})$.*

We also note that

(9.50) *$\mathfrak{E}_k^N(z, 0; p, q)$ belongs to $\mathfrak{N}_k\big(\Gamma(N), 0\big)$ if $k > 0$. In particular, $E_k(z)$ defined by (8.14e) belongs to $\mathfrak{N}_k\big(\Gamma(N), 0\big)$.*

In view of (8.14d) we may assume that $(N, p, q) = 1$. Then there exist relatively prime integers p_0 and q_0 such that $(p_0, q_0) - (p, q) \in N\mathbf{Z}^2$. We have $\mathfrak{E}_k^N(z, s; p, q) = \mathfrak{E}_k^N(z, s; p_0, q_0)$, and (8.14a) reduces the problem to $\mathfrak{E}_k^N(z, s; 0, 1)$. From (8.16) we see that

$$\varphi(N)\mathfrak{E}_k^N(z, s; 0, 1) = \sum_{\psi\in\Psi} E_k^N(z, s; \psi),$$

where Ψ is the set of all characters modulo N such that $\psi(-1) = (-1)^k$. Notice that $\varphi(N) = \#(\Psi) = 1$ and $k \in 2\mathbf{Z}$ if $N \le 2$. By (8.18) we obtain

$$\varphi(N)\mathfrak{E}_k^N(z, s; 0, 1) = 2\sum_{\psi\in\Psi} L_N(2s+k, \psi)E_k(z, s; \Gamma, \psi),$$

which together with (9.35) proves (9.50).

9.24. We note here one of the easiest cases of $\Phi_k(s, \Gamma)$ of (9.39a, b). Take $\Gamma = \Gamma(1)$ and $0 \le k \in 2\mathbf{Z}$. Let χ_0 denote the principal character. Then (8.18) shows that

$$2\zeta(2s+k)E_k\big(z, s; \Gamma(1), \chi_0\big) = E_k^1(z, s; \chi_0) = \mathfrak{E}_k^1(z, s; 0, 1).$$

The Fourier expansion of $\mathfrak{E}_k^1(z, s; 0, 1)$ is a special case of the formula given in [S07, p. 134]. Employing it, we obtain

$$E_k\big(z, s; \Gamma(1), \chi_0\big) = y^s + y^{1-k-s}\pi i^{-k}2^{2-k-2s}\frac{\Gamma(2s+k-1)\zeta(2s+k-1)}{\Gamma(s)\Gamma(s+k)\zeta(2s+k)}$$

+ nonconstant terms.

Put $\xi(s) = \pi^{-s/2}\Gamma(s/2)\zeta(s)$. Since $\Gamma(s)\Gamma(s-1/2) = \pi^{1/2}2^{2-2s}\Gamma(2s-1)$, we have

$$E_k\big(z,\, s;\, \Gamma(1),\, \chi_0\big) = y^s + \Phi_k\big(s,\, \Gamma(1)\big)y^{1-k-s} + \text{nonconstant terms} \tag{9.51}$$

$$\text{with } \Phi_k\big(s,\, \Gamma(1)\big) = i^{-k}\cdot\frac{\xi(2s+k-1)}{\xi(2s+k)}\cdot\frac{\Gamma(s+k/2)^2}{\Gamma(s)\Gamma(s+k)}.$$

Then the relation $\Phi_k\big(s,\, \Gamma(1)\big)\Phi_k\big(1-k-s,\, \Gamma(1)\big) = 1$ means the well-known equality $\xi(1-s) = \xi(s)$ combined with the fact that the map $s \mapsto 1-k-s$ transforms $\Gamma(s)\Gamma(s+k)\Gamma(s+k/2)^{-2}$ (which is a rational expression in s) into its inverse.

Lemma 9.25. *Given $f \in \mathscr{M}_k$, $\alpha \in G_k$, and $\sigma \in \mathrm{Aut}(\mathbf{C})$, there exists an element β of G_k such that $(f\|_k\alpha)^\sigma = f^\sigma\|_k\beta$.*

PROOF. We first prove the case $k \in \mathbf{Z}$. Since $SL_2(\mathbf{Q}) = P\Gamma(1)$, it is sufficient to prove the cases $\alpha \in \Gamma(1)$ and α is of the form $\alpha = \begin{bmatrix} 1 & b \\ 0 & 1 \end{bmatrix}$ or $\alpha = \begin{bmatrix} a & 0 \\ 0 & d \end{bmatrix}$. The first case is included in Lemma 7.6. The latter two cases can easily be verified. Now suppose $k \notin \mathbf{Z}$; let $f \in \mathscr{M}_k$ and $\gamma = (\alpha,\, p) \in G_k$. Then $f^2 \in \mathscr{M}_{2k}$ and we find $\beta \in SL_2(\mathbf{Q})$ such that $(f^2\|_{2k}\alpha)^\sigma = (f^2)^\sigma\|_{2k}\beta$. We easily see that $\big((f\|_k\gamma)^\sigma\big)^2 = \zeta(f^2\|_{2k}\alpha)^\sigma$ with a root of unity ζ. Therefore $(f\|_k\gamma)^\sigma = f^\sigma\|_k(\beta,\, q)$ with a suitable $(\beta,\, q) \in G_k$. This completes the proof.

Theorem 9.26. *For every weight $k > 0$ the space $\mathscr{E}_k$ is spanned by its $\mathbf{Q}$-rational elements.*

PROOF. This follows from (9.49) if $k = 1/2$. Since the cases $k = 3/2$ and $k = 2$ require special considerations, we first assume $k > 2$ or $k = 1$. Our task is to show that $\mathscr{E}_k$ for such a k is spanned by $\mathbf{Q}$-rational elements. By (9.23) and (9.24), $\mathscr{E}_k$ is spanned by $E_k\big(z,\, 0;\, \Gamma(N,\, N),\, \psi\big)\|_k\alpha$ for all choices of $(N,\, \psi)$ and $\alpha \in G_k$. By Theorem 9.23(ii) and (9.21a), such a function is $\mathbf{Q}_{\mathrm{ab}}$-rational. To obtain the desired result, it is sufficient to show that $\mathscr{E}_k$ is stable under the action of $\mathrm{Aut}(\mathbf{C})$ defined by (7.2). Indeed, assuming such a stability, take a $\mathbf{Q}_{\mathrm{ab}}$-rational element $f(z) = \sum_{\xi\in\mathbf{Q}} c(\xi)\mathbf{e}(\xi z)$ of $\mathscr{E}_k$. Let K be the field generated over $\mathbf{Q}$ by the $c(\xi)$. By Theorem 7.5(v), $[K : \mathbf{Q}]$ is finite. Put $G = \mathrm{Gal}(K/\mathbf{Q})$ and $g_b = \sum_{\sigma\in G}(bf)^\sigma$ for $b \in K$. Then g_b is a $\mathbf{Q}$-rational element of $\mathscr{E}_k$ and f is a finite K-linear combination of g_b for some b, and so $\mathscr{E}_k$ is spanned by $\mathbf{Q}$-rational elements. Thus our problem is to show that $\big\{E_k\big(z,\, 0;\, \Gamma(N,\, N),\, \psi\big)\|_k\alpha\big\}^\sigma$ belongs to $\mathscr{E}_k$ for every $\sigma \in \mathrm{Aut}(\mathbf{C})$. This is indeed so by Lemma 9.25, Theorem 8.15(i), and Theorem 8.15(iii). This completes the proof in the case $k > 2$ or $k = 1$.

Next suppose $k = 2$. By (9.46), $\mathscr{E}_2(\Gamma)$ consists of the holomorphic elements of $\mathfrak{E}_2(0, \Gamma)$, and the nonholomorphic elements belong to $\mathscr{N}_2^1$. Therefore, including such nonholomorphic elements, we observe that the space spanned by $E_2\big(z, 0; \Gamma(N, N), \psi\big)\|_2\alpha$ is stable under $\mathrm{Aut}(\mathbf{C})$. For this, in addition to Theorem 8.15(i), we need Lemma 9.25 for $f \in \mathscr{N}_2$. The statement in the two cases of the elements of P are easy, and so we have only to check the case $\alpha \in \Gamma(1)$. Then expressing f in the form $f = g + cE_2$ with $g \in \mathscr{M}_2$, $c \in \mathbf{C}$, and the function E_2 of (7.10), we easily obtain the desired result. Thus $\bigcup_\Gamma \mathfrak{E}_2(0, \Gamma)$ is stable under $\mathrm{Aut}(\mathbf{C})$, and its subset consisting of the holomorphic elements is also stable under $\mathrm{Aut}(\mathbf{C})$ as expected.

Finally suppose $k = 3/2$. We invoke the result of Pei in [P82, 84], in which a subset $\mathscr{F}$ of $\mathscr{E}_{3/2}$ with the following properties is given: (i) $\mathscr{F}$ consists of $\mathbf{Q}_{\mathrm{ab}}$-rational elements, and is stable under $\mathrm{Gal}(\mathbf{Q}_{\mathrm{ab}}/\mathbf{Q})$; (ii) $\mathscr{E}_{3/2}$ is spanned by the elements of the form $f\|_{3/2}\alpha$ with $f \in \mathscr{F}$ and $\alpha \in SL_2(\mathbf{Q})$. Therefore our argument in the case $k > 2$ is applicable and we obtain the desired result. This completes the proof.

CHAPTER IV

THE CORRESPONDENCE BETWEEN FORMS OF INTEGRAL AND HALF-INTEGRAL WEIGHT

10. Theta series of indefinite quadratic forms

In this section we will associate a certain theta function with an indefinite quadratic form, and prove its automorphy properties. In later sections we will employ the function in various ways. We consider the set $S_n(A)$ defined by (0.2) with $A = \mathbf{Z}$, $\mathbf{Q}$, or $\mathbf{R}$, and put $S[x] = {}^txSx$ and $S(x, y) = {}^txSy$ for $x, y \in \mathbf{C}^n$ and $S \in S_n(\mathbf{R})$. We begin with some easy facts.

Lemma 10.1. *For two elements S and P of $S_n(\mathbf{R}) \cap GL_n(\mathbf{R})$ the following three conditions are mutually equivalent:*

(i) *$P > 0$ and $PS^{-1}P = S$.*

(ii) *There exists an element A of $GL_n(\mathbf{R})$ such that $P = {}^tAA$ and $S = {}^tAI_{p,q}A$ with nonnegative integers p and q such that $p+q = n$, where $I_{p,q} = \mathrm{diag}[1_p, -1_q]$.* (We of course ignore 1_p or 1_q if p or q is 0.)

(iii) *There exists a direct sum decomposition $\mathbf{R}^n = W \oplus W'$ such that $S[x] \geq 0$ and $Px = Sx$ for $x \in W$, $S[y] \leq 0$ and $Py = -Sy$ for $y \in W'$, and $S(x, y) = 0$ for $x \in W$ and $y \in W'$.*

PROOF. Given P and S as in (i), take $B \in GL_n(\mathbf{R})$ so that $P = {}^tBB$, and put $T = {}^tB^{-1}SB^{-1}$. Then ${}^tT = T$ and $T^{-1} = BS^{-1} \cdot {}^tB = BP^{-1}SP^{-1} \cdot {}^tB = {}^tB^{-1}SB^{-1} = T$, and so the eigenvalues of T are ± 1. Therefore we can find an element C of $GL_n(\mathbf{R})$ such that ${}^tCC = 1_n$ and $C^{-1}TC = I_{p,q}$ with some p and q as in (ii). Putting $A = {}^tCB$, we obtain (ii). Next, given (p, q) and A as in (ii), let X resp. Y denote the subspace of $\mathbf{R}^n$ consisting of the elements of $\mathbf{R}^n$ whose last q resp. first p coordinates are 0. Let $W = A^{-1}X$ and $W' = A^{-1}Y$. Then we obtain (iii). Finally suppose W and W' are taken as in (iii); then we can find an element U of $GL_n(\mathbf{R})$ such that $UX = W$ and $UY = W'$. We see that ${}^tUSU = \mathrm{diag}[G, -H]$ with $0 < G \in S_p(\mathbf{R})$ and $0 < H \in S_q(\mathbf{R})$. Then ${}^tUPU = \mathrm{diag}[G, H]$, and so $P > 0$ and ${}^tUPS^{-1}PU = {}^tUPU({}^tUSU)^{-1} \cdot {}^tUPU = \mathrm{diag}[G, -H] = {}^tUSU$. Thus we obtain (i) and our proof is complete.

10.2. For $S \in S_n(\mathbf{R}) \cap GL_n(\mathbf{R})$ we put

G. Shimura, *Modular Forms: Basics and Beyond*, Springer Monographs in Mathematics, DOI 10.1007/978-1-4614-2125-2_4, © Springer Science+Business Media, LLC 2012

$$O(S) = \{\alpha \in GL_n(\mathbf{R}) \mid {}^t\alpha S\alpha = S\}, \tag{10.1}$$

$$\mathfrak{P}(S) = \{P \in S_n(\mathbf{R}) \mid P > 0,\ PS^{-1}P = S\}. \tag{10.2}$$

Then we can show that $\mathfrak{P}(S)$ is a symmetric space that is $O(S)$ modulo a compact subgroup as follows. First take $A \in GL_n(\mathbf{R})$ and nonnegative integers p and q so that $S = {}^tAI_{p,q}A$ and put $P_0 = {}^tAA$. By Lemma 10.1, $P_0 \in \mathfrak{P}(S)$. (Thus $\mathfrak{P}(S) \neq \emptyset$.) Put $K = O(S) \cap O(P_0)$. Since $O(P_0)$ is compact, K is a compact subgroup of $O(S)$. In view of Lemma 10.1, it is an easy exercise to show that $\alpha \mapsto {}^t\alpha P_0 \alpha$ for $\alpha \in O(S)$ gives a bijection of $K\backslash O(S)$ onto $\mathfrak{P}(S)$. (We do not need this fact in our later treatment, however.)

10.3. We fix two elements S and P of $S_n(\mathbf{R}) \cap GL_n(\mathbf{R})$ such that $P \in \mathfrak{P}(S)$. For $z = x + iy \in \mathfrak{H}$ and $\gamma \in SL_2(\mathbf{R})$ we put

$$R(z) = xS + iyP, \tag{10.3}$$

$$\sigma_\gamma = \begin{bmatrix} a_\gamma 1_n & b_\gamma S \\ c_\gamma S^{-1} & d_\gamma 1_n \end{bmatrix}. \tag{10.4}$$

Then clearly $R(z) \in \mathfrak{H}_n$ and

$$\sigma_\gamma = \begin{bmatrix} 1_n & 0 \\ 0 & S \end{bmatrix}^{-1} \begin{bmatrix} a_\gamma 1_n & b_\gamma 1_n \\ c_\gamma 1_n & d_\gamma 1_n \end{bmatrix} \begin{bmatrix} 1_n & 0 \\ 0 & S \end{bmatrix}. \tag{10.5}$$

Therefore we easily see that $\sigma_\gamma \in Sp(n, \mathbf{R})$ and the map $\gamma \mapsto \sigma_\gamma$ is a homomorphism of $SL_2(\mathbf{R})$ into $Sp(n, \mathbf{R})$. Moreover we have

$$R(z) - \overline{R(z)} = 2iyP, \tag{10.6}$$

$$\sigma_\gamma(R(z)) = R(\gamma(z)), \tag{10.7}$$

$$j(\sigma_\gamma, R(z)) = j_\gamma(z)^p \overline{j_\gamma(z)}^q, \tag{10.8}$$

where p and q are determined by S as in Lemma 10.1(ii). Formula (10.6) is obvious. To prove the last two formulas, take A as in Lemma 10.1(ii) and put $Z = \mathrm{diag}[z1_p, -\bar{z}1_q]$. Then we easily see that $R(z) = {}^tAZA$, and so for $\gamma = \begin{bmatrix} a & b \\ c & d \end{bmatrix}$ we have

$$\sigma_\gamma \begin{bmatrix} R(z) \\ 1_n \end{bmatrix} = \begin{bmatrix} a1_n & bS \\ cS^{-1} & d1_n \end{bmatrix} \begin{bmatrix} {}^tAZA \\ 1_n \end{bmatrix} = \begin{bmatrix} a \cdot {}^tAZA + b \cdot {}^tAI_{p,q}A \\ cA^{-1}I_{p,q}ZA + d1_n \end{bmatrix} = \begin{bmatrix} {}^tAXA \\ A^{-1}YA \end{bmatrix}$$

with $X = \mathrm{diag}[(az+b)1_p, -(a\bar{z}+b)1_q]$ and $Y = \mathrm{diag}[(cz+d)1_p, (c\bar{z}+d)1_q]$, and so

$$\sigma_\gamma \begin{bmatrix} R(z) \\ 1_n \end{bmatrix} = \begin{bmatrix} {}^tA \cdot \mathrm{diag}[\gamma(z)1_p - \overline{\gamma(z)}1_q]A \\ 1_n \end{bmatrix} A^{-1}YA$$

$$= \begin{bmatrix} R(\gamma(z)) \\ 1_n \end{bmatrix} A^{-1} \cdot \mathrm{diag}[j_\gamma(z)1_p, \overline{j_\gamma(z)}1_q]A.$$

Recalling formula (1.6), we obtain (10.7) and also

$$\mu(\sigma_\gamma, R(z)) = A^{-1} \cdot \mathrm{diag}[j_\gamma(z)1_p, \overline{j_\gamma(z)}1_q]A. \tag{10.9}$$

Taking the determinant, we obtain (10.8).

10.4. We now assume that $S \in S_n(\mathbf{Q})$. Then the map $\gamma \mapsto \sigma_\gamma$ sends $SL_2(\mathbf{Q})$ into $Sp(n, \mathbf{Q})$. We then consider the function $\varphi(u, Z; \lambda)$ of (4.48a) with $u \in \mathbf{C}^n$, $Z \in \mathfrak{H}_n$, and $\lambda \in \mathscr{L}(\mathbf{Q}^n)$. Put

$$\mathfrak{f}(u, z; \lambda) = \varphi\bigl(u, R(z); \lambda\bigr). \tag{10.10}$$

Here $z \in \mathfrak{H}$. More explicitly

$$\mathfrak{f}(u, z; \lambda) = \mathbf{e}\bigl((4iyP)^{-1}[u]\bigr) \sum_{\xi \in \mathbf{Q}^n} \lambda(\xi)\mathbf{e}\bigl(2^{-1}R(z)[\xi] + {}^t\xi u\bigr). \tag{10.11}$$

Put

$$M_S = \bigl\{\gamma \in SL_2(\mathbf{Q}) \bigm| \sigma_\gamma \in P_n\Gamma_n^\theta\bigr\} \tag{10.12}$$

with P_n of Lemma 2.2(iii) and Γ_n^θ of (4.6). Then $P_1 M_S = M_S$. Take positive integers r and s so that $rS \prec 2\mathbf{Z}$ and $sS^{-1} \prec 2\mathbf{Z}$. Then $\sigma_\gamma \in \Gamma_n^\theta$ if $\gamma \in \Gamma(r, s)$; see (8.11) for the notation. Also $\sigma_\iota = \mathrm{diag}[S, S^{-1}]\iota_n \in P_n\Gamma_n^\theta$. Thus

$$\Gamma(r, s) \subset M_S \quad \text{and} \quad \sigma_\iota \in M_S. \tag{10.12a}$$

If $\gamma \in M_S$ and $\sigma = \sigma_\gamma$, then from (10.7) and (4.49a) we obtain

$$\mathfrak{f}\bigl({}^tM_\gamma^{-1}u, \gamma(z); \lambda\bigr) = h_\sigma\bigl(R(z)\bigr)\mathfrak{f}(u, z; \lambda^\sigma), \tag{10.13}$$

where $M_\gamma = \mu\bigl(\sigma_\gamma, R(z)\bigr)$. Also, from (10.8) and (4.49b) we see that

$$h_\sigma\bigl(R(z)\bigr) = \kappa_\gamma(cz + d)^{p/2}(c\bar{z} + d)^{q/2} \quad \text{with} \quad \kappa_\gamma \in \mathbf{T}. \tag{10.14}$$

10.5. Let S, P, and $R(z)$ be as in §10.3 with $S \in S_n(\mathbf{Q})$. Put $V = \mathbf{R}^n$, $V_\mathbf{C} = \mathbf{C}^n$, and

$$V_\mathbf{C}^+ = \bigl\{x \in V_\mathbf{C} \bigm| Px = Sx\bigr\}, \quad V_\mathbf{C}^- = \bigl\{x \in V_\mathbf{C} \bigm| Px = -Sx\bigr\}. \tag{10.15}$$

We consider a $\mathbf{C}$-valued polynomial function χ on V given by

$$\chi(\xi) = \prod_{\rho \in \{\rho\}} ({}^t\rho S\xi)^{\ell_\rho} \prod_{\tau \in \{\tau\}} ({}^t\tau S\xi)^{m_\tau} \qquad (\xi \in V). \tag{10.16}$$

Here $\{\rho\}$ resp. $\{\tau\}$ is a finite subset of $V_\mathbf{C}^+$ resp. $V_\mathbf{C}^-$; $0 \le \ell_\rho \in \mathbf{Z}$, $0 \le m_\tau \in \mathbf{Z}$. We assume: $S(\rho, \rho') = 0$ if $\rho, \rho' \in \{\rho\}$ and $\rho \ne \rho'$; $S(\tau, \tau') = 0$ if $\tau, \tau' \in \{\tau\}$ and $\tau \ne \tau'$; $S[\rho] = 0$ if $\ell_\rho > 1$; $S[\tau] = 0$ if $m_\tau > 1$. We then put

$$f(z, \lambda) = f(z; \lambda, \chi) = \sum_{\xi \in V} \lambda(\xi)\chi(\xi)\mathbf{e}\bigl(2^{-1}R(z)[\xi]\bigr), \tag{10.17}$$

where $z \in \mathfrak{H}$ and $\lambda \in \mathscr{L}(\mathbf{Q}^n)$. We call $f(z, \lambda)$ a **theta series associated with** S. Take A as in Lemma 10.1(ii). Then we easily see that

(10.18) $Ax = I_{p,q}Ax$ if $x \in V_\mathbf{C}^+$ and $Ay = -I_{p,q}Ay$ if $y \in V_\mathbf{C}^-$;

(10.19) the last q resp. first p coordinates of Ax are 0 if $x \in V_\mathbf{C}^+$ resp. $x \in V_\mathbf{C}^-$.

We define a factor of automorphy $J_\alpha^{p,q}(z)$ as follows:

$$J_\alpha^{p,q}(z) = \begin{cases} j_\alpha(z)^{(p-q)/2}|j_\alpha(z)|^q & \text{if} \quad n \in 2\mathbf{Z}, \\ h_\alpha(z)^{p-q}|j_\alpha(z)|^q & \text{if} \quad n \notin 2\mathbf{Z}, \end{cases} \tag{10.20}$$

where $\alpha \in SL_2(\mathbf{Q})$ if $n \in 2\mathbf{Z}$ and $\alpha \in P_1\Gamma^\theta$ if $n \notin 2\mathbf{Z}$; h_α is defined in Theorem 4.12. Notice that $p - q - n \in 2\mathbf{Z}$. In the following theorem S and χ are fixed.

Theorem 10.6. *Let $\gamma \in SL_2(\mathbf{Q})$ if $n \in 2\mathbf{Z}$ and $\gamma \in P_1\Gamma^\theta$ if $n \notin 2\mathbf{Z}$. Then for $\lambda \in \mathscr{L}(\mathbf{Q}^n)$ we can define an element λ^γ of $\mathscr{L}(\mathbf{Q}^n)$ such that*

$$f\bigl(\gamma(z), \lambda\bigr) = j_\gamma(z)^\ell \overline{j_\gamma(z)}^m J_\gamma^{p,q}(z) f(z, \lambda^\gamma), \tag{10.21}$$

where $\ell = \sum_{\rho \in \{\rho\}} \ell_\rho$ and $m = \sum_{\tau \in \{\tau\}} m_\tau$. Moreover, λ^γ is independent of χ and $\{\gamma \in SL_2(\mathbf{Q}) \mid \lambda^\gamma = \lambda\}$ contains a congruence subgroup of $SL_2(\mathbf{Q})$.

PROOF. From (10.14) we see that $h_\sigma\bigl(R(z)\bigr)$ is $J_\gamma^{p,q}(z)$ times an element of $\mathbf{T}$. Therefore we can reformulate (10.13) in the form

$$\mathfrak{f}\bigl({}^tM_\gamma^{-1}u, \gamma(z); \lambda\bigr) = J_\gamma^{p,q}(z)\mathfrak{f}(u, z; \lambda^\gamma) \tag{10.22}$$

with a well-defined $\lambda^\gamma \in \mathscr{L}(\mathbf{Q}^n)$ for $\gamma \in M_S \cap P\Gamma^\theta$. Suppose $n \in 2\mathbf{Z}$; then $J_{\beta\gamma}^{p,q}(z) = J_\beta^{p,q}(\gamma z) J_\gamma^{p,q}(z)$ for every $\beta, \gamma \in SL_2(\mathbf{Q})$. By Lemma 2.2(iv), $SL_2(\mathbf{Q})$ is generated by P and ι. Since (10.22) is valid for $\gamma \in P$ and $\gamma = \iota$, we can define λ^γ by (10.22) for every $\gamma \in SL_2(\mathbf{Q})$ when $n \in 2\mathbf{Z}$. Next suppose $n \notin 2\mathbf{Z}$. Then a similar reasoning establishes (10.22) for every $\gamma \in SL_2(\mathbf{Q})$ if we replace $J_\gamma^{p,q}$ and λ^γ by $J' = j_\gamma(z)^{p/2}\overline{j_\gamma(z)}^{q/2}$ and some element $\lambda' \in \mathscr{L}(\mathbf{Q}^n)$. Here λ' depends on the choice of J'. Therefore, if we take $J' = J_\gamma^{p,q}(z)$ as defined by (10.20), then (10.22) is valid for $\gamma \in P\Gamma^\theta$ with a well-defined λ^γ when $n \notin 2\mathbf{Z}$. Thus (10.22) can be extended to $\gamma \in SL_2(\mathbf{Q})$ or $\gamma \in P\Gamma^\theta$; we call this extended formula (10.22).

We will derive (10.21) by applying some differential operators to (10.22) and putting $u = 0$. We first treat the case with $\chi(\xi) = ({}^t\rho S\xi)^\ell({}^t\tau S\xi)^m$, and will add a comment in a more general case at the end of the proof. Put $D_x = \sum_{i=1}^n (Sx)_i \partial/\partial u_i$ for $x \in V_\mathbf{C}$ and $g(u) = \mathbf{e}\bigl((4iyP)^{-1}[u] + {}^t\xi u\bigr)$. Then

$$(D_\rho^\ell D_\tau^m g)(0) = (2\pi i)^{\ell+m}\chi(\xi). \tag{10.23}$$

To prove this we first observe that $D_x B[u] = 2B(Sx, u)$ for every $B \in S_n(\mathbf{R})$. Then

$$(D_x g)(u) = 2\pi i\bigl\{S(\xi, x) + (2iyP)^{-1}(Sx, u)\bigr\}g(u),$$

$$(D_x^2 g)(u) = (2\pi i)^2\bigl\{S(\xi, x) + (2iyP)^{-1}(Sx, u)\bigr\}^2 g(u) + 2\pi i(2iyP)^{-1}[Sx]g(u).$$

Now $P^{-1}[Sx] = {}^t xPx = \pm {}^t xSx$ for $x \in V_{\mathbf{C}}^{\pm}$. Therefore, by our assumption $S[\tau] = 0$ if $m_\tau > 1$, we obtain $(D_\tau^m g)(0) = (2\pi i)^m S(\tau, \xi)^m$ for $0 \le m \in \mathbf{Z}$. Also, $D_\rho D_\tau g$ involves $P^{-1}(S\tau, S\rho)$, which equals ${}^t\tau P\rho = {}^t\tau S\rho = 0$. Thus we obtain (10.23), and consequently termwise differentiation of (10.11) gives

$$(D_\rho^\ell D_\tau^m)\{\mathfrak{f}(u, z; \lambda)\}_{u=0} = (2\pi i)^{\ell+m} f(z, \lambda). \tag{10.24}$$

On the other hand,

$$\begin{aligned}(D_\rho^\ell D_\tau^m)&\{\mathfrak{f}({}^t M_\gamma^{-1}u, \gamma z; \lambda)\} \\ &= j_\gamma(z)^{-\ell}\overline{j_\gamma(z)}^{-m}(D_\rho^\ell D_\tau^m \mathfrak{f})({}^t M_\gamma^{-1}u, \gamma z; \lambda).\end{aligned} \tag{10.25}$$

Indeed, in view of (10.9) and (10.19) we have

$$\begin{aligned}{}^t M_\gamma^{-1} S\rho &= {}^t A \cdot \mathrm{diag}[j_\gamma(z)1_p, \overline{j_\gamma(z)1_q}]^{-1} \cdot {}^t A^{-1} S\rho \\ &= {}^t A \cdot \mathrm{diag}[j_\gamma(z)1_p, \overline{j_\gamma(z)1_q}]^{-1} I_{p,q} A\rho = j_\gamma(z)^{-1} S\rho,\end{aligned}$$

and similarly ${}^t M_\gamma^{-1} S\tau = \overline{j_\gamma(z)}^{-1} S\tau$, and so (10.25) holds. Therefore, applying $D_\rho^\ell D_\tau^m$ to (10.22) and putting $u = 0$, we obtain

$$(2\pi i)^{\ell+m} j_\gamma(z)^{-\ell}\overline{j_\gamma(z)}^{-m} f(\gamma z, \lambda) = (2\pi i)^{\ell+m} J_\gamma^{p,q}(z) f(z, \lambda^\gamma),$$

which can be written in the form (10.21). When χ is defined in the most general form, we apply $\prod_\rho D_\rho^{\ell_\rho} \prod_\tau D_\tau^{m_\tau}$ to (10.22). In view of our assumption that ${}^t\rho S\rho' = {}^t\tau S\tau' = 0$, we obtain (10.21) in the general case.

Clearly λ^γ is independent of χ. The last assertion will be proven in §10.10.

Lemma 10.7. *Let the symbols be as in §10.5 and Theorem 10.6. Then $f(z, \lambda)$ is slowly increasing or rapidly decreasing at every cusp, locally uniformly in the parameters ρ and τ of (10.16), according as $\ell = m = 0$ or $\ell + m > 0$.*

PROOF. Let $\alpha \in SL_2(\mathbf{Q})$ and $\kappa = \ell + m + (p+q)/2$. Then by (10.21),

$$|\mathrm{Im}(\alpha z)^{\kappa/2} f(\alpha z, \lambda)| = |\mathrm{Im}(z)^{\kappa/2} f(z, \lambda^\alpha)|,$$

and so our task is to make an estimate of $|f(z, \mu)|$ for an arbitrary $\mu \in \mathscr{L}(\mathbf{Q}^n)$. Clearly it is sufficient to treat the case in which μ is the characteristic function of $L = \mathbf{Z}^n$. Let χ be a homogeneous polynomial function of $\xi \in V$ of degree d. Then we can find a positive constant C such that $|\chi(\xi)| \le CP[\xi]^{d/2}$ for every $\xi \in V$. Since $|\mathbf{e}(2^{-1}R(z)[\xi])| = \exp(-\pi y P[\xi])$, we have

$$\Bigl|\sum_{\xi \in L} \chi(\xi)\mathbf{e}(2^{-1}R(z)[\xi])\Bigr| \le C\sum_{\xi \in L} P[\xi]^{d/2} \exp\bigl(-\pi y P[\xi]\bigr).$$

Suppose $d = 0$. Then we see that $|f(z, \mu)| \le C'$ for $y > 1/2$ with a constant C'. Thus $f(z, \lambda)$ is slowly increasing at every cusp. Suppose $d > 0$. Since L is discrete in V and $\{\xi \in V \mid P[\xi] \le 1\}$ is compact, $\{\xi \in L \mid P[\xi] \le 1\}$ is a finite set, and so we can find a positive constant M such that $P[\xi] \geq M$ for $0 \neq \xi \in L$. Therefore if $d > 0$, we have

$$|f(z, \mu)| \leq C \exp(-\pi My/2) \sum_{\xi \in L} P[\xi]^{d/2} \exp(-\pi y P[\xi]/2) \leq C_0 \exp(-\pi My/2)$$

if $y > 1/2$ with a positive constant C_0. This shows that $f(z, \lambda)$ is rapidly decreasing at every cusp. Since we can take the same C when the parameters ρ and τ of (10.16) stay in compact sets, we obtain the desired local uniformity in ρ and τ.

Lemma 10.8. *For* $\gamma = \begin{bmatrix} a & b \\ c & d \end{bmatrix} \in \Gamma(2)$ *define* σ_γ *by (10.4). Suppose* $\sigma_\gamma \in \Gamma_n^\theta$ *and* $0 \leq d - 1 \in 2\mathbf{Z}$; *then*

$$h(\sigma_\gamma, R(z)) = \left(\frac{(-1)^p \det(bcS)}{d}\right) h_\gamma(z)^p \overline{h_\gamma(z)}^q \qquad (z \in \mathfrak{H}). \tag{10.26}$$

If $S \prec \mathbf{Z}$ *in particular,* $(-1)^p \det(bcS)$ *can be replaced by* $(-1)^q \det(S)$.

PROOF. Since $bS \prec \mathbf{Z}$, $cS^{-1} \prec \mathbf{Z}$, and $ad1_n - (bS)(cS^{-1}) = 1_{2n}$, we see that $\det(bS)$ is prime to d. By Theorem 4.7(1) we have $h(\sigma_\gamma, Z) = \kappa j(\sigma_\gamma, Z)^{1/2}$ with

$$\kappa = d^{-n/2} \sum_{x \in \mathbf{Z}^n/d\mathbf{Z}^n} \mathbf{e}(bS[x]/(2d)), \quad \lim_{Z \to 0} j(\sigma_\gamma, Z)^{1/2} > 0,$$

where Z is a variable on $\mathfrak{H}_n$. Since d is odd, putting $x = 2y$ with $y \in \mathbf{Z}^n/d\mathbf{Z}^n$, we obtain

$$\kappa = d^{-n/2} \sum_{y \in \mathbf{Z}^n/d\mathbf{Z}^n} \mathbf{e}(2bS[y]/d).$$

By Lemma 2.3 there exist an element α of $M_n(\mathbf{Z})$ and $r_i \in \mathbf{Z}$ such that ${}^t\alpha bS\alpha - \mathrm{diag}[r_1, \ldots, r_n] \prec d\mathbf{Z}$ and $\det(\alpha)$ is a positive integer prime to d. Clearly we may assume that the diagonal elements of ${}^t\alpha bS\alpha$ are $r_1, \ldots, r_n$. Then

$$\kappa = d^{-n/2} \prod_{i=1}^{n} \sum_{y=1}^{d} \mathbf{e}(2r_i y^2/d).$$

and $\det(\alpha)^2 \det(bS) - r_1 \cdots r_n \in d\mathbf{Z}$. By Theorem 2.6 we have

$$\kappa = \prod_{i=1}^{n} \varepsilon_d \left(\frac{2r_i}{d}\right) = \varepsilon_d^n \left(\frac{\det(2bS)}{d}\right) = \varepsilon_d^{-n} \left(\frac{2c}{d}\right)^n \left(\frac{\det(-bcS)}{d}\right).$$

Now $\mu(\sigma_\gamma, R(z))$ is given by (10.9). Therefore, by (4.40) we obtain

$$h(\sigma_\gamma, R(z)) = \varepsilon_d^{2q} \left(\frac{\det(-bcS)}{d}\right) h_\gamma(z)^p \overline{h_\gamma(z)}^q,$$

which gives (10.26). If $S \prec \mathbf{Z}$, we can replace $-bcS$ by S, since $bc + 1 \in d\mathbf{Z}$. This completes the proof.

Theorem 10.9. *Given* $0 < S \in S_n(\mathbf{Z})$, *put*

$$f(z) = \sum_{\xi \in \mathbf{Z}^n} \chi(\xi)\mathbf{e}(2^{-1}S[\xi]z) \qquad (z \in \mathfrak{H}) \tag{10.27}$$

with χ *as in (10.16) such that* $\{\tau\} = \emptyset$. *Then* $f \in \mathscr{M}_k$ *with* $k = \ell + n/2$, $\ell = \sum_{\rho \in \{\rho\}} \ell_\rho$, *and* $f \in \mathscr{S}_k$ *if* $\ell > 0$. *Moreover, if* $\gamma = \begin{bmatrix} a & b \\ c & d \end{bmatrix} \in \Gamma(1)$, $bS \prec 2\mathbf{Z}$, $cS^{-1} \prec 2\mathbf{Z}$, *and* $d > 0$, *then*

$$f(\gamma z) = \left(\frac{\det(S)}{d}\right) h_\gamma(z)^n j_\gamma(z)^\ell f(z). \tag{10.28}$$

PROOF. Since $bc \in 4\mathbf{Z}$, d is odd. Take $S = P$ and $\tau = \emptyset$ in Theorem 10.6; take also λ to be the characteristic function of $\mathbf{Z}^n$. Then our $f(z)$ is $f(z, \lambda)$, or rather, $f(z) = \mathfrak{f}(0,\, z;\, \lambda)$ with $\mathfrak{f}$ of (10.11). Therefore we have

$$\mathfrak{f}({}^tM_\gamma^{-1}u,\, \gamma z;\, \lambda) = h(\sigma_\gamma,\, zS)\mathfrak{f}(u,\, z;\, \lambda)$$

for $\gamma \in \Gamma^\theta$. Now $h(\sigma_\gamma, zS)$ is determined by Lemma 10.8. Then the application of D_ρ^ℓ produces $j_\gamma(z)^\ell$ as explained in the proof of Theorem 10.6, and so we obtain (10.28). Also, from (10.21) we see that f satisfies condition (3.4d). That $f \in \mathscr{S}_k$ if $\ell > 0$ can easily be seen. This proves our theorem.

The fact that a theta series of type (10.27) belongs to $\mathscr{M}_k$ was proved by Hecke and Schoeneberg when n is even. A general formula for both even and odd n was given in [S73a]. There is a paper cited in [S73a] that treated the case of odd n, but its proof is erroneous.

10.10. Let us now prove the last assertion of Theorem 10.6. Take an integer m so that $mS \prec \mathbf{Z}$. Suppose $\gamma = \begin{bmatrix} a & b \\ c & d \end{bmatrix} \in \Gamma(r,\, s)$ with $d > 0$ and positive even integers r and s such that $rs \in m\mathbf{Z}$ and $\sigma_\gamma \in \Gamma_n^\theta$. Put $bc = mt$. Since $mt + 1 = bc + 1 \in d\mathbf{Z}$, we have

$$\left(\frac{(-1)^p \det(bcS)}{d}\right) = \left(\frac{(-1)^p t^n \det(mS)}{d}\right) = \left(\frac{(-1)^q m^n \det(mS)}{d}\right) = 1$$

if $d - 1 \in 4m^n \det(mS)\mathbf{Z}$. By Lemma 10.8, we have $h(\sigma_\gamma, R(z)) = J_\gamma^{p,q}(z)$ for such a γ, that is, for γ in a sufficiently small congruence subgroup. Also, our proof of (10.21) shows that λ^γ does not depend on χ, that is, λ^γ is determined by (10.22). Since $h(\sigma_\gamma, R(z)) = J_\gamma^{p,q}(z)$ for γ as above, comparing (10.22) with (10.13), we have $\lambda^\sigma = \lambda^\gamma$ for $\sigma = \sigma_\gamma$, where λ^σ is determined by (4.49a). Therefore the last assertion of Theorem 10.6 follows from (4.49f).

10.11. If S of §10.3 has signature $(m,\, 2)$ with $0 < m \in \mathbf{Z}$ (that is, (p, q) above is $(m,\, 2)$), then the symmetric space associated with $\mathfrak{P}(S)$ in the sense of §10.2 is a noncompact hermitian symmetric space, and so has a complex structure. Here, however, without proving this in a precise form, let us merely show that $\mathfrak{P}(S)$ can be parametrized by some complex variables.

For that purpose, we take, instead of $S \in S(\mathbf{R}^n)$, a vector space V over $\mathbf{R}$ of dimension $m+2$, and take also an $\mathbf{R}$-bilinear symmetric form $S : V \times V \to \mathbf{R}$ that has signature $(m, 2)$. We put $V_{\mathbf{C}} = V \otimes_{\mathbf{R}} \mathbf{C}$ and extend S to a $\mathbf{C}$-valued $\mathbf{C}$-bilinear form on $V_{\mathbf{C}} \times V_{\mathbf{C}}$, using the same letter S. For $v \in V_{\mathbf{C}}$ we can define its complex conjugate $\bar{v} \in V_{\mathbf{C}}$ in an obvious way. We now put

$$\mathfrak{Y}(S) = \{v \in V_{\mathbf{C}} \mid S[v] = 0,\ S(v, \bar{v}) < 0\}, \tag{10.29}$$

where $S[v] = S(v, v)$. Taking $1_{m,2}$ as S, we see that $\mathfrak{Y}(S) \neq \emptyset$. Given $v \in \mathfrak{Y}(S)$, put $u = v + \bar{v}$, $u' = iv - i\bar{v}$, $W = \mathbf{R}u + \mathbf{R}u'$, and $W' = \{y \in V \mid S(y, W) = 0\}$. Since $S[u] < 0$, $S[u'] < 0$, and $S(u, u') = 0$, we see that $\dim(W) = 2$, $V = W \oplus W'$, S is negative definite on W, and S is positive definite on W'. Let $W_{\mathbf{C}} = W \otimes_{\mathbf{R}} \mathbf{C}$, Then $W_{\mathbf{C}} = \mathbf{C}v + \mathbf{C}\bar{v}$. Define $P_v : V \times V \to \mathbf{R}$ so that $P_v = -S$ on $W \times W$, $P_v = S$ on $W' \times W'$, and $P_v(W, W') = 0$. Then P_v is positive definite, and we have $W'_{\mathbf{C}} = V^+_{\mathbf{C}}$ and $W_{\mathbf{C}} = V^-_{\mathbf{C}}$ in the sense of (10.15) with respect to (S, P_v); also, we have

$$P_v[\xi] - S[\xi] = -4S(v, \bar{v})^{-1}|S(\xi, v)|^2 \quad \text{for every} \quad \xi \in V. \tag{10.30}$$

Indeed, given $\xi \in V$, we can find $c \in \mathbf{C}$ and $z \in W'$ such that $\xi = cv + \bar{c}\bar{v} + z$. Then $S(\xi, v) = \bar{c}S(v, \bar{v})$ and $P_v[\xi] - S[\xi] = -2S[cv + \bar{c}\bar{v}] = -4c\bar{c}S(v, \bar{v})$, and so we obtain (10.30). By Lemma 10.1, the matrices representing S and P_v with respect to an $\mathbf{R}$-basis of V are of the type described in (i) of that lemma.

11. Theta integrals

11.1. In the setting of §10.11 let us consider the special case $m = 1$ by taking

$$V = \{x \in M_2(\mathbf{R}) \mid \mathrm{tr}(x) = 0\}, \qquad S(x, y) = -2^{-1}\mathrm{tr}(xy) \qquad (x, y \in V). \tag{11.1}$$

Thus $S[\xi] = -a^2 - bc = \det(\xi)$ for $\xi = \begin{bmatrix} a & b \\ c & -a \end{bmatrix} \in V_{\mathbf{C}}$. We put

$$p(w) = \begin{bmatrix} w & -w^2 \\ 1 & -w \end{bmatrix}, \quad q(w) = \begin{bmatrix} 1 & -2w \\ 0 & -1 \end{bmatrix} = \frac{\partial p(w)}{\partial w} \qquad (w \in \mathfrak{H} \cup \bar{\mathfrak{H}}), \tag{11.2a}$$

$$[\xi, w] = 2S(\xi, p(w)) \qquad (\xi \in V). \tag{11.2b}$$

Then we have

$$[\xi, w] = cw^2 - 2aw - b \quad \text{if} \quad \xi = \begin{bmatrix} a & b \\ c & -a \end{bmatrix} \in V, \tag{11.3a}$$

$$\gamma p(w)\gamma^{-1} = j_\gamma(w)^2 p(\gamma w) \quad \text{if} \quad \gamma \in SL_2(\mathbf{R}), \tag{11.3b}$$

$$[\gamma^{-1}\xi\gamma, w] = j_\gamma(w)^2[\xi, \gamma w] \quad \text{if} \quad \gamma \in SL_2(\mathbf{R}). \tag{11.3c}$$

Formula (11.3a) is easy. As for the latter two formulas, we can easily verify that $p(w) = \begin{bmatrix} w \\ 1 \end{bmatrix} [w \quad 1]\iota$ and $\gamma^{-1} = \iota^{-1} \cdot {}^t\gamma\iota$ with $\iota = \begin{bmatrix} 0 & -1 \\ 1 & 0 \end{bmatrix}$ for $\gamma \in SL_2(\mathbf{R})$. Therefore

$$\gamma p(w)\gamma^{-1} = \gamma \begin{bmatrix} w \\ 1 \end{bmatrix} [w \quad 1] \cdot {}^t\gamma\iota = j_\gamma(w)^2 \begin{bmatrix} \gamma w \\ 1 \end{bmatrix} [\gamma w \quad 1]\iota,$$

which gives (11.3b). The left-hand side of (11.3c) equals $2S(\gamma^{-1}\xi\gamma,\, p(w)) = 2S(\xi,\, \gamma p(w)\gamma^{-1}) = 2j_\gamma(w)^2 S(\xi,\, p(\gamma w))$ by (11.3b), and so we obtain (11.3c).

Clearly $p(w) \in V_{\mathbf{C}}$ and $S[p(w)] = 0$. Also, a direct calculation shows

$$S(p(w),\, \overline{p(w)}) = 2^{-1}(w - \bar{w})^2 = -2\mathrm{Im}(w)^2 < 0, \tag{11.4}$$

and so $p(w) \in \mathfrak{Y}(S)$.

11.2. For $v = p(w)$ define P_v as in §10.10 and put $R_w(z) = xS + iyP_v$ for $z = x + iy \in \mathfrak{H}$. Then for $\xi \in V_{\mathbf{C}}$ we have

$$\begin{aligned} R_w(z)[\xi] &= S[\xi]z + iy(P_v[\xi] - S[\xi]) \\ &= \det(\xi)z + 2^{-1}iy \cdot \mathrm{Im}(w)^{-2}|[\xi,\, w]|^2, \end{aligned} \tag{11.5}$$

because of (10.30), (11.2b), and (11.4). We have seen that $V_{\mathbf{C}}^- = W_{\mathbf{C}} = \mathbf{C}v + \mathbf{C}\bar{v}$ in §10.11, and so $P_v v = -Sv$ and $P_v\bar{v} = -S\bar{v}$.

We now consider a function of the form

$$\Theta(z,\, w;\, \eta) = \mathrm{Im}(z)^{1/2}\mathrm{Im}(w)^{-2m} \sum_{\xi \in V} \eta(\xi)[\xi,\, \bar{w}]^m \mathbf{e}(2^{-1}R_w(z)[\xi]). \tag{11.6}$$

Here $0 \le m \in \mathbf{Z}$, $(z,\, w) \in \mathfrak{H} \times \mathfrak{H}$, and $\eta \in \mathscr{L}(V)$. This is $y^{1/2}$ times a special case of (10.17). Indeed, we have $S[p(\bar{w})] = 0$ and $[\xi,\, \bar{w}] = 2S(\xi,\, p(\bar{w}))$, and $P_v p(\bar{w}) = -Sp(\bar{w})$ as noted above, and so $\mathrm{Im}(w)^{-2m}[\xi,\, \bar{w}]^m$ is a special case of $\chi(\xi) = ({}^t\tau S\xi)^m$ of (10.16). Notice that $p = 1$ and $q = 2$ in the present case. We put

$$k = m + 1/2, \quad j_\gamma^k(z) = h_\gamma(z)j_\gamma(z)^m \qquad (\gamma \in \Gamma^\theta). \tag{11.7}$$

This is consistent with (5.1b). Then from Theorem 10.6 we obtain

$$\Theta(\gamma z,\, w;\, \eta) = \overline{j_\gamma^k(z)}\Theta(z,\, w;\, \eta) \quad \text{for every} \quad \gamma \in \Gamma \tag{11.8}$$

with a sufficiently small congruence subgroup Γ of Γ^θ. (The factor $y^{1/2}$ in (11.6) eliminates $|j_\gamma|$.) Also, from (11.3c) we easily obtain

$$\Theta(z,\, \beta w;\, \eta) = j_\beta(w)^{2m}\Theta(z,\, w;\, \eta^\beta) \quad \text{for every} \quad \beta \in GL_2^+(\mathbf{Q}), \tag{11.9}$$

where $\eta^\beta(\xi) = \eta(\beta\xi\beta^{-1})$. In the rest of this section we assume $m > 0$.

Theorem 11.3. (i) *Given* $f \in \mathscr{M}_k(\Gamma)$ *with* $k = m+1/2$, $0 < m \in \mathbf{Z}$, *and a congruence subgroup* Γ *for which* (11.8) *holds, put*

$$g(w) = \int_{\Phi} f(z)\Theta(z,\, w;\, \eta)y^k \mathbf{d}z, \qquad \Phi = \Gamma\backslash\mathfrak{H}, \tag{11.10}$$

where $\mathbf{d}z$ *is as in* (6.5). *Then the integral is convergent and* g *belongs to* $\mathscr{M}_{2m}$.

(ii) *In the setting of* (i) *suppose that* f *is a cusp form. Then* g *is a cusp form at least in the following two cases:* (a) $m > 1$; (b) $m = 1$ *and* $\langle f,\, \theta^*(z,\, \mu)\rangle = 0$ *for every* $\mu \in \mathscr{L}(\mathbf{Q})$, *where* θ^* *is as in* (4.51).

We call the expression on the right-hand side of (11.10) a **theta integral**. The proof of (i) will be completed in §11.7 and (ii) will be proven in §12.7.

11.4. In this subsection we assume that every integral is convergent. The convergence will be proven in the next subsection. Now we need to compute $(\partial/\partial\bar{w})\Theta(z,\, w;\, \eta)$. We have $(\partial/\partial\bar{w})[\xi,\, \bar{w}]^m = 2mS\bigl(\xi,\, q(\bar{w})\bigr)[\xi,\, \bar{w}]^{m-1}$, where $q(w) = \partial p(w)/\partial w$, as we defined in (11.2a). Writing simply p and q for $p(w)$ and $q(w)$, we obtain, from (11.5) and (10.30),

$$\begin{aligned}(\partial/\partial\bar{w})R_w(z)[\xi] &= iy(\partial/\partial\bar{w})P_v[\xi]\\ &= -4iy(\partial/\partial\bar{w})\bigl\{S(p,\, \bar{p})^{-1}S(\xi,\, p)S(\xi,\, \bar{p})\bigr\}\\ &= 4iyS(\xi,\, p)\bigl\{S(p,\, \bar{p})^{-2}S(p,\, \bar{q})S(\xi,\, \bar{p}) - S(p,\, \bar{p})^{-1}S(\xi,\, \bar{q})\bigr\}\\ &= iy\cdot \mathrm{Im}(w)^{-4}S(\xi,\, p)S(\xi,\, r) \quad \text{with}\end{aligned}$$

$$r = S(p,\, \bar{q})\bar{p} - S(p,\, \bar{p})\bar{q} = 2\mathrm{Im}(w)^2\bar{q} - 2i\mathrm{Im}(w)\bar{p}, \tag{11.11}$$

since $S(p,\, \bar{q}) = -2i\mathrm{Im}(w)$. Notice that

$$S(p,\, q) = S(p,\, r) = S(\bar{p},\, r) = 0. \tag{11.12}$$

Thus, putting $E = \mathbf{e}\bigl(2^{-1}R_w(z)[\xi]\bigr)$ for simplicity, we have

$$\begin{aligned}(\partial/\partial\bar{w})\Theta(z,\, w;\, \eta) &= -miy^{1/2}\mathrm{Im}(w)^{-2m-1}\sum_{\xi\in V}\eta(\xi)[\xi,\, \bar{w}]^m E\\ &\quad + 2my^{1/2}\mathrm{Im}(w)^{-2m}\sum_{\xi\in V}\eta(\xi)[\xi,\, \bar{w}]^{m-1}S(\xi,\, \bar{q})E\\ &\quad - (\pi/2)y^{3/2}\mathrm{Im}(w)^{-2m-4}\sum_{\xi\in V}\eta(\xi)[\xi,\, \bar{w}]^m[\xi,\, w]S(\xi,\, r)E.\end{aligned} \tag{11.13}$$

We need an auxiliary series H defined by

$$\begin{aligned}H(z,\, w;\, \eta) &= y^{3/2}\mathrm{Im}(w)^{-2m-2}\\ &\quad \cdot\sum_{\xi\in V}\eta(\xi)[\xi,\, \bar{w}]^{m-1}S(\xi,\, r)\mathbf{e}\bigl(2^{-1}R_w(z)[\xi]\bigr),\end{aligned} \tag{11.14}$$

and also the operator $\delta_{k-2} = (k-2)(2iy)^{-1} + \partial/\partial z$ on $\mathfrak{H}$ defined in (6.13b). Writing simply H for $H(z,\, w;\, \eta)$, we are going to prove

$$(\partial/\partial\bar{w})\Theta(z,\, w;\, \eta) = -2i\overline{\delta_{k-2}\bar{H}}. \tag{11.15}$$

Indeed, since $R_w(z) = xS + iyP_v$, we have, by (10.30),

$$\begin{aligned}(\partial/\partial\bar{z})R_w(z)[\xi] &= 2^{-1}(S - P_v)[\xi]\\ &= 2S(p,\bar{p})^{-1}|S(\xi,\, p)|^2 = -4^{-1}\mathrm{Im}(w)^{-2}[\xi,\, w][\xi,\, \bar{w}],\end{aligned}$$

and so

$$\begin{aligned}\overline{\delta_{k-2}\bar{H}} &= (2-k)(2iy)^{-1}H + (\partial/\partial\bar{z})H \\ &= (k-2)(i/2)y^{1/2}\mathrm{Im}(w)^{-2m-2}\sum_{\xi\in V}\eta(\xi)[\xi,\, \bar{w}]^{m-1}S(\xi,\, r)E\\ &\quad + (3i/4)y^{1/2}\mathrm{Im}(w)^{-2m-2}\sum_{\xi\in V}\eta(\xi)[\xi,\, \bar{w}]^{m-1}S(\xi,\, r)E\\ &\quad - (\pi i/4)y^{3/2}\mathrm{Im}(w)^{-2m-4}\sum_{\xi\in V}\eta(\xi)[\xi,\, \bar{w}]^{m}[\xi,\, w]S(\xi,\, r)E.\end{aligned} \tag{11.16}$$

The last line times $-2i$ equals the last line of (11.13). Substituting (11.11) into r of the first two terms of the right-hand side of (11.16) and multiplying by $-2i$, we obtain the first two terms on the right-hand side of (11.13). This proves (11.15). Now $g(w)$ is $\mu(\Phi)$ times $\langle \bar{f},\, \Theta(z,\, w;\, \eta)\rangle$, and so $\partial g/\partial\bar{w}$ is $\mu(\Phi)$ times

$$\begin{aligned}(\partial/\partial\bar{w})\langle \bar{f},\, \Theta(z,\, w;\, \eta)\rangle &= \langle \bar{f},\, (\partial/\partial\bar{w})\Theta(z,\, w;\, \eta)\rangle\\ &= -2i\langle \bar{f},\, \overline{\delta_{k-2}\bar{H}}\rangle = -2i\langle \delta_{k-2}\bar{H},\, f\rangle = -2i\langle \bar{H},\, \varepsilon f\rangle\end{aligned} \tag{11.17}$$

by (6.16). This is 0, since f is holomorphic. Thus $g(w)$ is holomorphic in w. By Lemma 10.7, $\Theta(z,\, w;\, \eta)$ is rapidly decreasing locally uniformly in w, and so the first equality of (11.17) is justified.

11.5. Since $m > 0$, $\Theta(z,\, w;\, \eta)$ is rapidly decreasing at every cusp as a function in z by Lemma 10.7. Every element of $\mathscr{M}_k$ is slowly increasing at every cusp, and so the integral of (11.10) is convergent. Similarly, H of (11.14) is a special case of (10.17), since $r \in V_{\mathbf{C}}^-$ and $S(\bar{p},\, r) = 0$, and so by Lemma 10.7, H is rapidly decreasing at every cusp. Also, as can be seen from (11.13) or (11.16), $(\partial/\partial\bar{w})\Theta$ and $\delta_{k-2}\bar{H}$ are rapidly decreasing at every cusp. Therefore all the inner products appearing in (11.17) are meaningful; also the last equality of (11.17) can be justified, since the conditions stated in Theorem 6.8 are satisfied.

To study the nature of g beyond its holomorphy, we first put $g(w) = g(w,\, \eta)$. From (11.9) we obtain $g(w,\, \eta)\|_{2m}\,\beta = g(w,\, \eta^\beta)$ for every $\beta\in SL_2(\mathbf{Q})$. Put $\Delta = \{\beta \in SL_2(\mathbf{Q}) \mid \eta^\beta = \eta\}$. Then Δ is a congruence subgroup and $g\|_{2m}\,\delta = g$ for every $\delta \in \Delta$. Therefore, in view of Lemma 6.4, to show that $g \in \mathscr{M}_{2m}$, it is sufficient to show that g is slowly increasing at every cusp. For that purpose, we take T and E as in (6.10a, b) As shown there, $\Gamma\backslash\mathfrak{H}$ is covered by $\bigcup_{\varepsilon\in E}\varepsilon T$, and so the integral over Φ is majorized by $\sum_{\varepsilon\in E}\bigl|\int_{\varepsilon T}\bigr|$.

Thus, putting $f_\varepsilon(z) = j_\varepsilon(z)^{-k}f(\varepsilon z)$, our question is reduced to the estimate of $\int_T f_\varepsilon(z)\Theta(z,\, w;\, \eta)y^k\mathbf{d}z$, or rather, to the problem of showing that

$$h(w) = \int_T f(z)\Theta(z,\, w;\, \eta)y^k\mathbf{d}z$$

as a function of w satisfies (6.9a) (with $2m$ and h in place of k and f there) for every $f \in \mathscr{M}_k$. Fixing f, let $\alpha \in SL_2(\mathbf{Q})$. Then

$$\begin{aligned}\mathrm{Im}(\alpha w)^m h(\alpha w) &= \mathrm{Im}(w)^m \int_T f(z)j_\alpha(w)^{-2m}\Theta(z,\, \alpha w;\, \eta)y^k\mathbf{d}z \\ &= \mathrm{Im}(w)^m \int_T f(z)\Theta(z,\, w;\, \eta^\alpha)y^k\mathbf{d}z\end{aligned}$$

with η^α as in (11.9). Our task is to make an estimate of $\Theta(z,\, w;\, \zeta)$ for $\zeta \in \mathscr{L}(V)$, when $\mathrm{Im}(w)$ is sufficiently large. It is sufficient to treat the case in which ζ is the characteristic function of a lattice L in V. Then Θ is invariant under $w \mapsto w + 2p$ with a positive number p. Thus our estimate will be made under the condition

(11.18) $$|\mathrm{Re}(w)| \le p \quad \text{and} \quad \mathrm{Im}(w) > q$$

with a positive constant q. Changing the notation, write P_w for P_v of (11.5). For $\xi = \begin{bmatrix} a & b \\ c & -a \end{bmatrix} \in V$ put $\|\xi\| = (a^2 + b^2 + c^2)^{1/2}$. In §11.7 we will prove

(11.19) $$P_w[\xi] \ge 8^{-1}\mathrm{Im}(w)^{-2}\|\xi\|^2 \quad \text{under (11.18)},$$

(11.20) $$\big|[\xi,\, w]\big|^2 \le A \cdot \mathrm{Im}(w)^4\|\xi\|^2 \quad \text{under (11.18)},$$

with a constant A depending only on p and a sufficiently large q.

We need two more easy facts. The first one is:

(11.21) *For $0 \le m \in \mathbf{Z}$ there is a constant B_m depending only on m such that $\sum_{N=1}^{\infty} N^m e^{-tN} \le B_m t^{-m-1}$ for $0 < t \le 1$.*

Indeed, $\sum_{N=1}^{\infty} N^m x^N = xF_m(x)(1-x)^{-m-1}$ with a polynomial F_m of degree $\mathrm{Max}\{m-1,\, 0\}$. This can be obtained by applying $x \cdot d/dx$ successively to $x/(1-x) = \sum_{N=1}^{\infty} x^N$. Putting $x = e^{-t}$ and observing that $t/2 \le 1 - e^{-t}$ for $0 < t \le 1$, we obtain the desired inequality.

Next, for $L = \mathbf{Z}^n$ and $0 < N \in \mathbf{Z}$ we have

(11.22) $\#\{\xi \in L \,|\, \mathrm{Max}_{1 \le \nu \le n}|\xi_\nu| = N\} \le C_n N^{n-1}$ *with a constant C_n depending only on n.*

The proof is left to the reader, as it is completely elementary.

11.6. For ξ in the set of (11.22) we have $N \le \|\xi\|^2 \le nN^2$, and so for $0 < m \in \mathbf{Z}$ and $y > 1/2$ we have

$$\begin{aligned}(11.23)\quad \sum_{\xi\in L}\|\xi\|^m \exp\bigl(-yt\|\xi\|^2\bigr) &\le \exp(-yt/2)\sum_{\xi\in L}\|\xi\|^m \exp\bigl(-yt\|\xi\|^2/2\bigr)\\ &\le C_n\exp(-yt/2)\sum_{N=1}^{\infty} N^{n-1}(n^{1/2}N)^m\exp(-ytN/2)\\ &= n^{m/2}C_n\exp(-yt/2)\sum_{N=1}^{\infty} N^{m+n-1}\exp(-tN/4)\\ &\le 4^{m+n}n^{m/2}C_nB_{m+n-1}t^{-m-n}\exp(-yt/2)\end{aligned}$$

for $0 < t \le 4$, by (11.21). We apply this to $\Theta(z, w; \zeta)$ with the characteristic function of $L = V \cap M_2(\mathbf{Z})$ as ζ. (Thus $n = 3$.) Since $\bigl|\mathbf{e}(2^{-1}R_w(z)[\xi])\bigr| = \exp\bigl(-\pi y P_w[\xi]\bigr)$, we have, by (11.6), (11.19), and(11.20),

$$\begin{aligned}\bigl|\Theta(z, w; \zeta)\bigr| &\le y^{1/2}\mathrm{Im}(w)^{-2m}\sum_{\xi\in L}\bigl|[\xi, w]\bigr|^m \exp\bigl(-\pi y P_w[\xi]\bigr)\\ &\le A^{m/2}y^{1/2}\sum_{\xi\in L}\|\xi\|^m\exp\bigl(-(\pi/8)y\mathrm{Im}(w)^{-2}\|\xi\|^2\bigr).\end{aligned}$$

Applying the estimate of (11.23) to this, we find that, under (11.18),

$$\bigl|\Theta(z, w; \zeta)\bigr| \le Dy^{1/2}\mathrm{Im}(w)^{2m+6}\exp\bigl(-(\pi/16)y\mathrm{Im}(w)^{-2}\bigr)$$

with a constant D that depends only on m. (We take z in T, and so $y > 1/2$.) Now $f(z)$ is bounded on T and $k = m + 1/2$, and so

$$\mathrm{Im}(w)^m\left|\int_T f(z)\Theta(z, w; \zeta)y^k\mathbf{d}z\right| \le E\cdot\mathrm{Im}(w)^{3m+6}\int_{1/2}^{\infty} e^{-ay}y^{m-1}dy$$

with $a = (\pi/8)\mathrm{Im}(w)^{-2}$ and a constant E. Replacing $\int_{1/2}^{\infty}$ by $\int_0^{\infty}$ and noting that $\int_0^{\infty} e^{-ay}y^{m-1}dy = \Gamma(m)a^{-m}$, we see that

$$\mathrm{Im}(w)^m\left|\int_T f(z)\Theta(z, w; \zeta)y^k\mathbf{d}z\right| \le F\cdot\mathrm{Im}(w)^{5m+6}$$

with a constant F, which proves that g of (11.10) is slowly increasing at every cusp, and so $g \in \mathscr{M}_{2m}$ as stated in Theorem 11.3(i).

11.7. It remains to prove (11.19) and (11.20). By (11.5) we have

$$(11.24)\qquad P_w[\xi] = \det(\xi) + 2^{-1}\mathrm{Im}(w)^{-2}\bigl|[\xi, w]\bigr|^2.$$

Put $u = \mathrm{Re}(w)$ and $v = \mathrm{Im}(w)$. Then a direct calculation shows that

$$(*)\quad \bigl|[\xi, w]\bigr|^2 = c^2|w|^4 + 4a^2|w|^2 + b^2 - 4acu|w|^2 - 2bc(u^2 - v^2) + 4abu,$$

and so

$$\begin{aligned}2v^2P_w[\xi] &= \bigl|[\xi, w]\bigr|^2 - 2(a^2 + bc)v^2\\ &= c^2|w|^4 + 4a^2|w|^2 + b^2 - 2a^2v^2 - 2bcu^2 - 4acu|w|^2 + 4abu.\end{aligned}$$

We are assuming (11.18). Thus $|u| \le p$, and so $-2bcu^2 \ge -2|2^{-1}b \cdot 2p^2c| \ge -(4^{-1}b^2 + 4p^4c^2)$. Similarly $-4acu|w|^2 \ge -(a^2 + 4p^2c^2)|w|^2$ and $4abu \ge -(8p^2a^2 + 2^{-1}b^2)$. Therefore

$$2v^2 P_w[\xi] \ge a^2(3|w|^2 - 2v^2 - 8p^2) + 4^{-1}b^2 + c^2\bigl(|w|^4 - 4p^2|w|^2 - 4p^4\bigr).$$

We easily see that the right-hand side is $\ge 4^{-1}(a^2 + b^2 + c^2)$ for sufficiently large v, which proves (11.19).

As for (11.20), viewing $[\xi,\, w]$ as the inner product of the vectors $(c,\, a,\, b)$ and $(w^2,\, -2w,\, -1)$, we find that

$$\bigl|[\xi,\, w]\bigr|^2 \le \|\xi\|^2\bigl(|w|^4 + 4|w|^2 + 1\bigr),$$

from which we obtain (11.20). This completes the proof of Theorem 11.3(i).

12. Main theorems on the correspondence

12.1. We need nonholomorphic modular forms involving the Hermite polynomial $H_n(x)$ defined by

$$H_n(x) = (-1)^n \exp(x^2/2)(d/dx)^n \exp(-x^2/2) \qquad (0 \le n \in \mathbf{Z}). \tag{12.1}$$

This is a polynomial of degree n with coefficients in $\mathbf{Z}$. We easily obtain

$$(-\sqrt{c})^n H_n(\sqrt{c}\,x) = \exp(cx^2/2)(d/dx)^n \exp(-cx^2/2) \qquad (c > 0). \tag{12.1a}$$

Here are some basic formulas on H_n, in which $H_n'(x) = (d/dx)H_n(x)$:

$$H_0(x) = 1, \qquad H_1(x) = x, \tag{12.2}$$

$$H_n(-x) = (-1)^n H_n(x), \tag{12.3}$$

$$H_{n+1}(x) = xH_n(x) - H_n'(x), \tag{12.4}$$

$$H_n'(x) = nH_{n-1}(x), \tag{12.5}$$

$$(x+iy)^n = \sum_{k=0}^{n} \binom{n}{k} i^k H_k(y) H_{n-k}(x), \tag{12.6}$$

$$\int_0^\infty H_n(\sqrt{4\pi y}\,a)\mathbf{e}(ia^2y)y^{(s/2)-1}dy = 2^{-n/2}(2\pi)^{-s/2}a^{-s}\Gamma\bigl((s-n)/2\bigr)\prod_{\nu=1}^{n}(s-\nu) \quad (\mathrm{Re}(s) > n,\ a > 0). \tag{12.7}$$

The first three are easy; (12.5) can be derived from (12.4) by induction on n. As for (12.6), putting $E = e^{-z\bar{z}/2}$, we have $(x+iy)^n E = (-2\partial/\partial\bar{z})^n E = \sum_{k=0}^{n} \binom{n}{k} (-i\partial/\partial y)^k(-\partial/\partial x)^{n-k}E$, which gives the desired result. The case $n = 0$ of (12.7) is

$$\int_0^\infty \mathbf{e}(ia^2y)y^{(s/2)-1}dy = \Gamma(s/2)(2\pi)^{-s/2}a^{-s} \qquad (\mathrm{Re}(s) > 0),$$

which follows from (8.2). Applying d^n/da^n to this, we obtain, by (12.1a),

$$(-1)^n \int_0^\infty H_n(\sqrt{4\pi y}\,a)\mathbf{e}(ia^2y)(4\pi y)^{n/2}y^{(s/2)-1}dy$$
$$= (2\pi)^{-s/2}a^{-s-n}\Gamma(s/2)\prod_{\nu=1}^{n}(1-s-\nu).$$

Substituting $s-n$ for s, we obtain (12.7).

The nonholomorphic form we need is

$$\theta_n(z,\,\lambda) = (4\pi y)^{-n/2}\sum_{\xi\in\mathbf{Q}}\lambda(\xi)H_n(\sqrt{4\pi y}\,\xi)\mathbf{e}(\xi^2z/2), \tag{12.8}$$

where $z \in \mathfrak{H}, \lambda \in \mathscr{L}(\mathbf{Q})$, and $y = \mathrm{Im}(z)$. Since $H_n(x)$ is a polynomial in x of degree n, the sum is convergent. Clearly

$$\theta_0(z,\,\lambda) = \sum_{\xi\in\mathbf{Q}}\lambda(\xi)\mathbf{e}(\xi^2z/2) = \theta(0,\,z;\,\lambda), \tag{12.9a}$$

$$\theta_1(z,\,\lambda) = \sum_{\xi\in\mathbf{Q}}\lambda(\xi)\xi\mathbf{e}(\xi^2z/2) = \theta^*(z,\,\lambda), \tag{12.9b}$$

where $\theta(0,\,z;\,\lambda)$ is as in (4.48b) with $n = 1$ and θ^* as in (4.51), and so $\theta_n(z,\,\lambda)$ is holomorphic in z for $n \le 1$. From (4.50b) and (4.52) we obtain

$$\theta_0(\alpha z,\,\lambda) = h_\alpha(z)\theta_0(z,\,\lambda^\alpha) \quad \text{for every} \quad \alpha\in P\Gamma^\theta, \tag{12.9c}$$

$$\theta_1(\alpha z,\,\lambda) = h_\alpha(z)j_\alpha(z)\theta_1(z,\,\lambda^\alpha) \quad \text{for every} \quad \alpha\in P\Gamma^\theta. \tag{12.9d}$$

Consequently $\theta_0(z,\,\lambda)$ belongs to $\mathscr{M}_{1/2}$ and $\theta_1(z,\,\lambda)$ to $\mathscr{S}_{3/2}$, as we already observed in §§5.1 and 5.8. Now we have:

Lemma 12.2. *Let the symbols $h_\alpha(z)$ and λ^α be defined as in Theorem 4.12 in the one-dimensional case. Then for every $\alpha \in P\Gamma^\theta$ we have*

$$\theta_n\big(\alpha(z),\,\lambda\big) = h_\alpha(z)j_\alpha(z)^n\theta_n(z,\,\lambda^\alpha). \tag{12.10}$$

PROOF. We first prove

$$(\pi i)^{-1}\delta_k\theta_n(z,\,\lambda) = \theta_{n+2}(z,\,\lambda), \qquad k = n+1/2, \tag{12.10a}$$

where δ_k is the operator of (6.13b). To prove this, we put

$$K_n(z) = K_n(\xi,\,z) = (4\pi y)^{-n/2}H_n(\sqrt{4\pi y}\,\xi)\mathbf{e}(\xi^2z/2), \tag{12.11}$$

where ξ is fixed. Then our task is to prove that

$$(\pi i)^{-1}\delta_kK_n(z) = K_{n+2}(z), \tag{12.12}$$

which implies (12.10a). To make our formulas short, put $Y = \sqrt{4\pi y}$. Then

$$(\pi i)^{-1}(\partial/\partial z)K_n(z)$$
$$= Y^{-n}\mathbf{e}(\xi^2z/2)\big\{nY^{-2}H_n(Y\xi) - \xi Y^{-1}H_n'(Y\xi) + \xi^2H_n(Y\xi)\big\}.$$

Since $\delta_k = k(2iy)^{-1} + \partial/\partial z$, we have

$$(\pi i)^{-1}\delta_k K_n(z) = -2kY^{-2}K_n(z) + (\pi i)^{-1}(\partial/\partial z)K_n(z)$$
$$= Y^{-n-2}\mathbf{e}(\xi^2 z/2)\{-(n+1)H_n(Y\xi) - \xi Y H_n'(Y\xi) + \xi^2Y^2H_n(Y\xi)\}.$$

By (12.4) we have $Y\xi H_{n+1}(Y\xi) = Y^2\xi^2H_n(Y\xi) - Y\xi H_n'(Y\xi)$, and by (12.5), $H_{n+1}'(Y\xi) = (n+1)H_n(Y\xi)$. Therefore

$$(\pi i)^{-1}\delta_k K_n(z) = Y^{-n-2}\mathbf{e}(\xi^2 z/2)\{Y\xi H_{n+1}(Y\xi) - H_{n+1}'(Y\xi)\}$$
$$= Y^{-n-2}\mathbf{e}(\xi^2 z/2)H_{n+2}(Y\xi)$$

by (12.4). This proves (12.12), and (12.10a) as well. Recall the operator δ_r^m defined by $\delta_r^m = \delta_{r+2m-2}\cdots\delta_{r+2}\delta_r$ in (6.13d). Let $n = 2m+\nu$ and $r = \nu + 1/2$ with $\nu = 0$ or 1 and $0 < m \in \mathbf{Z}$. Then from (12.10a) we obtain

(12.12a) $$(\pi i)^{-m}\delta_r^m\theta_\nu(z,\,\lambda) = \theta_n(z,\,\lambda).$$

Applying $(\pi i)^{-m}\delta_r^m$ to (12.9c, d) and employing (6.14c), we obtain (12.10). This completes the proof.

12.3. This subsection concerns a formula for the Fourier transform of K_n. We do not need this result in our later discussion, however. We have $K_0(\xi,\,z) = \mathbf{e}(\xi^2z/2)$ and $K_1(\xi,\,z) = \xi\mathbf{e}(\xi^2z/2)$, and so from (12.12) we obtain, for $0 \le m \in \mathbf{Z}$ and $\nu = 0$ or 1,

(12.13) $$K_{2m+\nu}(\xi,\,z) = (\pi i)^{-m}\delta_{\nu+1/2}^m K_\nu(\xi,\,z) = (\pi i)^{-m}\delta_{\nu+1/2}^m\xi^\nu\mathbf{e}(\xi^2z/2).$$

Now we have an integral formula

(12.14) $$\int_{\mathbf{R}} K_\nu(\xi,\,z)\mathbf{e}(-\xi\eta)d\xi = (-iz)^{-1/2}z^{-\nu}K_\nu\big(\eta,\,\iota(z)\big) \qquad (\eta\in\mathbf{R}),$$

where $\iota = \begin{bmatrix} 0 & -1 \\ 1 & 0 \end{bmatrix}$. Indeed, if $\nu = 0$, this is merely (4.30) with $n = 1$. Applying $\partial/\partial\eta$ to it, we obtain the case $\nu = 1$. Then applying $(\pi i)^{-m}\delta_{\nu+1/2}^m$ to (12.14), from (12.13) and (6.14c) we obtain

(12.15) $$\int_{\mathbf{R}} K_{2m+\nu}(\xi,\,z)\mathbf{e}(-\xi\eta)d\xi = (-iz)^{-1/2}z^{-m-\nu}K_{2m+\nu}\big(\eta,\,\iota(z)\big)$$
$$(\nu = 0 \text{ or } 1,\ 0 \le m \in \mathbf{Z},\ \eta\in\mathbf{R}).$$

12.4. Fixing a positive integer N divisible by 4, we put, for simplicity,

(12.16) $$\Gamma_N^0 = \Gamma(2,\,N/2) = \big\{\gamma\in\Gamma(1)\,\big|\,b_\gamma\in 2\mathbf{Z},\ c_\gamma\in 2^{-1}N\mathbf{Z}\big\}.$$

In addition, we will be considering $\Gamma_0(N/2)$. We also take a half-integral weight k and a character ψ of $(\mathbf{Z}/N\mathbf{Z})^\times$ (which may be imprimitive) such that

(12.17) $$k = m + 1/2 \quad\text{and}\quad \psi(-1) = (-1)^m, \quad 0 < m \in \mathbf{Z}.$$

We then consider an element f of $\mathscr{S}_k(2,\,N/2;\,\psi)$ (see §8.6), that is, an element f of $\mathscr{S}_k$ such that

(12.18) $$f\|_k\gamma = \psi(d_\gamma)f \quad\text{for every}\quad \gamma\in\Gamma_N^0.$$

Next, we define $\eta \in \mathscr{L}(V)$ with V of §11.1 by

$$\eta\left(\begin{bmatrix} a & b \\ c & -a \end{bmatrix}\right) = \omega(a)\omega(2c/N)\eta_0(b), \tag{12.19}$$

where ω is the characteristic function of $\mathbf{Z}$ and

$$\eta_0(b) = \begin{cases} N^{-1} \sum_{t \in \mathbf{Z}/N\mathbf{Z}} \psi(t)\mathbf{e}(-bt/2) & \text{if } Nb \in 2\mathbf{Z}, \\ 0 & \text{otherwise,} \end{cases} \tag{12.19a}$$

with $\psi(t) = 0$ for t not prime to N. Since $\eta_0(b)$ depends only on $b \pmod{2\mathbf{Z}}$, we see that $\eta \in \mathscr{L}(V)$. Moreover, a simple calculation shows that

$$\eta(\beta\alpha\beta^{-1}) = \psi(d_\beta^2)\eta(\alpha) \quad \text{for every} \quad \beta \in \Gamma_0(N/2). \tag{12.20}$$

Let us simply write $\Theta(z, w)$ for $\Theta(z, w; \eta)$ of (11.6) with this η. Then from (12.20) and (11.9) we obtain

$$\Theta(z, \beta w)j_\beta(w)^{-2m} = \psi(d_\beta^2)\Theta(z, w) \quad \text{for every} \quad \beta \in \Gamma_0(N/2). \tag{12.21}$$

We have several aims: one is to prove Theorem 11.3(ii); the other is to make preparations for the proof of our main theorem, Theorem 12.8 below. We thus take f as above and define g by (11.10) with $\Gamma = \{\gamma \in \Gamma_N^0 \mid a_\gamma - 1 \in N\mathbf{Z}\}$. Then from Theorem 11.3(i) and (12.21) we see that $g \in \mathscr{M}_{2m}(N/2, \psi^2)$. Put

$$g(w) = \sum_{\xi \in \mathbf{Z}} c(\xi)\mathbf{e}(\xi w). \tag{12.22}$$

We first note that $g(ir) = \sum_{\xi \in \mathbf{Z}} c(\xi)e^{-2\pi\xi r}$ for $0 < r \in \mathbf{R}$. (We make these choices of f and Γ for some definite reason, but actually our calculation is valid for an arbitrary $f \in \mathscr{S}_k$ and a suitable Γ, with some obvious modifications, about which we will be more explicit in our later discussion.) In view of (8.2) we have

$$\int_0^\infty g(ir)r^{m+s-1}dr = (2\pi)^{-m-s}\Gamma(s+m) \sum_{0<\xi \in \mathbf{Z}} c(\xi)\xi^{-s-m}, \tag{12.23}$$

provided $c(0) = 0$ and the right-hand side is convergent, which is the case if $\mathrm{Re}(s) > m + 1$, because of the estimate of $c(\xi)$ given in Lemma 6.2(ii). Also, if $c(0) \neq 0$, the integral of (12.23) is divergent for large $\mathrm{Re}(s)$, since $|g(ir)| > |c(0)|/2$ for sufficiently large r. In the following subsection we will show that the integral is indeed convergent for sufficiently large $\mathrm{Re}(s)$, and so $c(0) = 0$ and (12.23) holds for $\mathrm{Re}(s) > m + 1$.

12.5. For $\xi = \begin{bmatrix} a & b \\ c & -a \end{bmatrix} \in V$ and $w = ir$ we have, by (11.3a) and (11.5),

$$[\xi, w] = -cr^2 - 2air - b, \quad \bigl|[\xi, w]\bigr|^2 = 4a^2r^2 + b^2 + c^2r^4 + 2bcr^2,$$
$$R_w(z)[\xi] = (-a^2 - bc)z + (iy/2)(4a^2 + b^2r^{-2} + 2bc + c^2r^2)$$

$$= -a^2\bar{z} - bcx + (iy/2)(c^2r^2 + b^2r^{-2}),$$

and so

$$y^{(m-1)/2}(\sqrt{\pi}\, r)^m \overline{\Theta(z,\, ir)} = (-1)^m \sum_{(a,b,c)\in\mathbf{Q}^3} \omega(a)\overline{\eta_0(b)}\omega(2c/N)$$
$$\cdot\, (\pi y)^{m/2}(cr + br^{-1} + 2ai)^m \mathbf{e}\big((a^2/2)z + (bc/2)x + (iy/4)(c^2r^2 + b^2r^{-2})\big).$$

By (12.6) we have

$$(\pi y)^{m/2}(cr + br^{-1} + 2ai)^m$$
$$= \sum_{n=0}^{m} \binom{m}{n} i^{m-n} H_n\big(\sqrt{\pi y}(cr + br^{-1})\big) H_{m-n}\big(\sqrt{4\pi y}\cdot a\big).$$

Therefore, putting

(12.24) $$A_n(z,\, r) = \sum_{(b,c)\in\mathbf{Q}^2} \overline{\eta_0(b)}\omega(2c/N) H_n\big(\sqrt{\pi y}(cr + br^{-1})\big)$$
$$\cdot\, \mathbf{e}\big((bc/2)x + (iy/4)(c^2r^2 + b^2r^{-2})\big),$$

(12.25) $$\tau_\ell(z) = y^{-\ell/2} \sum_{a\in\mathbf{Z}} H_\ell\big(\sqrt{4\pi y}\cdot a\big)\mathbf{e}(a^2z/2),$$

we obtain

(12.26) $$y^{(m-1)/2}(\sqrt{\pi}\, r)^m \overline{\Theta(z,\, ir)}$$
$$= (-1)^m \sum_{n=0}^{m} \binom{m}{n} i^{m-n} y^{(m-n)/2} \tau_{m-n}(z) A_n(z,\, r).$$

Notice that $\tau_\ell(z)$ is $(4\pi)^{\ell/2}$ times $\theta_\ell(z,\, \omega)$ of (12.8), and it is identically equal to 0 if ℓ is odd, because of (12.3).

We now need a formula

(12.27) $$i^{-n}2^{1/2}\pi^{-n/2}(y/r^2)^{(n+1)/2} A_n(z,\, r)$$
$$= \sum_{(c,d)\in T} \bar{\psi}(d)(c\bar{z} + d)^n \mathbf{e}\big((ir^2/(4y))|cz + d|^2\big),$$

where $T = \big\{(c,\, d) \in 2^{-1}N\mathbf{Z}\times\mathbf{Z} \,\big|\, d\mathbf{Z} + N\mathbf{Z} = \mathbf{Z}\big\}$. To prove this, we first note

(12.28) $$\int_{\mathbf{R}} \exp\big(-\pi p(u+q)^2\big)\mathbf{e}(-uv)du = p^{-1/2}\mathbf{e}(qv)\exp(-\pi v^2/p)$$

for $0 < p \in \mathbf{R}$, $q \in \mathbf{R}$, and $v \in \mathbf{R}$. This follows easily from (4.28). Given $z = x + iy \in \mathfrak{H}$, $0 < t \in \mathbf{R}$, and $c \in \mathbf{R}$, take $p = t^2/y$ and $q = cx$ in (12.28) and multiply the result by $\exp(-\pi y c^2 t^2)$. Then we obtain

(12.29) $$\int_{\mathbf{R}} \exp\big(-\pi t^2|cz + u|^2/y\big)\mathbf{e}(-uv)du$$
$$= (\sqrt{y}/t)\mathbf{e}\big(cxv + (iy/2)(c^2t^2 + v^2t^{-2})\big).$$

This is the special case $n = 0$ of

(12.30) $$(\sqrt{2\pi}\, i)^n \int_{\mathbf{R}} \exp(-\pi t^2 y^{-1}|cz + u|^2)(c\bar{z} + u)^n \mathbf{e}(-uv)du$$

$$= (\sqrt{y}/t)^{n+1} H_n\big(\sqrt{2\pi y}(ct + vt^{-1})\big)\mathbf{e}\big(cxv + (iy/2)(c^2t^2 + v^2t^{-2})\big)$$
$$(0 \le n \in \mathbf{Z},\ z = x + iy \in \mathfrak{H},\ c \in \mathbf{R},\ v \in \mathbf{R},\ 0 < t \in \mathbf{R}).$$

The case $n > 0$ of (12.30) can be proved by induction on n, by applying $2\pi ic\bar{z} - \partial/\partial v$ to (12.29) and employing (12.4).

Before proceeding further, we consider a variation of the Poisson summation formula:

$$\sum_{d\in\mathbf{Z}} \bar{\psi}(d)h(d) = \sum_{b\in(2/N)\mathbf{Z}} \overline{\eta_0(b)}\hat{h}(b/2) \tag{12.31}$$

for a function h on $\mathbf{R}$. Indeed, if $f(x) = h(Nx)$, then $\hat{f}(x) = N^{-1}\hat{h}(x/N)$. By (4.27) with $n = 1$ and y/N in place of r we have $\sum_{m\in\mathbf{Z}} h(Nm + y) = \sum_{m\in\mathbf{Z}} f(N^{-1}y + m) = \sum_{n\in\mathbf{Z}} \mathbf{e}(ny/N)\hat{f}(n) = N^{-1}\sum_{n\in\mathbf{Z}} \mathbf{e}(ny/N)\hat{h}(n/N)$. Multiply this by $\bar{\psi}(y)$, take the sum over $1 \le y \le N$, and replace n by $Nb/2$; then we obtain (12.31).

Now put $h(u) = (c\bar{z} + u)^n\mathbf{e}\big((ir^2/4y)|cz + u|^2\big)$. Then $\hat{h}$ is given by (12.30) with $r^2/2$ in place of t^2, and (12.31) combined with summation over $c \in 2^{-1}N\mathbf{Z}$ gives the desired (12.27).

12.6. Let $C_n(z, r)$ denote the function of (12.27). Combining (11.10) with (12.26), we obtain

$$g(ir) = i^m \sum_{n=0}^{m} \binom{m}{n} 2^{-1/2}\pi^{(n-m)/2} \cdot \int_{\Phi} f(z)\overline{\tau_{m-n}(z)C_n(z, r)}r^{n-m+1}y^{k-n}\mathbf{d}z. \tag{12.32}$$

Employing the right-hand side of (12.27), we have

$$\int_0^\infty \overline{C_n(z, r)}r^{s+n}dr = \sum_{(c,d)\in T}\int_0^\infty \psi(d)(cz + d)^n \exp\big(-(\pi r^2/2y)|cz + d|^2\big)r^{s+n}dr. \tag{12.33}$$

Since $r^{s+n}dr = 2^{-1}(r^2)^{(s+n-1)/2}d(r^2)$, the last sum equals

$$2^{-1}(2y/\pi)^{(s+n+1)/2}\Gamma\big((s + n + 1)/2\big)\sum_{(c,d)\in T}\psi(d)(cz + d)^n|cz + d|^{-s-n-1}$$
$$= 2^{-1}(2/\pi)^{(s+n+1)/2}\Gamma\big((s + n + 1)/2\big)E_{-n}^{N/2}\big(z, (s + n + 1)/2; \psi\big)$$

with E of (8.15). (The definition of T shows that d is always nonzero. Our calculation becomes invalid if $m = 0$ and the term with $c = d = 0$ appears.) We observe that the integral of (12.33) is absolutely convergent for sufficiently large $\mathrm{Re}(s)$ for every n. Indeed, we have

$$\int_0^\infty |C_n(z, r)|r^{n+\sigma}dr \le 2^{-1}(2/\pi)^{(\sigma+n+1)/2}y^{n/2} \cdot \Gamma\big((\sigma + n + 1)/2\big)E_0^{N/2}\big(z, (\sigma + 1)/2; \psi_0\big)$$

for sufficiently large $\sigma \in \mathbf{R}$, where ψ_0 is the trivial character modulo N. Thus from (12.32) and (12.33) we obtain, at least formally,

$$(12.34) \quad \int_0^\infty g(ir)r^{m-1+s}dr = i^m \sum_{n=0}^m \binom{m}{n} 2^{(s+n)/2-1}\pi^{-(s+m+1)/2}$$
$$\cdot \Gamma((s+n+1)/2)\int_\Phi f(z)\overline{\tau_{m-n}(z)}E_{-n}^{N/2}(z,\,(s+n+1)/2;\,\psi)y^{k-n}\mathbf{d}z.$$

By (8.18) the last integral over Φ can be written

$$(12.35) \quad 2L_N(s+1,\,\bar\psi)\int_\Phi f(z)\overline{\tau_{m-n}(z)}E_{-n}(z,\,(s+n+1)/2;\,\Gamma,\,\psi)y^{k-n}\mathbf{d}z.$$

By (12.12a) we have $\tau_\ell = (4\pi)^{\ell/2}\theta_\ell$ and $\theta_\ell = (\pi i)^{-t}\delta_r^t\theta_\nu$, where $\ell = 2t+\nu$, $r = \nu + 1/2$, $\nu = 0$ or 1, and so θ_ℓ is slowly increasing at every cusp by Lemma 6.10. By Theorem 8.12(ii) the same is true with $E_{-n}(z,\,s;\,N,\,\psi)$ for every s where the function is finite. Since f is a cusp form, the integral over Φ in (12.34) is convergent for every such s. We can say the same for the integral of (12.35) at least for sufficiently large $\mathrm{Re}(s)$. Replacing every factor of the integrand by its absolute value, we find that $\int_0^\infty |g(ir)|r^{\sigma+m-1}dr$ is convergent for sufficiently large σ, and so we can justify our formal calculation, and at the same time we have proved that $c(0) = 0$. But this does not necessarily mean that g is a cusp form.

12.7. Let us now prove Theorem 11.3(ii). By (11.9), for $\alpha \in SL_2(\mathbf{Q})$ we have

$$(g\|_{2m}\,\alpha)(w) = \int_\Phi f(z)\Theta(z,\,w;\,\eta^\alpha)\mathbf{d}z.$$

We see that η^α is a finite $\mathbf{C}$-linear combination of functions of the form

$$\eta\left(\begin{bmatrix} a & b \\ c & -a \end{bmatrix}\right) = \omega_1(a)\omega_2(b)\omega_3(c)$$

with $\omega_i \in \mathscr{L}(\mathbf{Q})$. We now repeat the calculation of §12.5 with this η in place of η of (12.19) and an arbitrary $f \in \mathscr{S}_k(\Gamma)$. Then (12.26) is true with A_n for which $\overline{\eta_0(b)}\omega(2c/N)$ is replaced by $\overline{\omega_2(b)}\omega_3(c)$, and with τ_ℓ replaced by $\theta_\ell(z,\,\omega_1)$. To carry out the calculation, we first have to modify (12.31) as follows. Put $\omega'(b) = \omega_2(2b)$ and take positive rational numbers K and M so that ω' is essentially a function on $M\mathbf{Z}/K\mathbf{Z}$. If $f(x) = h(x/K)$, then $\hat f(x) = K\hat h(Kx)$, and so, by (4.27a),

$$K\sum_{n\in\mathbf{Z}}\hat h(Kn+Kr) = \sum_{n\in\mathbf{Z}}\hat f(n+r) = \sum_{m\in\mathbf{Z}}\mathbf{e}(-mr)f(m) = \sum_{m\in\mathbf{Z}}\mathbf{e}(-mr)h(m/K).$$

We can find a finite subset R of $\mathbf{Q}$ such that $M\mathbf{Z} = \bigsqcup_{r\in R}(K\mathbf{Z}+Kr)$. Then

$$\sum_{b\in\mathbf{Q}}\overline{\omega_2(2b)}\hat h(b) = \sum_{r\in R}\sum_{n\in\mathbf{Z}}\overline{\omega'(Kn+Kr)}\hat h(Kn+Kr)$$

$$= \sum_{r\in R} \overline{\omega'(Kr)} \sum_{n\in \mathbf{Z}} \hat{h}(Kn+Kr).$$

Therefore,

$$\sum_{b\in \mathbf{Q}} \overline{\omega_2(2b)}\hat{h}(b) = \sum_{m\in \mathbf{Z}} \zeta(m)h(m/K) \quad \text{with} \quad \zeta(m) = K^{-1}\sum_{r\in R} \overline{\omega'(Kr)}\mathbf{e}(-mr).$$

Employing this instead of (12.31) we obtain the modification of (12.27) whose right-hand side is

$$\sum_{(c,d)\in \mathbf{Q}^2} \xi(c,\, d)(c\bar{z}+d)^n \mathbf{e}\big((ir^2/(4y))|cz+d|^2\big)$$

with some $\xi \in \mathscr{L}(\mathbf{Q}^2)$. We eventually find that

$$(g\|_{2m}\,\alpha)(ir) = \sum_{j\in J}\sum_{n=0}^{m} \int_{\Phi} f(z)\overline{\tau_{j,m-n}(z)C_{j,n}(z,\, r)}r^{n-m+1}y^{k-n}\mathbf{d}z,$$

where J is a finite set of indices, $\tau_{j,\ell}(z) = \theta_\ell(z,\, \lambda_{j,\ell})$ with $\lambda_{j,\ell} \in \mathscr{L}(\mathbf{Q})$, and

$$C_{j,n}(z,\, r) = \sum_{(c,d)\in \mathbf{Q}^2} \xi_{j,n}(c,\, d)(c\bar{z}+d)^n \mathbf{e}\big((ir^2/y)|cz+d|^2\big)$$

with $\xi_{j,n} \in \mathscr{L}(\mathbf{Q}^2)$. If $n > 0$, we see that $\int_0^\infty \overline{C_{j,n}(z,\, r)}r^{s+n}dr$ is convergent for sufficiently large $\mathrm{Re}(s)$, but if $n = 0$, we have to be careful, since $C_{j,0}(z,\, r)$ has the constant term $\xi_{j,0}(0,\, 0)$. Thus put

$$B = \sum_{j\in J} \xi_{j,0}(0,\, 0)\int_{\Phi} f(z)\overline{\tau_{j,m}(z)}y^k\mathbf{d}z.$$

Then our previous argument shows that

$$\int_0^\infty \big|(g\|_{2m}\,\alpha)(ir) - Br^{1-m}\big|r^\sigma dr$$

is convergent for sufficiently large σ. If $m > 1$, this implies that $g\|_{2m}\,\alpha$ has zero constant term. If $m = 1$ and $\langle f,\, \theta^*(z,\, \eta)\rangle = 0$ for every $\eta \in \mathscr{L}(\mathbf{Q})$, then $B = 0$, since $\tau_{j,1}$ is $\theta_1(z,\, \omega_1)$ with $\omega_1 \in \mathscr{L}(\mathbf{Q})$ and $\theta^* = \theta_1$. Therefore we see again that $g\|_{2m}\,\alpha$ has zero constant term. This proves Theorem 11.3(ii).

We now state the first main theorem on the correspondence $\mathscr{S}_k \to \mathscr{M}_{2m}$.

Theorem 12.8. *Let* $f(z) = \sum_{\xi=1}^\infty \lambda(\xi)\mathbf{e}(\xi z/2) \in \mathscr{S}_k(2,\, N/2;\, \psi)$ *with* $k = m+1/2$, $0 < m \in \mathbf{Z}$, $0 < N \in 4\mathbf{Z}$ *and a character* ψ *modulo* N *such that* $\psi(-1) = (-1)^m$. *Given a square-free positive integer* t, *define a character* χ_t *modulo* tN *by*

$$\chi_t(n) = \psi(n)\left(\frac{t}{n}\right) \qquad (n \in \mathbf{Z}).$$

Then the following assertions hold:

(i) *There exists an element* $g_t(w) = \sum_{n=0}^\infty c_t(n)\mathbf{e}(nw)$ *of* $\mathscr{M}_{2m}(N,\, \psi^2)$ *such that* $c_t(0) = 0$ *and*

$$\sum_{n=1}^\infty c_t(n)n^{-s} = L_N(s-m+1,\, \chi_t)\sum_{\xi=1}^\infty \lambda(t\xi^2)\xi^{-s}. \tag{12.36}$$

(ii) *The function* g_t *can be obtained as a theta integral (11.10)* with a suitable $\eta \in \mathscr{L}(V)$ and $f(tz)$ in place of f there.

(iii) *The function* g_t *is a cusp form if* $m > 1$ *or if* $m = 1$ *and* $\langle f, \theta_1(z, \mu)\rangle = 0$ *for every* $\mu \in \mathscr{L}(\mathbf{Q})$*, where* θ_1 *is as in (12.9b).*

PROOF. We return to §12.6 with f as in our theorem. We took a sufficiently small congruence subgroup Γ of Γ_N^0 and put $\Phi = \Gamma\backslash\mathfrak{H}$. We can replace Γ by a larger subgroup of Γ_N^0 provided the integral of (12.35) is meaningful, though the value of the integral is multiplied by an element of $\mathbf{Q}^\times$ that depends on the choice of the group. As for τ_ℓ, we have $\theta_\ell = (\pi i)^{-t}\delta_r^t\theta_\nu$ as above, and so $\tau_\ell(\gamma z) = h_\gamma(z)j_\gamma(z)^\ell\tau_\ell(z)$ for every $\gamma \in \Gamma(2, 2)$ by (4.20) and (4.47). Therefore, in view of (8.16a) we can take $\Gamma = \Gamma_N^0 = \Gamma(2, N/2)$.

We now calculate the integral over Φ of (12.35) with this Γ. We put $f(z) = \sum_{\xi=1}^\infty \lambda(\xi)\mathbf{e}(\xi z/2)$ and $\ell = m - n$. Since τ_ℓ is given by (12.25), we have

$$f(z)\overline{\tau_\ell(z)} = \sum_{\xi=1}^{\infty}\sum_{a\in\mathbf{Z}} \lambda(\xi)e^{\pi i(\xi-a^2)x}e^{-\pi(\xi+a^2)y}y^{-\ell/2}H_\ell\big(\sqrt{4\pi y}\cdot a\big),$$

and so

$$\int_0^2 f(z)\overline{\tau_\ell(z)}dx = 4\sum_{a=1}^{\infty}\lambda(a^2)e^{-2\pi a^2 y}y^{-\ell/2}H_\ell\big(\sqrt{4\pi y}\cdot a\big).$$

Let $q = m - 1 + (s-n)/2$. Then, employing (12.7) we obtain, for $\ell = m - n$,

$$(*)\quad \int_0^\infty\int_0^2 f(z)\overline{\tau_\ell(z)}dx\, y^q dy = \sum_{a=1}^{\infty}4\lambda(a^2)\int_0^\infty e^{-2\pi a^2 y}y^{-\ell/2}H_\ell\big(\sqrt{4\pi y}\cdot a\big)y^q dy$$

$$= 2^{2-\ell/2}(2\pi)^{-(s+m)/2}\prod_{\nu=1}^{\ell}(s+m-\nu)\Gamma\big((s+n)/2\big)\sum_{a=1}^{\infty}\lambda(a^2)a^{-s-m}.$$

Let $\Xi = \big\{x+iy \in \mathfrak{H}\,\big|\, 0 \le x < 2\big\}$. Then Ξ gives $(P\cap\Gamma_N^0)\backslash\mathfrak{H}$, and as observed in §9.3, $\bigsqcup_{\gamma\in R}\gamma\Phi$ represents Ξ, where $R = (P\cap\Gamma_N^0)\backslash\Gamma_N^0$. Now

$$(f\bar{\tau}_\ell y^{q+2})\circ\gamma = \psi(d_\gamma)j_\gamma^n|j_\gamma|^{-s-n-1}f\bar{\tau}_\ell y^{q+2}$$

for every $\gamma \in \Gamma$, and so

$$\int_0^\infty\int_0^2 f\bar{\tau}_\ell dx\, y^q dy = \int_\Xi f\bar{\tau}_\ell y^{q+2}\mathbf{d}z = \sum_{\gamma\in R}\int_\Phi (f\bar{\tau}_\ell y^{q+2})\circ\gamma\,\mathbf{d}z$$

$$= \int_\Phi f\bar{\tau}_\ell y^{q+2}\sum_{\gamma\in R}\psi(d_\gamma)j_\gamma^n|j_\gamma|^{-s-n-1}\mathbf{d}z$$

$$= \int_\Phi f(z)\overline{\tau_\ell(z)}E_{-n}(z, (s+n+1)/2; \Gamma, \psi)y^{k-n}\mathbf{d}z,$$

which is the integral of (12.35). If we replace $\psi(d_\gamma)j_\gamma^n|j_\gamma|^{-s-n-1}$ by its absolute value, then E_{-n} is replaced by $E_0\big(z, (\sigma+1)/2; \Gamma\big)$ with $\sigma = \mathrm{Re}(s)$, which is finite and slowly increasing at every cusp for sufficiently large σ. Since f is a cusp form, our calculation is justified for sufficiently large $\mathrm{Re}(s)$. Combining this result with (12.23) and (12.34), we find that

$$(2\pi)^{-m-s}\Gamma(s+m)\sum_{\xi=1}^{\infty}c(\xi)\xi^{-s-m}$$
$$= i^m\sum_{n=0}^{m}\binom{m}{n}2^{(s+n)/2-1}\pi^{-(s+m+1)/2}\Gamma((s+n+1)/2)$$
$$\cdot\, 2L_N(s+1,\,\psi)\int_0^{\infty}\int_0^2 f\overline{\tau}_{m-n}dx\, y^q dy,$$

where $q = m-1+(s-n)/2$. Notice that for $\ell = m-n$ we have

$$\Gamma((s+n)/2)\Gamma((s+n+1)/2)\prod_{\nu=1}^{\ell}(s+m-\nu) = 2^{1-s-n}\pi^{1/2}\Gamma(s+m).$$

Also, $\displaystyle\sum_{n=0}^{m}\binom{m}{n} = 2^m$. Thus, employing $(*)$, we finally obtain

$$\sum_{\xi=1}^{\infty}c(\xi)\xi^{-s-m} = 2^{m+3}i^m L_N(s+1,\,\psi)\sum_{a=1}^{\infty}\lambda(a^2)a^{-s-m},$$

since $(2\pi)^{-s-m}\Gamma(s+m)$ appearing on both sides can be cancelled. Thus $i^{-m}2^{-m-3}g$ gives g_1 of our theorem, proving the case $t=1$.

To prove the case with $t>1$, we put $f_t(z) = f(tz) = \sum_{\xi=1}^{\infty}\mu(\xi)\mathbf{e}(\xi z/2)$. By Lemma 8.17(i), $f_t \in \mathscr{S}_k(2,\, tN/2,\, \chi_t)$ with χ_t as above. Applying our result in the case $t=1$ to f_t, we find an element $h(w) = \sum_{n=1}^{\infty}C(n)\mathbf{e}(nw) \in \mathscr{M}_{2m}(2tN,\, \psi^2)$ such that

$$(**)\qquad \sum_{n=1}^{\infty}C(n)n^{-s} = \sum_{n=1}^{\infty}\chi_t(a)a^{m-1-s}\sum_{\xi=1}^{\infty}\mu(\xi^2)\xi^{-s}.$$

Since $f_t(z) = \sum_{\xi=1}^{\infty}\lambda(\xi)\mathbf{e}(t\xi z/2)$, we see that $\mu(\xi^2)\neq 0$ only if $t|\xi$, in which case $\mu(\xi^2) = \lambda(t\eta^2)$ with $\eta = \xi/t$. Thus $\sum_{\xi=1}^{\infty}\mu(\xi^2)\xi^{-s} = t^{-s}\sum_{\eta=1}^{\infty}\lambda(t\eta^2)\eta^{-s}$, and so from $(**)$ we see that $C(n)\neq 0$ only if $t|n$. Put $g_t(w) = h(w/t)$, Then $g_t(w) = \sum_{n=1}^{\infty}c_t(n)\mathbf{e}(nw)$ with $c_t(n) = C(tn)$, and by Lemma 8.17(i), $g_t \in \mathscr{M}_{2m}(t,\, 2N;\, \psi^2)$. Also, we have (12.36). It remains to prove that $g_t \in \mathscr{M}_{2m}(1,\, 2N;\, \psi^2)$. By Lemma 8.18, $\Gamma_0(2N)$ is generated by $\Gamma(t,\, 2N)$ and $\begin{bmatrix}1 & 1\\ 0 & 1\end{bmatrix}$, Since $g_t(w+1) = g_t(w)$, we obtain the desired fact. This proves (i) and (ii), and (iii) as well, by virtue of Theorem 11.3(ii).

The above theorem excludes the case $k=1/2$. We can actually determine $\mathscr{M}_{1/2}$ completely as follows.

Theorem 12.9. *The space $\mathscr{M}_{1/2}$ is spanned by the series $\theta_0(az,\, \lambda)$ with $0 < a \in \mathbf{Q}$ and $\lambda \in \mathscr{L}(Q)$, where θ_0 is as in (12.9a), that is,*

$$\theta_0(az,\, \lambda) = \sum_{\xi \in \mathbf{Q}} \lambda(\xi)\mathbf{e}(a\xi^2 z/2).$$

After some preliminary observations in §12.10 and two lemmas, the proof will be completed in §12.13.

12.10. Throughout this subsection we put $k = 1/2$. Thus $m = 0$. Define g by (11.10) with $f \in \mathscr{S}_k(\Gamma)$. By Lemma 10.7, Θ as a function of z is slowly increasing at every cusp, locally uniformly in w. Since f is a cusp form, (11.10) is convergent, and so g is meaningful. Besides, the estimate of $\Theta(z,\, w;\, \zeta)$ in §11.6 is valid in the case $m = 0$, if we ignore the constant term, which is $y^{1/2}$. Therefore g is slowly increasing at every cusp.

For α in V of (11.1), $s \in \mathbf{C}$, and $w \in \mathfrak{H}$ put

$$\kappa[\alpha,\, w,\, s] = \mathrm{Im}(w)^{2s}|[\alpha,\, w]|^{-2s}, \tag{12.37}$$

where $[\alpha,\, w]$ is as in (11.3a). Let L_0 be the operator of (6.13c) with $k = 0$. Then

$$L_0\kappa[\alpha,\, w,\, s] = 2s(1-2s)\kappa[\alpha,\, w,\, s] - 16s^2\det(\alpha)\kappa[\alpha,\, w,\, s+1]. \tag{12.38}$$

To make our calculation easier, we note that $\kappa[\alpha,\, \gamma w,\, s] = \kappa[\gamma^{-1}\alpha\gamma,\, w,\, s]$ for every $\gamma \in SL_2(\mathbf{R})$, which follows from (6.3) and (11.3c), and so

$$L_0\big(\kappa[\gamma^{-1}\alpha\gamma,\, w,\, s]\big) = (L_0\kappa)[\alpha,\, \gamma w,\, s]$$

by (6.14d). Therefore it is sufficient to verify (12.38) when $\alpha = \mathrm{diag}[-a,\, a]$, which can be done easily.

Put $f(z) = \sum_{\xi \in \mathbf{Q}} \lambda(\xi)\mathbf{e}(\xi/2)$ and define two infinite series P and Q by

$$P(w,\, s) = \sum_{\alpha \in V} \eta(\alpha)\lambda\big(-\det(\alpha)\big)\kappa[\alpha,\, w,\, s], \tag{12.39a}$$

$$Q(w,\, s) = \sum_{\alpha \in V} \eta(\alpha)\lambda\big(-\det(\alpha)\big)\det(\alpha)\kappa[\alpha,\, w,\, s+1], \tag{12.39b}$$

where $w \in \mathfrak{H}$, $s \in \mathbf{C}$, and $\eta \in \mathscr{L}(\mathbf{Q})$. Since $\lambda(0) = 0$, the sums are over α such that $\det(\alpha) < 0$. These series are convergent for sufficiently large $\mathrm{Re}(s)$, provided $\mathrm{Im}(w) > c$ with a constant c that depends on η. Indeed, by Lemma 6.2(iii), $|\lambda(-\det(\alpha))| = O(|\det(\alpha)|^{1/4})$, and so our task is to prove the convergence of

$$\sum_{\alpha \in V,\, \det(\alpha)<0} \eta(\alpha)|\det(\alpha)|^{a}|[\alpha,\, w]|^{-2s} \tag{12.40}$$

under those conditions on s and w, for any $a \in \mathbf{R}$, ≥ 0. We may assume that η is a characteristic function of a lattice in V. For $\alpha = \begin{bmatrix} a & b \\ c & -a \end{bmatrix} \in V$ put $\|\alpha\| = (a^2+b^2+c^2)^{1/2}$. Then, by (11.24) and (11.19) we have

$$|[\alpha,\, w]|^2 = 2\mathrm{Im}(w)^2\big(P_w[\alpha] - \det(\alpha)\big) \geq 2\mathrm{Im}(w)^2 P_w[\alpha] \geq 4^{-1}\|\alpha\|^2,$$

since $\det(\alpha) < 0$. Notice also that $|\det(\alpha)| \le \|\alpha\|^2$. Therefore (12.39) is majorized by a constant times $\sum_{0\ne\alpha\in L} \|\alpha\|^{2a-2s}$ for $\mathrm{Re}(s) > 0$ with a lattice L in V. Thus we obtain the desired convergence. Consequently P and Q define holomorphic functions in s for $\mathrm{Re}(s) > b$ with some $b \in \mathbf{R}$ when $\mathrm{Im}(w) > c$ with a positive constant c.

Lemma 12.11. *The series $P(w, s)$ and $Q(w, s)$ can be continued as meromorphic functions in s to the whole s-plane with at most a simple pole at $s = 1$. The residues of P and Q at $s = 1$ are $Ag(w)$ and $(-A/8)g(w)$, respectively, where A is a nonzero constant and g is the function of (11.10) defined with the present f, η, $k = 1/2$, and $m = 0$.*

PROOF. We prove this lemma by finding integral expressions for these series. We take a sufficiently large even positive integer N, and replace Γ by $\Gamma(N)$, or rather, put $\Gamma = \Gamma(N)$. We also put $\Gamma_\infty = \Gamma \cap P$, $\Psi = \Gamma_\infty \backslash \mathfrak{H}$, and $\Phi = \Gamma\backslash\mathfrak{H}$. Let Θ_0 denote $\Theta(z, w; \eta)$ of (11.6) with $m = 0$. We have

$$f(z)\Theta_0(z) = y^{1/2} \sum_{\xi\in\mathbf{Q}} \sum_{\alpha\in V} \lambda(\xi)\eta(\alpha)\mathbf{e}\Big(2^{-1}\big(\xi + \det(\alpha)\big)z + (iy/4)v^{-2}|[\alpha, w]|^2\Big)$$

in view of (11.5), where $y = \mathrm{Im}(z)$ and $v = \mathrm{Im}(w)$, and so

$$\int_0^N f(z)\Theta_0(z)dx = Ny^{1/2} \sum_{\alpha\in V} \eta(\alpha)\lambda\big(-\det(\alpha)\big)\mathbf{e}\big((iy/4)v^{-2}|[\alpha, w]|^2\big).$$

Since Ψ is represented by $\{x + iy \mid y > 0,\ 0 \le x < N\}$, we have, using (8.2),

$$\begin{aligned}\int_\Psi f\Theta_0 y^{s+1/2}\mathbf{d}z &= \int_0^\infty \Big\{\int_0^N f\Theta_0 dx\Big\} y^{s-3/2}dy \\ &= N\int_0^\infty \sum_{\alpha\in V} \eta(\alpha)\lambda\big(-\det(\alpha)\big)\exp\big(-y(\pi/2)v^{-2}|[\alpha, w]|^2\big)y^{s-1}dy \\ &= N(2/\pi)^s\Gamma(s)P(w, s)\end{aligned}$$

for $\mathrm{Re}(s) > b$ and $\mathrm{Im}(w) > c$ with some b and c. As observed in §9.13, Ψ is represented by $\bigsqcup_{\gamma\in R} \gamma\Phi$ with $R = \Gamma_\infty\backslash\Gamma$. By (11.8) we have $(f\Theta_0)\circ\gamma = |j_\gamma(z)| f\Theta_0$, and so

$$\begin{aligned}\int_\Psi f\Theta_0 y^{s+1/2}\mathbf{d}z &= \sum_{\gamma\in R} \int_\Phi \big(f\Theta_0 y^{s+1/2}\big)\circ\gamma\mathbf{d}z \\ &= \int_\Phi f\Theta_0 \sum_{\gamma\in R} |j_\gamma(z)|^{-2s} y^{s+1/2}\mathbf{d}z = \int_\Phi f\Theta_0 E_0(z, s; \Gamma)y^{1/2}\mathbf{d}z,\end{aligned}$$

where E_0 is the function of (9.12) with $k = 0$. Our calculation is justified for the same reason as in the proof of Theorem 12.8. Thus

$$N(2/\pi)^s\Gamma(s)P(w, s) = \int_\Phi f\Theta_0 E_0(z, s; \Gamma)y^{1/2}\mathbf{d}z. \tag{12.41}$$

Now $E_0(z, s; \Gamma)$ can be continued to a meromorphic function on the whole s-plane in the sense of Theorem 9.9 with a simple pole at $s = 1$ with a positive number, say ρ, as the residue. Besides, Lemma 9.8 shows that if $(s - s_0)^b E_0$ is finite at $s = s_0$, then $(s - s_0)^b E_0$ is slowly increasing at every cusp locally uniformly in s. Since f is a cusp form, the last integral over Φ defines a meromorphic function in s on the whole $\mathbf{C}$. In addition, it has at most a simple pole at $s = 1$ with residue

$$\rho \int_{\Phi} f(z)\Theta_0(z, w; \eta) y^{1/2} \mathbf{d}z = \rho g(w).$$

Therefore from (12.41) we see that $P(w, s)$ has at most a simple pole at $s = 1$ with residue $(2N)^{-1}\pi\rho g(w)$. This proves our assertion on P with $A = (2N)^{-1}\pi\rho$.

Next, to deal with $Q(w, s)$, we take

$$(\pi i)^{-1}\partial f/\partial z = \sum_{\xi \in \mathbf{Q}} \xi\lambda(\xi)\mathbf{e}(\xi z/2)$$

in place of f. By the same technique as before we easily find that

$$(*) \qquad \int_{\Psi} (\pi i)^{-1}(\partial f/\partial z)\Theta_0 y^{s+3/2} \mathbf{d}z = -N(2/\pi)^{s+1}\Gamma(s+1)Q(w, s).$$

We consider the operator δ_k of (6.13b) with $k = 1/2$. Then $(\pi i)^{-1}\partial f/\partial z = (\pi i)^{-1}\delta_{1/2} f + (4\pi y)^{-1} f$, and so the last integral over Ψ equals

$$(**) \qquad (\pi i)^{-1} \int_{\Psi} (\delta_{1/2} f)\Theta_0 y^{s+3/2} \mathbf{d}z + (4\pi)^{-1} \int_{\Psi} f\Theta_0 y^{s+1/2} \mathbf{d}z.$$

Since $(\Theta_0 \delta_{1/2} f) \circ \gamma = j_\gamma^2 |j_\gamma| \Theta_0 \delta_{1/2} f$ for every $\gamma \in \Gamma$, the previous technique used for P produces

$$\int_{\Psi} (\delta_{1/2} f)\Theta_0 y^{s+3/2} \mathbf{d}z = \int_{\Phi} (\delta_{1/2} f)\Theta_0 E_{-2}(z, s+1; \Gamma) y^{1/2} \mathbf{d}z.$$

By Lemma 6.10, $\delta_{1/2} f$ is rapidly decreasing at every cusp. Thus, for the same reason as for P, we see that $Q(w, s)$ can be continued to a meromorphic function on the whole s-plane. By Theorem 9.9, $E_{-2}(z, s+1; \Gamma)$ is finite at $s = 1$. Therefore from (12.41), $(*)$, and $(**)$ we see that $Q(w, s)$ has at most a simple pole at $s = 1$ with residue $-(16N)^{-1}\pi\rho g(w)$, and our proof of Lemma 12.9 is complete.

Lemma 12.12. *Define* g *by* (11.10) *with* $f \in \mathscr{S}_k$, $k = 1/2$. *Suppose that* $\langle f, \theta_0(z, \lambda)\rangle = 0$ *for every* $\lambda \in \mathscr{L}(\mathbf{Q})$, *where* θ_0 *is as in (12.9a). Then* $g = 0$.

PROOF. We have shown in §12.10 that g is slowly increasing at every cusp. Let the notation be as in §12.10 and Lemma 12.11. By (12.38) we have $L_0 P(w, s) = 2s(1 - 2s)P(w, s) - 16s^2 Q(w, s)$. Taking the residues of P and Q at $s = 1$, from Lemma 12.9 we obtain $AL_0 g = -2Ag + (16A/8)g = 0$, and so $L_0 g = 0$. Thus g belongs to $\mathfrak{A}_0(0)$, and has an expansion

$$g(u+iv) = a + cv + \sum_{0\neq h\in r\mathbf{Z}} b_h W_k(hv,\, 0)\mathbf{e}(hu)$$

with $a,\, c,\, b_h \in \mathbf{C}$; see (9.10) and (9.12a). By Lemma 9.2(ii) the last sum is $O(e^{-Bv})$ with $B > 0$ as $v \to \infty$. Now our calculation of §12.7 is valid in the present case with $m = 0$ under our assumption that $\langle f,\, \theta_0(z,\, \lambda)\rangle = 0$. In particular, with $m = n = 0$ we see that $\int_0^\infty |g(ir)|r^\sigma dr < \infty$ for a sufficiently large σ, which shows that $a = c = 0$. Since this is so for $g \circ \alpha$ in place of g for every $\alpha \in SL_2(\mathbf{Q})$, we see that g is a cusp form. By Lemma 9.2(iv), $\mathfrak{S}_0(0) = \mathscr{S}_0 = \{0\}$. Thus $g = 0$.

12.13. We now prove Theorem 12.9. In this proof we put $k = 1/2$. We know that $\theta_0(az,\, \lambda)$ belongs to $\mathscr{M}_k$. We have shown in (9.45a) that $\mathscr{M}_k = \mathscr{S}_k \oplus \mathscr{E}_k$, and also in (9.49) that $\mathscr{E}_k$ is spanned by some series of the type $\theta_0(az,\, \lambda)$. Therefore it is sufficient to show that every element f of $\mathscr{S}_k$ orthogonal to all series of the type $\theta_0(az,\, \lambda)$ is 0. Take such an f. Then we can find $q \in \mathbf{Q},\, > 0$, such that $f(qz)$ belongs to $\mathscr{S}_k(\Gamma')$ with $\Gamma' = \{\gamma \in \Gamma_N^0 \,|\, a_\gamma - 1 \in N\mathbf{Z}\}$, where $0 < N \in 4\mathbf{Z}$ and Γ_N^0 is as in (12.16). Put $f_0(z) = f(qz)$. For each character χ of $(\mathbf{Z}/N\mathbf{Z})^\times$ such that $\chi(-1) = 1$ put $f_\chi = \sum_{\gamma\in R} \chi(a_\gamma) f_0\|_k\,\gamma$, where $R = \Gamma_N^0/\{\pm 1\}\Gamma'$. Then $f_\chi \in \mathscr{S}_k(2,\, N/2;\, \chi)$ and $\#\{\chi\} f_0 = \sum_{\chi\in\{\chi\}} f_\chi$, where $\{\chi\}$ is the set of all such characters χ. Our aim is to show that $f_\chi = 0$ for every χ. For a square-free positive integer t put $f_{\chi,t}(z) = f_\chi(tz)$. By Lemma 8.17(i), $f_{\chi,t} \in \mathscr{S}_k(2,\, tN/2;\, \psi_t)$ with $\psi_t(a) = \chi(a)\left(\dfrac{t}{a}\right)$. Moreover, our assumption on f implies that $\langle f_{\chi,t},\, \theta_0(z,\, \lambda)\rangle = 0$ for every $\lambda \in \mathscr{L}(\mathbf{Q})$. Therefore, by Lemma 12.12, the theta integral of $f_{\chi,t}$ is 0. Our calculations in §§12.5 and 12.6 and the proof of Theorem 12.8 are valid in the present case with $f_{\chi,t}$ in place of f, and so the vanishing of the theta integral of $f_{\chi,t}$ means that (**) in the proof of Theorem 12.8 is 0, and so $\sum_{\xi=1}^\infty \mu(\xi^2)\xi^{-s} = 0$ if $f_{\chi,t}(z) = \sum_{\xi=1}^\infty \mu(\xi)\mathbf{e}(\xi z/2)$. The argument of the last part of the proof of Theorem 12.8 shows that $\sum_{\xi=1}^\infty \lambda(t\xi^2)\xi^{-s} = 0$ if $f_\chi(z) = \sum_{\xi=1}^\infty \lambda(\xi)\mathbf{e}(\xi z/2)$. Since this is so for every square-free positive integer t, we obtain $f_\chi = 0$ as expected. This completes the proof.

12.14. Theorem 12.8(i) was essentially given in [S73a, p. 458] in a somewhat different form, and later in [S87] it was generalized to the case of forms with respect to congruence subgroups of $SL_2(F)$ with a totally real algebraic number field F. The difference between Theorem 12.8(i) and the corresponding statement in [S73a] is caused by our choice of j_α^k in (5.1b) and of h_γ in (4.40). Our choice is more natural than $(h'_\delta)^{2k}$ employed in [S73a], where h'_δ is as in (5.6). Also, generalizations of Theorem 12.8(ii, iii) and Theorem 12.9 in the case of $SL_2(F)$ were given in [S87]. To show that $g_t(w) = \sum_{m=1}^\infty c_t(m)\mathbf{e}(mw)$

of Theorem 12.8(i) is a form of weight $2m$, we employed in [S73a] the characterization of such a form by the functional equations of $\sum_{m=1}^{\infty} \chi(t)c_t(m)\mathbf{e}(mw)$ for all Dirichlet characters χ. Such a characterization originated in Hecke [H36], and Weil, inspired by my idea on the modularity of **Q**-rational elliptic curves (as he implied in 1967 and 1986), proved the characterization of the forms of $\mathscr{S}_\nu(\Gamma_0(N))$, $0 < \nu \in \mathbf{Z}$. It is easy to extend it to the forms of $\mathscr{S}_\nu(N, \psi)$ with an arbitrary character ψ modulo N, and it was this characterization that I employed in [S73a].

The methods of [S87], which we followed in Sections 11 and 12, were completely different. I calculate explicitly the theta integral (11.10), which I believe, gives a shorter proof and better results. In fact, similar integrals had been investigated by a few researchers, but their methods required that the weight be sufficiently large. I found that this difficulty was avoidable by using the operators ε and δ_{k-2}, and proving an equality of type (11.15). It seems that there is a conceptual explanation of such an equality.

In any case, in the intervening years, no small number of authors published papers on the correspondence, as can be seen from the references of [S87]. However, their connection with our main theorems is not so clear-cut, and so we included in the references of the present book only those which may be called truly relevant. The reader who is interested in those works can check the papers listed at the end of [S87] and compare them with our theorems. I may be allowed to say that not every paper there is reliable, and some have serious gaps.

The main theorem of [S73a] was formulated for eigenfunctions of Hecke operators of half-integral weight, but we stated Theorem 12.8 without such operators. We will discuss them in the next section. We note that a few examples of the correspondence $f \mapsto g$ and also examples of the dimensions of $\mathscr{S}_k(\Gamma_N^0)$ and $\mathscr{S}_{2m}(\Gamma_0(N/2))$ are given in [S73a, Section 4].

13. Hecke operators

13.1. We first introduce the notion of Hecke algebra in an abstract setting. We say that two subgroups D and D' of a group are **commensurable** if $D \cap D'$ is of finite index in D and in D'. We now fix a multiplicative group $\mathscr{G}$ and a subgroup D of $\mathscr{G}$, and assume that $\alpha D \alpha^{-1}$ is commensurable with D for every $\alpha \in \mathscr{G}$. It is an easy exercise to show that $\#(D\alpha D/D) = [D : D \cap \alpha D \alpha^{-1}]$ and $\#(D\backslash D\alpha D) = [D : D \cap \alpha^{-1} D \alpha]$ for every $\alpha \in \mathscr{G}$. Also we have

(13.0) *If $\#(D\alpha D/D) = \#(D\backslash D\alpha D)$, then there exists a set $\{\zeta_i\}_{i \in I}$ such that* $D\alpha D = \bigsqcup_{i \in I} D\zeta_i = \bigsqcup_{i \in I} \zeta_i D$.

Indeed, let $D\alpha D = \bigsqcup_{i \in I} D\xi_i = \bigsqcup_{i \in I} \eta_i D$. Since $\xi_i \in D\eta_i D$, we have $\xi_i = \delta_i \eta_i \varepsilon_i$ with δ_i, $\varepsilon_i \in D$. Then we obtain the desired result with $\zeta_i = \delta_i^{-1}\xi_i$.

Let $\mathfrak{R}$ denote the vector space over $\mathbf{Q}$ consisting of all formal finite sums $\sum c_\alpha D\alpha D$ with $c_\alpha \in \mathbf{Q}$ and $\alpha \in \mathscr{G}$. We introduce a law of multiplication $\mathfrak{R} \times \mathfrak{R} \to \mathfrak{R}$ which makes $\mathfrak{R}$ an associative algebra as follows. Given $u = D\alpha_0 D$ and $v = D\beta_0 D$ with $\alpha_0,\ \beta_0 \in \mathscr{G}$, take coset decompositions

$$u = D\alpha_0 D = \bigsqcup\nolimits_{\alpha\in A} D\alpha, \quad v = D\beta_0 D = \bigsqcup\nolimits_{\beta\in B} D\beta. \tag{13.1}$$

We have $D\alpha_0 D\beta_0 D = \bigcup_{\beta\in B} D\alpha_0 D\beta = \bigcup_{\alpha\in A,\, \beta\in B} D\alpha\beta$, so that

$$D\alpha_0 D\beta_0 D = \bigsqcup\nolimits_{\xi\in X} D\xi D \tag{13.1a}$$

with a finite set X. We then define the product $u \cdot v$ to be the element of $\mathfrak{R}$ given by

$$u \cdot v = \sum_w \mu(u\cdot v,\, w)w, \tag{13.2}$$

where the sum is extended over all the different $w = D\xi D \subset D\alpha_0 D\beta_0 D$, and

$$\mu(u\cdot v,\, w) = \#\big\{(\alpha,\, \beta) \in A \times B \,\big|\, D\alpha\beta = D\xi\big\}. \tag{13.2a}$$

To make this definition meaningful, we have to show that the right-hand side is independent of the choice of A, B and ξ. Once this is done, we extend the map $(u,\, v) \mapsto u \cdot v$ to a $\mathbf{Q}$-bilinear map of $\mathfrak{R} \times \mathfrak{R}$ into $\mathfrak{R}$. Though this was done in [S71], we present here a simpler proof.

Thus our task is to show that the above law is well-defined and associative. For this purpose, we first consider the vector space $\mathfrak{M}$ over $\mathbf{Q}$ consisting of all formal finite sums $\sum_\gamma c_\gamma D\gamma$ with $c_\gamma \in \mathbf{Q}$ and $\gamma \in \mathscr{G}$. Let $u = D\alpha_0 D = \bigsqcup_{\alpha\in A} D\alpha$. Clearly $D\alpha_0 D\gamma = \bigsqcup_{\alpha\in A} D\alpha\gamma$ and this set depends only on u and $D\gamma$. Therefore, if we let u act on $\mathfrak{M}$ by the rule

$$u \cdot \sum_\gamma c_\gamma D\gamma = \sum_\gamma \sum_{\alpha\in A} c_\gamma D\alpha\gamma,$$

then this is well defined independently of the choice of A and γ. We can also let $\mathscr{G}$ act on $\mathfrak{M}$ on the right by putting

$$\Big(\sum_\gamma c_\gamma D\gamma\Big)\xi = \sum_\gamma c_\gamma D\gamma\xi \qquad (\xi \in \mathscr{G}).$$

We can view $\mathfrak{R}$ as a subspace of $\mathfrak{M}$ by identifying $D\alpha_0 D$ with $\sum_{\alpha\in A} D\alpha$. Then it is easy to see that $\mathfrak{R}$ as a subspace of $\mathfrak{M}$ consists of the elements $x \in \mathfrak{M}$ such that $x\delta = x$ for every $\delta \in D$. We now restrict the map $\mathfrak{R} \times \mathfrak{M} \to \mathfrak{M}$ to $\mathfrak{R} \times \mathfrak{R}$. Since $(u \cdot x)\delta = u \cdot (x\delta)$, we see that $u \cdot x \in \mathfrak{R}$ if $x \in \mathfrak{R}$. Thus we obtain a $\mathbf{Q}$-bilinear map $\mathfrak{R} \times \mathfrak{R} \to \mathfrak{R}$. From our definition we see that if $u,\ v,\ A,\ B$ are as in (13.1), then

$$u \cdot v = \sum_{\alpha\in A} \sum_{\beta\in B} D\alpha\beta. \tag{13.3}$$

The right-hand side, being an element of $\mathfrak{R}$, can be written $\sum_{\xi\in X} m_\xi D\xi D$ with $0 \le m_\xi \in \mathbf{Z}$ and X of (13.1a). Then clearly $m_\xi = \mu(u\cdot v,\, w)$ for $w = D\xi D$. Thus (13.3) coincides with (13.2), and so (13.2) is well defined.

Now, for $x = D\gamma$ we have

$$u \cdot (v \cdot x) = u \cdot \sum_{\beta \in B} D\beta\gamma = \sum_{\alpha \in A} \sum_{\beta \in B} D\alpha\beta\gamma = \sum_{\xi \in X} m_\xi D\xi D\gamma = (u \cdot v) \cdot x.$$

By linearity we obtain $u \cdot (v \cdot y) = (u \cdot v) \cdot y$ for every $y \in \mathfrak{R}$, which proves the associativity of the algebra $\mathfrak{R}$. Taking α_0 or β_0 to be 1, we see that $D = D1D$ is the identity element of $\mathfrak{R}$.

We denote $\mathfrak{R}$ by $\mathfrak{R}(D, \mathscr{G})$ and call it the **Hecke algebra** of $(D, \mathscr{G})$. We define a $\mathbf{Q}$-linear map $\deg : \mathfrak{R}(D, \mathscr{G}) \to \mathbf{Q}$ by

$$\deg\Bigl(\sum_\alpha c_\alpha D\alpha D\Bigr) = \sum_\alpha c_\alpha \#(D\backslash D\alpha D). \tag{13.4}$$

Let $\mathscr{A}$ be a subset of $\mathscr{G}$ containing D which is closed under multiplication. Then we easily see that the $\mathbf{Q}$-linear span of $D\alpha D$ for all $\alpha \in \mathscr{A}$ is a subalgebra of $\mathfrak{R}(D, \mathscr{G})$. We denote this subalgebra by $\mathfrak{R}(D, \mathscr{A})$ and call it the **Hecke algebra** of $(D, \mathscr{A})$. As shown in [S71, Propositions 3.3 and 3.8], we have

(13.4a) $\deg(xy) = \deg(x)\deg(y)$ for every $x, y \in \mathfrak{R}(D, \mathscr{G})$,

(13.4b) *If $\mathscr{G}$ has an anti-automorphism $\alpha \mapsto \alpha^*$ such that $D^* = D$ and $(D\alpha D)^* = D\alpha D$ for every $\alpha \in \mathscr{A}$, then $\mathfrak{R}(D, \mathscr{A})$ is commutative.*

13.2. The symbol $\mathfrak{R}\bigl(\Gamma, GL_2^+(\mathbf{Q})\bigr)$ is meaningful for every congruence subgroup Γ of $\Gamma(1)$. We now take Γ of a special type and replace $GL_2^+(\mathbf{Q})$ by a smaller set as follows:

$$\Gamma = \bigl\{\gamma \in \Gamma(1) \bigm| a_\gamma \in \mathfrak{h},\ b_\gamma \in t\mathbf{Z},\ c_\gamma \in N\mathbf{Z}\bigr\}.$$

Here $0 < t \in \mathbf{Z}$, $0 < N \in \mathbf{Z}$, $t|N$, and $\mathfrak{h}$ is a subgroup of $(\mathbf{Z}/tN\mathbf{Z})^\times$; we use the same letter $\mathfrak{h}$ for the inverse image of $\mathfrak{h}$ under the natural map $\mathbf{Z} \to \mathbf{Z}/tN\mathbf{Z}$. (In fact, we are interested only in the two special cases $\mathfrak{h} = \{1\}$ and $\mathfrak{h} = (\mathbf{Z}/tN\mathbf{Z})^\times$.) We then consider $\mathfrak{R}(\Gamma, \Xi)$ with

$$\Xi = \bigl\{\alpha \in M_2(\mathbf{Z}) \bigm| \det(\alpha) > 0,\ a_\alpha \in \mathfrak{h},\ b_\alpha \in t\mathbf{Z}, c_\alpha \in N\mathbf{Z}\bigr\}.$$

For $0 < n \in \mathbf{Z}$ we denote by $T'(n)$ the sum of all different $\Gamma\alpha\Gamma$ with $\alpha \in \Xi$ such that $\det(\alpha) = n$. Also, for two positive integers a and d such that $a|d$ and $(a, N) = 1$ we denote by $T'(a, d)$ the element $\Gamma\xi\Gamma$ of $\mathfrak{R}(\Gamma, \Xi)$ with $\xi \in \Xi \cap \Gamma(1)\mathrm{diag}[a, d]\Gamma(1)$. Such a ξ exists and $\Gamma\xi\Gamma$ is uniquely determined independently of the choice of ξ; see [S71, Proposition 3.32].

Lemma 13.3. (i) *The algebra $\mathfrak{R}(\Gamma, \Xi)$ is a polynomial ring over $\mathbf{Q}$ of the elements of the following two types:*

(1) $T'(p)$ for all prime factors p of N;

(2) $T'(1, p)$ and $T'(p, p)$ for all prime numbers p not dividing N.

(ii) *Every* $\Gamma\xi\Gamma$ *with* $\xi \in \Xi$ *can be expressed as a product* $T'(m)T'(a,\, d)$ *with* $m|N^\infty$, $a|d$, d *prime to* N.

(iii) $T'(m)T'(n) = T'(mn)$ *if* $m|N^\infty$ *and* $n|N^\infty$.

(iv) $T'(\ell m) = T'(\ell)T'(m)$ *if* ℓ *is prime to* m.

(v) $\mathfrak{R}(\Gamma,\, \Xi)$ *is generated over* $\mathbf{Q}$ *by the* $T'(n)$ *for all positive integers* n.

This is [S71, Theorem 3.34]. Here we write $m|N^\infty$ when the prime factors of m divide N. In fact, these are formulated only when $\mathfrak{h}$ is a subgroup of $(\mathbf{Z}/N\mathbf{Z})^\times$, but all the statements are valid for Γ and Ξ given as above, as can be seen by verifying various statements of [S71], Proposition 3.32, in particular, for such Γ and Ξ, which, if tedious, is straightforward. For this, see the errata of the paperback edition of [S71] in 1994.

We note that

(13.5) *If* $m|N^\infty$, *then* $T'(m) = \Gamma\sigma\Gamma$ *with* $\sigma = \mathrm{diag}[1,\, m]$.

This can be obtained by taking $\Gamma\sigma\Gamma$ as $\Gamma\xi\Gamma$ of (ii).

We also note a basic formula

$$T'(m)T'(n) = \sum_d d \cdot T'(d,\, d)T'(mn/d^2), \tag{13.6a}$$

where d runs over all positive divisors of $(m,\, n)$ prime to N. We also have an equality of formal Dirichlet series with coefficients in $\mathfrak{R}(\Gamma,\, \Xi)$:

$$\sum_{\alpha\in\Gamma\backslash\Xi/\Gamma} (\Gamma\alpha\Gamma)\det(\alpha)^{-s} = \sum_{n=1}^{\infty} T'(n)n^{-s} \tag{13.6b}$$

$$= \prod_{p|N} \bigl[1 - T'(p)p^{-s}\bigr]^{-1} \prod_{p\nmid N} \bigl[1 - T'(p)p^{-s} + T'(p,\, p)p^{1-2s}\bigr]^{-1}.$$

These are [S71, (3.3.6), (3.3.8)].

13.4. We now fix a half-integral weight $k > 0$ and consider $\mathfrak{R}(\Delta,\, G_k)$ with the group G_k and its congruence subgroup Δ in the sense of §9.5. For $\gamma \in \Gamma^\theta$ we define an element $\ell(\gamma)$ of G_k by $\ell(\gamma) = (\gamma,\, j_\gamma^k)$. Then ℓ is an injective homomorphism of Γ^θ into G_k.

We also fix a congruence subgroup Γ of $\Gamma(2)$ and put $\Delta = \ell(\Gamma)$. Taking a positive integer e, we consider $\Delta\xi\Delta$ with $\xi = (\alpha,\, e^k)$, $\alpha = \mathrm{diag}[e^{-1},\, e]$, $0 < e \in \mathbf{Q}$. We have then

$$\ell(\delta)\xi = \xi\ell(\gamma) \quad \text{if} \quad \delta = \alpha\gamma\alpha^{-1} \ \text{ with } \gamma \in \alpha^{-1}\Gamma\alpha \cap \Gamma. \tag{13.7}$$

Indeed, from (4.40) we see that $h_\delta(\alpha z) = h_\gamma(z)$, and so $j_\delta^k(\alpha z) = j_\gamma^k(z)$, and we can easily verify (13.7).

Lemma 13.5. (i) *With* Γ, Δ, ξ, *and* α *as above, for* $\{\xi_\nu\} \subset \Delta\xi\Delta$ *we have* $\Delta\xi\Delta = \bigsqcup_\nu \Delta\xi_\nu$ *if and only if* $\Gamma\alpha\Gamma = \bigsqcup_\nu \Gamma\mathrm{pr}(\xi_\nu)$, *and consequently* $\#(\Delta\backslash\Delta\xi\Delta) = \#(\Gamma\backslash\Gamma\alpha\Gamma)$.

(ii) *Let* $\eta = (\beta, f^k)$ *with* $\beta = \mathrm{diag}[f^{-1}, f]$, $0 < f \in \mathbf{Q}$. *Suppose* $\Gamma\alpha\Gamma \cdot \Gamma\beta\Gamma = \Gamma\alpha\beta\Gamma$. *Then* $\Delta\xi\Delta \cdot \Delta\eta\Delta = \Delta\xi\eta\Delta$.

PROOF. We can take $\xi_\nu = \xi\delta_\nu$ with $\delta_\nu = \ell(\gamma_\nu)$ with $\gamma_\nu \in \Gamma$. Suppose $\Delta\xi\Delta = \bigsqcup_\nu \Delta\xi_\nu$. Then $\Gamma\alpha\Gamma = \bigcup_\nu \Gamma\mathrm{pr}(\xi_\nu)$. Suppose $\Gamma\alpha\gamma_1 = \Gamma\alpha\gamma_2$. Then $\gamma_2\gamma_1^{-1} \in \alpha^{-1}\Gamma\alpha \cap \Gamma$, and so $\ell(\alpha\gamma_2\gamma_1^{-1}\alpha^{-1})\xi = \xi\ell(\gamma_2\gamma_1^{-1})$ by (13.7). Therefore $\xi_2\xi_1^{-1} = \xi\delta_2\delta_1^{-1}\xi^{-1} = \ell(\alpha\gamma_2\gamma_1^{-1}\alpha^{-1}) \in \ell(\Gamma) = \Delta$, and so $\Delta\xi_2 = \Delta\xi_1$. This shows that $\Gamma\alpha\Gamma = \bigsqcup_\nu \Gamma\mathrm{pr}(\xi_\nu)$ and $\#(\Delta\backslash\Delta\xi\Delta) = \#(\Gamma\backslash\Gamma\xi\Gamma)$. Conversely, suppose $\Gamma\alpha\Gamma = \bigsqcup_\nu \Gamma\alpha\gamma_\nu$. Since $\Gamma\alpha\gamma_\nu = \mathrm{pr}(\Delta\xi_\nu)$, we have $\Delta\xi_\nu \neq \Delta\xi_\mu$ if $\Gamma\alpha\gamma_\nu \neq \Gamma\alpha\gamma_\mu$. Also, since $\#(\Delta\backslash\Delta\xi\Delta) = \#(\Gamma\backslash\Gamma\xi\Gamma)$, we obtain $\Delta\xi\Delta = \bigsqcup_\nu \Delta\xi_\nu$. This proves (i). To prove (ii), put $\Gamma\alpha\Gamma = \bigsqcup_{i\in I} \Gamma\alpha_i$, $\Gamma\beta\Gamma = \bigsqcup_{j\in J} \Gamma\beta_j$, and $\Gamma\alpha\beta\Gamma = \bigsqcup_{h\in H} \Gamma\varepsilon_h$. By (i) we have $\Delta\xi\Delta = \bigsqcup_{i\in I} \Delta\xi_i$, $\Delta\eta\Delta = \bigsqcup_{j\in J} \Delta\eta_j$, and $\Delta\xi\eta\Delta = \bigsqcup_{h\in H} \Delta\zeta_h$ with ξ_i, η_j, and ζ_h such that $\mathrm{pr}(\xi_i) = \alpha_i$, $\mathrm{pr}(\eta_j) = \beta_j$, and $\mathrm{pr}(\zeta_h) = \varepsilon_h$. If $\Gamma\alpha\Gamma \cdot \Gamma\beta\Gamma = \Gamma\alpha\beta\Gamma$, then there is only one (i, j) such that $\Gamma\alpha_i\beta_j = \Gamma\alpha\beta$, and so there is only one (i, j) such that $\Delta\xi_i\eta_j = \Delta\xi\eta$. Thus $\Delta\xi\Delta \cdot \Delta\eta\Delta = \Delta\xi\eta\Delta$. This proves (ii).

Lemma 13.6. *The symbols* Γ *and* Δ *being as in* §13.4, *let* $\xi_m = (\alpha_m, m^k)$ *with* $\alpha_m = \mathrm{diag}[m^{-1}, m]$, $0 < m \in \mathbf{Z}$, *and* $\mathfrak{T}_m = \Delta\xi_m\Delta$. *Then* $\mathfrak{T}_m\mathfrak{T}_n = \mathfrak{T}_{mn}$ *if either* $m|N^\infty$ *or* m *is prime to* n.

PROOF. Put $\alpha = m\alpha_m$ and $\beta = n\alpha_n$. Our task is to show that $\Gamma\alpha\Gamma \cdot \Gamma\beta\Gamma = \Gamma\alpha\beta\Gamma$ in $\mathfrak{R}(\Gamma, GL_2^+(\mathbf{Q}))$. Indeed, if that were so, then $\Gamma\alpha_m\Gamma \cdot \Gamma\alpha_n\Gamma = \Gamma\alpha_{mn}\Gamma$, which combined with Lemma 13.5(ii) proves that $\mathfrak{T}_m\mathfrak{T}_n = \mathfrak{T}_{mn}$. Now $\Gamma\alpha\Gamma$, $\Gamma\beta\Gamma$, and $\Gamma\alpha\beta\Gamma$ are terms of $T'(m^2)$, $T'(n^2)$, and $T'(m^2n^2)$. Let $\varepsilon \in \Gamma\alpha\Gamma$, $\delta \in \Gamma\beta\Gamma$, and $\varepsilon\delta \in \Gamma(1)\mathrm{diag}[a, d]\Gamma(1)$ with positive integers a and d such that $a|d$. Suppose m is prime to n; take a prime factor p of a. If $p|m$, then $\beta \in SL_2(\mathbf{Z}_p)$, and so $\alpha \prec p\mathbf{Z}$, a contradiction. Thus $p\nmid m$, and similarly $p\nmid n$. Therefore, $a = 1$, and consequently $\varepsilon\delta \in \Gamma\alpha\beta\Gamma$, that is, $\Gamma\alpha\Gamma\beta\Gamma \subset \Gamma\alpha\beta\Gamma$. Therefore, $\Gamma\alpha\Gamma \cdot \Gamma\beta\Gamma = c\Gamma\alpha\beta\Gamma$ with $0 < c \in \mathbf{Z}$. Since $T'(m^2)T'(n^2) = T'(m^2n^2)$ by Lemma 13.3(iv), we see that $c = 1$, and so $\Gamma\alpha\Gamma \cdot \Gamma\beta\Gamma = \Gamma\alpha\beta\Gamma$.

Next suppose $m|N^\infty$; then $\Gamma\alpha\Gamma = T'(m^2)$ by (13.5). From (13.5) and Lemma 13.3(ii) we see that $\Gamma\beta\Gamma = \Gamma\sigma\Gamma \cdot \Gamma\tau\Gamma$, $\Gamma\sigma\Gamma = T'(\ell^2)$, $\ell|N^\infty$, $\tau = \mathrm{diag}[a, d]$, $a|d$ with d prime to N. Since $\beta = \mathrm{diag}[1, n^2]$, we see that $a = 1$, and so $\tau = \mathrm{diag}[1, h^2]$ with h such that $n = \ell h$. Taking $\Gamma\alpha\beta\Gamma$ in place of $\Gamma\beta\Gamma$, we have similarly by Lemma 13.3(ii) $\Gamma\alpha\beta\Gamma = T'(m^2\ell^2)\Gamma\tau\Gamma$. By Lemma 13.3(iii), $T'(m^2\ell^2) = T'(m^2)T'(\ell^2)$, and so $\Gamma\alpha\beta\Gamma = \Gamma\alpha\Gamma \cdot \Gamma\beta\Gamma$. This completes the proof.

13.7. So far we have used the symbol $f\|_k\xi$ for $\xi \in SL_2(\mathbf{R})$ or $\xi \in G_k$. When $k \in \mathbf{Z}$, we extend this to $\xi \in GL_2^+(\mathbf{R})$. Namely, for $\xi \in GL_2^+(\mathbf{R})$ and a function f on $\mathfrak{H}$ we put

(13.8a) $$f\|_k\xi = f\|_k\{\det(\xi)^{-1/2}\xi\}.$$

This is consistent with what we defined in [S71, §2.1]. Then (6.12) is valid for $\alpha \in GL_2^+(\mathbf{Q})$.

Let us now discuss the action of a Hecke algebra on modular forms of integral or half-integral weight. We first note

(13.8b) $\#(\Gamma\backslash\Gamma\alpha\Gamma) = \#(\Gamma\alpha\Gamma/\Gamma)$ *for every congruence subgroup* Γ *and* $\alpha \in GL_2^+(\mathbf{Q})$.

Indeed, we have $\Gamma \cap \{\pm 1\} = \Gamma \cap \alpha\Gamma\alpha^{-1} \cap \{\pm 1\}$ and

$$\mu(\Gamma\backslash\mathfrak{H})[\Gamma : \Gamma \cap \alpha\Gamma\alpha^{-1}] = \mu\big((\Gamma \cap \alpha\Gamma\alpha^{-1})\backslash\mathfrak{H}\big),$$

which equals $\mu\big((\alpha^{-1}\Gamma\alpha \cap \Gamma)\backslash\mathfrak{H}\big)$, since the measure on $\mathfrak{H}$ is invariant under the action of α^{-1}. Thus $[\Gamma : \Gamma \cap \alpha\Gamma\alpha^{-1}] = [\Gamma : \Gamma \cap \alpha^{-1}\Gamma\alpha]$, which gives (13.8b).

Let us first consider the case $k \in \mathbf{Z}$. Let Γ be a congruence subgroup of $\Gamma(1)$ and let $\alpha \in GL_2^+(\mathbf{Q})$. We take a decomposition $\Gamma\alpha\Gamma = \bigsqcup_\nu \Gamma\alpha_\nu$. Then we put $f|[\Gamma\alpha\Gamma]_k = \sum_\nu f\|_k\alpha_\nu$ for $f \in \mathscr{M}_k(\Gamma)$. We easily see that $f|[\Gamma\alpha\Gamma]_k \in \mathscr{M}_k(\Gamma)$. We call $[\Gamma\alpha\Gamma]_k$ a **Hecke operator.** We have, for $f \in \mathscr{M}_k(\Gamma)$ and $g \in \mathscr{S}_k(\Gamma)$,

(13.9) $$\langle f|[\Gamma\alpha\Gamma]_k,\, g\rangle = \langle f,\, g|[\Gamma\alpha^{-1}\Gamma]_k\rangle.$$

Indeed, by (13.8b) and (13.0), we can put $\Gamma\alpha\Gamma = \bigsqcup_{i\in I}\Gamma\alpha_i = \bigsqcup_{i\in I}\alpha_i\Gamma$ with some α_i. Then $\Gamma\alpha^{-1}\Gamma = \bigsqcup_{i\in I}\Gamma\alpha_i^{-1}$, and so by (6.12),

$$\langle f|[\Gamma\alpha\Gamma]_k,\, g\rangle = \sum_{i\in I}\langle f\|_k\alpha_i,\, g\rangle = \sum_{i\in I}\langle f,\, g\|_k\alpha_i^{-1}\rangle = \langle f,\, g|[\Gamma\alpha^{-1}\Gamma]_k\rangle,$$

which proves (13.9).

There is a traditional definition of Hecke operators acting on $\mathscr{M}_k(N, \psi)$. To be specific, take $t = 1$ and $\mathfrak{h} = (\mathbf{Z}/N\mathbf{Z})^\times$ in §13.2. Then $\Gamma = \Gamma_0(N)$ and $\Xi = \big\{\alpha \in M_2(\mathbf{Z}) \,\big|\, \det(\alpha) > 0,\ (a_\alpha, N) = 1,\ c_\alpha \in N\mathbf{Z}\big\}$. Let $\Gamma\alpha\Gamma = \bigsqcup_\nu \Gamma\alpha_\nu$ with $\alpha \in \Xi$ and $\det(\alpha) = q$. Then for $f \in \mathscr{M}_k(N, \psi)$ we put

$$f|[\Gamma\alpha\Gamma]_{k,\psi} = q^{k/2-1}\sum_\nu \psi\big(a(\alpha_\nu)\big)f\|_k\alpha_\nu,$$

and denote by $T'(n)_{k,\psi}$ resp. $T'(a, d)_{k,\psi}$ the sum of $[\Gamma\alpha\Gamma]_{k,\psi}$ for all $\Gamma\alpha\Gamma$ involved in $T'(n)$ resp. $T'(a, d)$. We have

(13.9a) $\psi\big(\det(\alpha)\big)\langle f[\Gamma\alpha\Gamma]_{k,\psi},\, g\rangle = \langle f,\, g|[\Gamma\alpha\Gamma]_{k,\psi}\rangle$ *if* $\det(\alpha)$ *is prime to* N *and* $f, g \in \mathscr{S}_k(N, \psi)$, *and consequently,* $r[\Gamma\alpha\Gamma]_{k,\psi}$ *with any* r *such that* $\bar{r}^2 = \psi\big(\det(\alpha)\big)$ *is a self-adjoint operator.*

To prove this, denote by $\xi \mapsto \xi^\iota$ the main involution of $M_2(\mathbf{Q})$, that is, $\begin{bmatrix} a & b \\ c & d\end{bmatrix}^\iota = \begin{bmatrix} d & -b \\ -c & a\end{bmatrix}$. Then $f\|_k\xi^{-1} = f\|_k\xi^\iota$, since $\xi^{-1} = \det(\xi)^{-1}\xi^\iota$. Now let $\alpha \in \Xi$ and $\det(\alpha) = q$ with q prime to N. As shown above, we can put $\Gamma\alpha\Gamma = \bigsqcup_\nu \Gamma\alpha_\nu = \bigsqcup_\nu \alpha_\nu\Gamma$ with suitable α_ν. Clearly $\Gamma^\iota = \Gamma$ and $\alpha^\iota \in \Xi$. Besides, $\Gamma\alpha\Gamma = \Gamma\alpha^\iota\Gamma$, since $\Gamma\alpha\Gamma = \Xi \cap \Gamma(1)\alpha\Gamma(1)$, which follows from [S71, Proposition 3.32(i)]. (Γ, Ξ here correspond to Γ', Δ' there.) Thus $\Gamma\alpha\Gamma = \Gamma\alpha^\iota\Gamma = \bigsqcup_\nu \Gamma\alpha_\nu^\iota$. Also, for $\xi \in \Xi$ with $\det(\xi) = q$ we have $a(\xi)d(\xi) - q \in N\mathbf{Z}$. Therefore

$$\begin{aligned}
&\psi(q)\langle f|[\Gamma\alpha\Gamma]_{k,\psi},\ g\rangle = q^{k/2-1}\big\langle \textstyle\sum_\nu \psi\big(a(\alpha_\nu)\big) f\|_k\alpha_\nu,\ \psi(q)g\big\rangle \\
&= q^{k/2-1}\big\langle f,\ \textstyle\sum_\nu \bar\psi\big(a(\alpha_\nu)\big)\psi(q)g\|_k\alpha_\nu^\iota\big\rangle = q^{k/2-1}\big\langle f,\ \textstyle\sum_\nu \psi\big(d(\alpha_\nu)\big)g\|_k\alpha_\nu^\iota\big\rangle \\
&= \big\langle f,\ g|[\Gamma\alpha\Gamma]_{k,\psi}\big\rangle,
\end{aligned}$$

since $d(\alpha_\nu) = a(\alpha_\nu^\iota)$. This proves (13.9a).

If $\alpha = q1_2$ with q prime to N, then $T'(q,\,q) = \Gamma\alpha\Gamma$, and so we have

(13.9b) $\qquad T'(q,\,q)_{k,\,\psi} = \psi(q)q^{k-2}$ *if* q *is prime to* N.

Let $0 \neq f(z) = \sum_{n=0}^\infty a(n)\mathbf{e}(nz) \in \mathscr{M}_k(N,\,\psi)$. Suppose $f|T'(p)_{k,\,\psi} = c_p f$ with $c_p \in \mathbf{C}$ for every prime number p; then $f|T'(n)_{k,\,\psi} = c_n f$ with $c_n \in \mathbf{C}$ for every $n \in \mathbf{Z},\ > 0$, and

(13.9c)
$$\sum_{n=1}^\infty a(n)n^{-s} = a(1)\sum_{n=1}^\infty c_n n^{-s} = a(1)\prod_p \big[1 - c_p p^{-s} + \psi(p)p^{k-1-2s}\big]^{-1};$$

see [S71, Theorem 3.43]. We call such an f a **Hecke eigenform,** and say that f is **normalized** if $a(1) = 1$, so that $a(n) = c_n$. Somewhat more generally, we have $f|T'(p)_{k,\,\psi} = c_p f$ for every $p \nmid r$ with a positive integer r if and only if the following equality holds:

(13.9d)
$$\sum_{(n,r)=1} a(n)n^{-s} = a(1)\prod_{p\nmid r}\big[1 - c_p p^{-s} + \psi(p)p^{k-1-2s}\big]^{-1}.$$

Taking $f = g$ in (13.9a), we obtain

(13.10) *If* $0 \neq f \in \mathscr{S}_k(N,\,\psi)$, $f|T'(n)_{k,\,\psi} = c_n f$, *and* n *is prime to* N, *then* $\psi(n)\bar c_n = c_n$.

13.8. Next suppose $k \notin \mathbf{Z}$. We put $\Delta = \ell(\Gamma)$ with a congruence subgroup Γ of $\Gamma(2)$. Let $\Delta\xi\Delta = \bigsqcup_\nu \Delta\xi_\nu$ with $\xi \in G_k$. Given $f \in \mathscr{M}_k(\Delta)$, we put

(13.11)
$$f|[\Delta\xi\Delta]_k = \sum_\nu f\|_k\xi_\nu,$$

where $(f\|_k\xi)(z) = q(z)^{-1}f(\alpha z)$ if $\xi = (\alpha,\,q)$; see (9.19). We easily see that $f|[\Delta\xi\Delta]_k \in \mathscr{M}_k(\Delta)$. Extending this $\mathbf{Q}$-linearly to $\mathfrak{R}(\Delta,\,G_k)$, we can let $\mathfrak{R}(\Delta,\,G_k)$ act on $\mathscr{M}_k(\Delta)$. Clearly $f|[\Delta\xi\Delta]_k \in \mathscr{S}_k(\Delta)$ if $f \in \mathscr{S}_k(\Delta)$.

Let us now consider the case $\xi = (\alpha,\,e^k)$ with $\alpha = \mathrm{diag}[e^{-1},\,e]$, $0 < e \in \mathbf{Q}$. Then

$$\text{(13.12a)} \qquad \#(\Delta\backslash\Delta\xi\Delta) = \#(\Delta\xi\Delta/\Delta),$$

$$\text{(13.12b)} \qquad \langle\, f\|[\Delta\xi\Delta]_k,\, g\,\rangle = \langle\, f,\, g\|[\Delta\xi^{-1}\Delta]_k\,\rangle.$$

Indeed, $\#(\Delta\backslash\Delta\xi\Delta) = \#(\Gamma\backslash\Gamma\alpha\Gamma)$ by Lemma 13.5. Similarly, $\#(\Delta\xi\Delta/\Delta) = \#(\Delta\backslash\Delta\xi^{-1}\Delta) = \#(\Gamma\backslash\Gamma\alpha^{-1}\Gamma) = \#(\Gamma\alpha\Gamma/\Gamma)$, and so we obtain (13.12a) from (13.8b). Therefore we can prove (13.12b) in the same manner as for (13.9).

We now consider $\mathscr{M}_k(2,\, N/2;\, \psi)$ of (8.11b) and (7.6) with a character ψ modulo N, and define an operator $\mathfrak{T}_m^{\psi}$ on that space as follows. Let $\Delta = \ell(\Gamma)$ with $\Gamma = \Gamma(2,\, N/2)$ and let $\Delta\xi_m\Delta = \bigsqcup_\nu \Delta\eta_\nu$ with ξ_m as in Lemma 13.6, that is, $\xi_m = (\alpha_m,\, m^k)$ with $\alpha_m = \mathrm{diag}[m^{-1},\, m]$. We see that if $\beta \in \Gamma\cdot\mathrm{diag}[1,\, n]\Gamma$ with $n \in \mathbf{Z}$, then a_β is prime to N. Therefore the a-entry of $m \cdot \mathrm{pr}(\eta_\nu)$ is prime to N. Thus for $f \in \mathscr{M}_k(2,\, N/2;\, \psi)$ we put

$$\text{(13.13)} \qquad f|\mathfrak{T}_m^{\psi} = m^{k-2} \sum_\nu \psi(a_\nu) f\|_k \eta_\nu,$$

where a_ν is the a-entry of $m \cdot \mathrm{pr}(\eta_\nu)$. We can easily verify that $f|\mathfrak{T}_m^{\psi}$ is well defined and belongs to $\mathscr{M}_k(2,\, N/2;\, \psi)$. Hereafter we make the convention that any product of numbers or symbols involving a factor $\psi(x)$ with $x|N^\infty$ means 0.

Theorem 13.9. *Let* $f(z) = \sum_{m=0}^{\infty} \lambda(m)\mathbf{e}(mz/2) \in \mathscr{M}_k(2,\, N/2;\, \psi)$ *and let* $(f|\mathfrak{T}_p^{\psi})(z) = \sum_{n=0}^{\infty} b(n)\mathbf{e}(nz/2)$ *with a prime number* p. *Then for* $0 \le n \in \mathbf{Z}$ *we have*

$$\text{(13.13a)} \qquad b(n) = \lambda(p^2 n) + \psi(p)\left(\frac{n}{p}\right)p^{k-3/2}\lambda(n) + \psi(p^2)p^{2k-2}\lambda(n/p^2),$$

where we understand that $\lambda(n/p^2) = 0$ *if* $p^2 \nmid n$.

PROOF. We have $\xi_p = (\alpha_p,\, p^k)$ with $\alpha_p = \mathrm{diag}[p^{-1},\, p]$, and so $\Gamma\alpha\Gamma = T'(1,\, p^2)$ with $\alpha = p\alpha_p = \mathrm{diag}[1,\, p^2]$.

Suppose $p|N$; then $\deg T'(p^2) = p^2$ and $\Gamma\alpha\Gamma = \bigsqcup_\nu \Gamma\beta_\nu$ with $\beta_\nu = \begin{bmatrix} 1 & 2\nu \\ 0 & p^2 \end{bmatrix}$, $0 \le \nu < p^2$; see [S71, Proposition 3.33]; notice that t in that proposition is 2. Thus $\Gamma\alpha_p\Gamma = \bigsqcup_\nu \Gamma\gamma_\nu$ with $\gamma_\nu = \begin{bmatrix} p^{-1} & 2p^{-1}\nu \\ 0 & p \end{bmatrix}$. We have $\gamma_\nu = \alpha_p\varepsilon_\nu$ with $\varepsilon_\nu = \begin{bmatrix} 1 & 2\nu \\ 0 & 1 \end{bmatrix}$. Let $\eta_\nu = (\gamma_\nu,\, p^k)$. Then $\ell(\varepsilon_\nu) = (\varepsilon_\nu,\, 1)$ and $\eta_\nu = (\alpha_p,\, p^k)(\varepsilon_\nu,\, 1) \in \Delta\xi_p\Delta$. By Lemma 13.5, $\Delta\xi_p\Delta = \bigsqcup_\nu \Delta\eta_\nu$. Thus

$$\begin{aligned} f|\mathfrak{T}_p^{\psi}(z) &= p^{-2} \sum_\nu f\big((z + 2\nu)/p^2\big) \\ &= p^{-2} \sum_{m=0}^{\infty} \lambda(m)\mathbf{e}(mz/2p^2) \sum_{\nu=0}^{p^2-1} \mathbf{e}(m\nu/p^2) = \sum_{n=0}^{\infty} \lambda(p^2 n)\mathbf{e}(nz/2). \end{aligned}$$

This gives the desired formula for $b(n)$ when $p|N$.

Next suppose $p \nmid N$. Then $\deg T'(1,\, p^2) = p^2 + p$, and $\Gamma\backslash\Gamma\alpha_p\Gamma$ can be given by the following elements:

$$\beta_\nu = \begin{bmatrix} p^{-1} & 0 \\ 0 & p \end{bmatrix}\begin{bmatrix} 1 & 2\nu \\ 0 & 1 \end{bmatrix} \qquad (0 \le \nu < p^2),$$

$$\gamma_h = \begin{bmatrix} 1 & 2h/p \\ 0 & 1 \end{bmatrix} = \begin{bmatrix} 1 & 0 \\ psN/2 & 1 \end{bmatrix}\begin{bmatrix} p^{-1} & 0 \\ 0 & p \end{bmatrix}\begin{bmatrix} p & 2h \\ -sN/2 & q \end{bmatrix} \quad (0 < h < p),$$

$$\delta = \begin{bmatrix} p & 0 \\ 0 & p^{-1} \end{bmatrix} = \begin{bmatrix} p^2 & -2t \\ N/2 & d \end{bmatrix}\begin{bmatrix} p^{-1} & 0 \\ 0 & p \end{bmatrix}\begin{bmatrix} p^2 d & 2t \\ -N/2 & 1 \end{bmatrix}.$$

Here for each h we take integers s and q so that $shN + qp = 1$, noting that hN is prime to p. As for δ, we take d and t so that $p^2 d + tN = 1$. To avoid ambiguity, we can choose positive q and d. Define elements β_ν^*, γ_h^*, and δ^* of G_k by

$$(13.14) \qquad \beta_\nu^* = (\beta_\nu,\, p^k), \quad \gamma_h^* = \left(\gamma_h,\, \varepsilon_p^{-1}\left(\frac{-h}{p}\right)\right), \quad \delta^* = (\delta,\, p^{-k}).$$

Then these belong to $\Delta\xi_p\Delta$. This is clear for β_ν^*. To treat γ_h^*, put $\sigma = \begin{bmatrix} 1 & 0 \\ psN/2 & 1 \end{bmatrix}$ and $\tau = \begin{bmatrix} p & 2h \\ -sN/2 & q \end{bmatrix}$. Then $\ell(\sigma) = (\sigma,\, j_\sigma^k)$ and $\ell(\tau) = \left(\tau,\, j_\tau(z)^k \varepsilon_q^{-1}\left(\frac{-sN}{q}\right)\right)$ with the branch of $j_\tau(z)^k$ such that $\lim_{z\to 0} j_\tau^k(z) > 0$, by (4.40), (4.41a), and (5.1b). Since $4|N$, we have $\left(\frac{-sN}{q}\right) = \left(\frac{-sN}{p}\right) = \left(\frac{-h}{p}\right)$ and $\varepsilon_q = \varepsilon_p$, and so $\ell(\sigma)\xi_p\ell(\tau) = \left(\gamma_h,\, \varepsilon_p^{-1}\left(\frac{-h}{p}\right)\right)$. As for δ, we note that $\varepsilon_d = 1$ and $\left(\frac{N}{d}\right) = 1$, and we obtain $\delta^* = (\delta,\, p^{-k}) \in \Delta\xi_p\Delta$ in a similar way. By Lemma 13.5, $\Delta\backslash\Delta\xi_p\Delta$ can be given by the elements of (13.14). Thus

$$\begin{aligned} f|\mathfrak{T}_p^\psi &= p^{-2}\sum_{\nu=0}^{p^2-1} f\big((z+2\nu)/p^2\big) \\ &\quad + \varepsilon_p p^{k-2}\psi(p)\sum_{h=1}^{p-1}\left(\frac{-h}{p}\right) f(z+2h/p) + \psi(p^2)p^{2k-2} f(p^2 z) \\ &= p^{-2}\sum_{m=0}^{\infty}\lambda(m)\mathbf{e}(mz/2p^2)\sum_{\nu=0}^{p^2-1}\mathbf{e}(m\nu/p^2) \\ &\quad + \varepsilon_p p^{k-2}\psi(p)\sum_{m=0}^{\infty}\lambda(m)\mathbf{e}(mz/2)\sum_{h=1}^{p-1}\left(\frac{-h}{p}\right)\mathbf{e}(mh/p) \\ &\quad + \psi(p^2)p^{2k-2}\sum_{m=0}^{\infty}\lambda(m)\mathbf{e}(mp^2 z/2). \end{aligned}$$

The last sum on the next to last line is $\left(\frac{-m}{p}\right)$ times the Gauss sum of (2.4b), and so

$$f|\mathfrak{T}_p^\psi = \sum_{n=0}^{\infty}\lambda(np^2)\mathbf{e}(nz/2) + p^{k-3/2}\psi(p)\sum_{m=0}^{\infty}\lambda(m)\left(\frac{m}{p}\right)\mathbf{e}(mz/2)$$

$$+\psi(p^2)p^{2k-2}\sum_{m=0}^{\infty}\lambda(m)\mathbf{e}(mp^2z/2).$$

Therefore we obtain the formula for $b(n)$ as stated in our theorem.

Theorem 13.10. (i) *Let f be a form as in Theorem 13.9, t a positive integer, and p a prime number. Put $\mu = k - 1/2$ and $\chi_t(a) = \psi(a)\left(\frac{t}{a}\right)$ for a prime to N. Suppose that $f|\mathfrak{T}_p^\psi = \omega_p f$ with $\omega_p \in \mathbf{C}$ and that $p|N$ or $p^2 \nmid t$. Then*

$$(13.15)\qquad \sum_{n=1}^{\infty}\lambda(tn^2)n^{-s}$$
$$=\sum_{p\nmid n}\lambda(tn^2)n^{-s}\cdot\left[1-\chi_t(p)p^{\mu-1-s}\right]\left[1-\omega_p p^{-s}+\psi(p)^2p^{2\mu-1-2s}\right]^{-1}.$$

(ii) *Suppose further that t has no nontrivial square factor prime to rN with a fixed positive integer r and that $f|\mathfrak{T}_p^\psi = \omega_p f$ with $\omega_p \in \mathbf{C}$ for every prime number $p\nmid r$. Then*

$$L_r(s-\mu+1,\chi_t)\sum_{0<n\in\mathbf{Z},(n,r)=1}\lambda(tn^2)n^{-s}$$
$$=\lambda(t)\prod_{p\nmid r}\left[1-\omega_p p^{-s}+\psi(p)^2p^{2\mu-1-2s}\right]^{-1}.$$

PROOF. Take p as in (i). By Theorem 13.9, if $p\nmid n$, we have

$$(*)\qquad \omega_p\lambda(tn^2)=\lambda(tp^2n^2)+\chi_t(p)p^{\mu-1}\lambda(tn^2),$$
$$(**)\quad \omega_p\lambda(tp^{2m}n^2)=\lambda(tp^{2m+2}n^2)+\psi(p)^2p^{2k-2}\lambda(tp^{2m-2}n^2)\quad(0<m\in\mathbf{Z}).$$

Put $H_n(X)=\sum_{m=0}^{\infty}\lambda(tp^{2m}n^2)X^m$ with an indeterminate X. Adding X times $(*)$ and X^{m+1} times $(**)$ for all $m>0$, we obtain

$$\omega_pXH_n(X)=H_n(X)-\lambda(tn^2)+\chi_t(p)p^{\mu-1}\lambda(tn^2)X+\psi(p)^2p^{2k-2}X^2H_n(X),$$

and so we have, for $p\nmid n$,

$$H_n(X)\left[1-\omega_pX+\psi(p)^2p^{2k-2}X^2\right]=\lambda(tn^2)\left[1-\chi_t(p)p^{\mu-1}X\right].$$

Since $\sum_{n=1}^{\infty}\lambda(tn^2)n^{-s}=\sum_{p\nmid n}H_n(p^{-s})n^{-s}$, we obtain (i). Then (ii) follows immediately from (i).

Theorem 13.11. *Let $f(z)=\sum_{m=1}^{\infty}\lambda(m)\mathbf{e}(mz/2)\in\mathscr{S}_k(2,N/2;\psi)$ and let $\mu=k-1/2$. Suppose that $f|\mathfrak{T}_p^\psi=\omega_pf$ with $\omega_p\in\mathbf{C}$ for every prime number $p\nmid r$ with a fixed positive integer r. Then there is an element $g(z)=\sum_{n=1}^{\infty}c(n)\mathbf{e}(nz)\in\mathscr{M}_{2\mu}(N,\psi^2)$ that is an eigenform of $T'(p)_{\mu,\psi^2}$ for every prime number p, such that $c(p)=\omega_p$ for every $p\nmid r$, where $T'(p)_{\mu,\psi^2}$ is the Hecke operator defined on $\mathscr{M}_{2\mu}(N,\psi^2)$. Moreover we have*

$$(13.16)\qquad L_r(s-\mu+1,\,\chi_t)\sum_{0<n\in\mathbf{Z},(n,r)=1}\lambda(tn^2)n^{-s} = \lambda(t)\sum_{0<n\in\mathbf{Z},(n,r)=1}c(n)n^{-s}.$$

PROOF. Let Y be the set of all $h \in \mathscr{M}_k(2,\,N/2;\,\psi)$ such that $h|\mathfrak{T}_p^{\psi} = \omega_p h$ for every $p \nmid r$. From Lemma 13.6 we easily see that the operators $\mathfrak{T}_p^{\psi}$ for all p form a commutative ring, and so Y is stable under it. Therefore by Lemma 2.12, Y contains an element that is an eigenform of $\mathfrak{T}_p^{\psi}$ for all p. This means that replacing f by a suitable element, we may assume that $f|\mathfrak{T}_p^{\psi} = \omega_p f$ with $\omega_p \in \mathbf{C}$ for every p. Take n so that $\lambda(n) \neq 0$ and put $n = tm^2$ with a squarefree t; define g_t by (12.36). Take $r = 1$ in Theorem 13.10(ii) and compare it with (12.36). Then we see that $\lambda(t) \neq 0$, and the function $g = \lambda(t)^{-1}g_t$ gives the desired eigenform in $\mathscr{M}_{2\mu}(N,\,\psi^2)$. (The letter μ here is m in Theorem 12.8. As for the eigenforms in $\mathscr{M}_{2\mu}$, basic principles are explained in [S71, Theorem 3.43].)

Lemma 13.12. *Let p be a prime number not dividing N. Then*

$$(13.17)\qquad \psi(p)^2\langle f|\mathfrak{T}_p^{\psi},\,g\rangle = \langle f,\,g|\mathfrak{T}_p^{\psi}\rangle$$

for every $f,\,g \in \mathscr{S}_k(2,\,N/2;\,\psi)$, and consequently $\bar{\psi}(p)\mathfrak{T}_p^{\psi}$ is a self-adjoint operator.

PROOF. We can put $\Gamma\alpha_p\Gamma = \bigsqcup_\nu \Gamma\alpha_\nu = \bigsqcup_\nu \alpha_\nu\Gamma$ with some $\{\alpha_\nu\}$ and $\Delta\xi_p\Delta = \bigsqcup_\nu \Delta\eta_\nu$ with η_ν such that $\mathrm{pr}(\eta_\nu) = \alpha_\nu$. Then $\Gamma\alpha_p^{-1}\Gamma = \bigsqcup_\nu \Gamma\alpha_\nu^{-1}$, and so $\Delta\xi_p^{-1}\Delta = \bigsqcup_\nu \Delta\eta_\nu^{-1}$ by Lemma 13.5(i). Now $\xi_p^{-1} = (\alpha_p^{-1},\,p^{-k}) = \delta^*$ with δ^* of (13.14), which belongs to $\Delta\xi_p\Delta$. Thus $\Delta\xi_p\Delta = \Delta\xi_p^{-1}\Delta = \bigsqcup_\nu \Delta\eta_\nu^{-1}$. Let a_ν resp. a'_ν be the a-entry of $p\alpha_\nu$ resp. $p\alpha_\nu^{-1}$. Then a'_ν is the d-entry of $p\alpha_\nu$. Observing that the b-entry resp. c-entry of $p\alpha_\nu$ is divisible by 2 resp. $N/2$, we obtain $a_\nu a'_\nu - p^2 \in N\mathbf{Z}$, and so $\psi(a'_\nu) = \psi(p^2)\bar{\psi}(a_\nu)$. Thus

$$\begin{aligned} p^{2-k}\langle f|\mathfrak{T}_p^{\psi},\,g\rangle &= \langle \textstyle\sum_\nu \psi(a_\nu)f\|_k\eta_\nu,\,g\rangle = \langle f,\,\textstyle\sum_\nu \bar{\psi}(a_\nu)g\|_k\eta_\nu^{-1}\rangle \\ &= \bar{\psi}(p^2)\langle f,\,\textstyle\sum_\nu \psi(a'_\nu)g\|_k\eta_\nu^{-1}\rangle = \bar{\psi}(p^2)p^{2-k}\langle f,\,g|\mathfrak{T}_p^{\psi}\rangle, \end{aligned}$$

which proves our lemma.

Lemma 13.13. (i) *Given $f \in \mathscr{S}_k(N,\,\psi)$, $\neq 0$, with $k \in \mathbf{Z}$, and $\sigma \in \mathrm{Aut}(\mathbf{C})$, suppose $f|T'(m)_{k,\psi} = c_m f$ with $c_m \in \mathbf{C}$ for a fixed m. Then $c_m \in \overline{\mathbf{Q}}$, $f^\sigma \in \mathscr{S}_k(N,\,\psi^\sigma)$, and $f^\sigma|T'(m)_{k,\psi^\sigma} = c_m^\sigma f^\sigma$.*

(ii) *Given $f \in \mathscr{S}_k(2,\,N/2;\,\psi)$, $\neq 0$, with $k \notin \mathbf{Z}$, and $\sigma \in \mathrm{Aut}(\mathbf{C})$, suppose $f|\mathfrak{T}_p^{\psi} = \omega_p f$ with $\omega_p \in \mathbf{C}$ for a fixed prime number p. Then $\omega_p \in \overline{\mathbf{Q}}$, $f^\sigma \in \mathscr{S}_k(2,\,N/2,\,\psi^\sigma)$, and $f^\sigma|\mathfrak{T}_p^{\psi^\sigma} = \omega_p^\sigma f^\sigma$.*

PROOF. In Theorem 7.5(iia) we showed that $f^\sigma \in \mathscr{S}_k(N,\,\psi^\sigma)$ if $k \in \mathbf{Z}$. By Theorem 7.5(iib), $\mathscr{S}_k(N,\,\psi) = Z\otimes_{\overline{\mathbf{Q}}}\mathbf{C}$ with $Z = \mathscr{S}_k(N,\,\psi)\cap\mathscr{M}_k(\mathbf{Q}_{\mathrm{ab}})$. Also, by Theorem 7.5(iii), $T'(m)_{k,\psi}$ sends Z into itself, and so its characteristic roots are algebraic. Thus $c_m \in \overline{\mathbf{Q}}$. Let $f(z) = \sum_{n=1}^{\infty} \lambda(n)\mathbf{e}(nz)$ and $f|T'(m)_{k,\psi} = \sum_{n=1}^{\infty} \mu(n)\mathbf{e}(nz)$. Then

$$\mu(n) = \sum_{d|(m,n)} \psi(d)d^{k-1}\lambda(mn/d^2); \tag{13.18}$$

see [S71, (3.5.12)]. Applying σ to this equality, we see that $\bigl(f|T'(m)_{k,\psi}\bigr)^{\sigma} = f^{\sigma}|T'(m)_{k,\psi^{\sigma}}$, and so we obtain (i). We can similarly prove (ii). The only difference is that we employ (13.13a) instead of (13.18).

13.14. In [S73a] and also in a few later papers of ours we formulated Hecke operators of half-integral weight in terms of $\Gamma_0(N)$ and the factor of automorphy $h'_{\delta}(z)^{\kappa}$ with h'_{δ} of (5.6) and odd $\kappa \in \mathbf{Z}$. The reason why we have changed the formulation as we have done in the present book is that in this way we can generalize the theory naturally to the Hilbert modular case, whereas we cannot do so with the old formulation.

CHAPTER V

THE ARITHMETICITY OF CRITICAL VALUES OF DIRICHLET SERIES

14. The theory on $SL_2(\mathbf{Q})$ for integral weight

14.1. The purpose of this section is to show that many of the main results of the previous section on the forms of half-integral weight have analogues in the case of integral weight, and to discuss their consequences. Thus in this section k denotes a positive integral weight, unless otherwise stated. The principal idea is to consider Hecke operators within $SL_2(\mathbf{Q})$. We fix a positive integer N and a character ψ modulo N, and define $\mathscr{M}_k(N, \psi)$ and $\mathscr{S}_k(N, \psi)$ as in (8.11d). Put $\Gamma = \Gamma_0(N)$ and take a prime number p. We take a decomposition $\Gamma\alpha_p\Gamma = \bigsqcup_{\beta\in B}\Gamma\beta$ with $\alpha_p = \mathrm{diag}[p^{-1}, p]$. Then for $f \in \mathscr{M}_k(N, \psi)$ we define a function $f|T_p^\psi$ on $\mathfrak{H}$ by

$$f|T_p^\psi = p^{k-2}\sum_{\beta\in B}\psi(pa_\beta)f\|_k\beta. \tag{14.1}$$

We easily see that $f|T_p^\psi$ is well defined and belongs to $\mathscr{M}_k(N, \psi)$, and also that $T_p^\psi = T'(1, p^2)_{k,\psi}$; however, we employ the symbol T_p^ψ, since that will suggest a similarity to the operator $\mathfrak{T}_p^\psi$ of (13.13) in the case of half-integral weight, though we will later find that $\bar{\psi}(p)T_p^\psi$ is a more natural object when $p\nmid N$. Thus, from (13.9a) we obtain

$$\psi(p^2)\langle f|T_p^\psi, g\rangle = \langle f, g|T_p^\psi\rangle \quad \textit{if} \quad p\nmid N \tag{14.2}$$

for every $f, g \in \mathscr{S}_k(N, \psi)$, which implies that $\bar{\psi}(p)T_p^\psi$ is a self-adjoint operator. This is an analogue of (13.17).

We keep the convention that any product of numbers or symbols involving a factor $\psi(x)$ with $x|N^\infty$, $N > 1$, means 0.

Lemma 14.2. *Let* $f(z) = \sum_{m=0}^\infty \lambda(m)\mathbf{e}(mz) \in \mathscr{M}_k(N, \psi)$ *and let* $f|T_p^\psi = \sum_{n=0}^\infty b(n)\mathbf{e}(nz)$ *with a prime number* p. *Then for* $0 \le n \in \mathbf{Z}$ *we have*

$$b(n) = \lambda(p^2n) + \psi(p)p^{k-2}\big(\delta(n/p)p - 1\big)\lambda(n) + \psi(p^2)p^{2k-2}\lambda(n/p^2),$$

where $\delta(x) = 1$ *or* 0 *according as* $x \in \mathbf{Z}$ *or* $x \notin \mathbf{Z}$, *and we understand that* $\lambda(n/p^2) = 0$ *if* $p^2 \nmid n$.

G. Shimura, *Modular Forms: Basics and Beyond*, Springer Monographs in Mathematics, DOI 10.1007/978-1-4614-2125-2_5, © Springer Science+Business Media, LLC 2012

PROOF. Suppose $p \nmid N$; then we can take B of (14.1) to be the set consisting of the following $p^2 + p$ elements:

$$\begin{bmatrix} p^{-1} & p^{-1}j \\ 0 & p \end{bmatrix} \quad (0 \le j < p^2), \qquad \begin{bmatrix} 1 & p^{-1}h \\ 0 & 1 \end{bmatrix} \quad (0 < h < p), \qquad \begin{bmatrix} p & 0 \\ 0 & p^{-1} \end{bmatrix}.$$

We then make calculations similar to and simpler than what was done in the proof of Theorem 13.9. We have

$$f|T_p^{\psi} = p^{-2}\sum_{j=0}^{p^2-1} f\bigl((z+j)/p^2\bigr)$$
$$+ p^{k-2}\psi(p)\sum_{h=1}^{p-1} f(z+p^{-1}h) + \psi(p^2)p^{2k-2}f(p^2z).$$

Since $\sum_{h=1}^{p-1}\mathbf{e}(mh/p)$ equals $p-1$ or -1 according as $p|m$ or $p \nmid m$, we obtain

$$f|T_p^{\psi} = \sum_{n=0}^{\infty}\lambda(np^2)\mathbf{e}(nz) + (p-1)p^{k-2}\psi(p)\sum_{n=0}^{\infty}\lambda(np)\mathbf{e}(npz)$$
$$- p^{k-2}\psi(p)\sum_{p\nmid n}\lambda(n)\mathbf{e}(nz) + \psi(p^2)p^{2k-2}\sum_{n=0}^{\infty}\lambda(n)\mathbf{e}(np^2z).$$

Clearly $\sum_{p\nmid n}\lambda(n)\mathbf{e}(nz) = \sum_{n=0}^{\infty}\{\lambda(n)\mathbf{e}(nz)-\lambda(np)\mathbf{e}(npz)\}$, and so we obtain the formula for $b(n)$ as stated above when $p\nmid N$. If $p|N$, the terms involving $\psi(p)$ don't appear, and so $b(n) = \lambda(p^2n)$. This completes the proof.

Theorem 14.3. *Let f be a form as in Lemma 14.2, r a positive integer, and t a positive integer with no nontrivial square factor prime to rN. Suppose $f|T_p^{\psi} = \omega_p f$ with $\omega_p \in \mathbf{C}$ for every prime number p not dividing r. Then*

$$L_{tr}(2s-2k+2,\,\psi^2)\cdot\sum_{0<n\in\mathbf{Z},\,(n,\,r)=1}\lambda(tn^2)n^{-s}$$
$$= \lambda(t)L_{tr}(s-k+1,\,\psi)\cdot\prod_{p\nmid r}\bigl[1-\zeta_p p^{-s}+\psi(p)^2p^{2k-2-2s}\bigr]^{-1},$$

where $\zeta_p = \omega_p - \psi(p)p^{k-2}(p-1)$.

PROOF. Our argument is similar to the proof of Theorem 13.10. To make our formulas short, put $q = \psi(p)p^{k-2}$. Take a prime number p not dividing r. Then for $p \nmid n$ we have, by Lemma 14.2,

$$(*) \qquad \omega_p\lambda(tn^2) = \lambda(tp^2n^2) + q\bigl(\delta(t/p)p-1\bigr)\lambda(tn^2),$$
$$(**) \qquad \omega_p\lambda(tp^{2m}n^2) = \lambda(tp^{2m+2}n^2) + q(p-1)\lambda(tp^{2m}n^2)$$
$$+ \psi(p)^2p^{2k-2}\lambda(tp^{2m-2}n^2) \qquad (0 < m \in \mathbf{Z}).$$

Put $\varepsilon = 1-\delta(t/p)$ and $K_n(X) = \sum_{m=0}^{\infty}\lambda(tp^{2m}n^2)X^m$ with an indeterminate X. Adding X times $(*)$ and X^{m+1} times $(**)$ for all $m > 0$, we obtain

$$\omega_pXK_n = K_n - \lambda(tn^2) + q(p-1)XK_n - qp\varepsilon\lambda(tn^2)X + \psi(p)^2p^{2k-2}X^2K_n,$$

and so we have $K_n\big[1-\zeta_p X+\psi(p)^2p^{2k-2}X^2\big]=\lambda(tn^2)(1+qp\varepsilon X)$ with ζ_p defined as in our theorem. Thus

$$K_n=\lambda(tn^2)(1+qp\varepsilon X)\big[1-\zeta_p X+\psi(p)^2p^{2k-2}X^2\big]^{-1}.$$

We have therefore

$$\sum_{(n,\,r)=1}\lambda(tn^2)n^{-s}=\sum_{(n,\,pr)=1}K_n(p^{-s})n^{-s}$$
$$=\sum_{(n,\,pr)=1}\lambda(tn^2)n^{-s}\cdot(1+q\varepsilon p^{1-s})\big[1-\zeta_p p^{-s}+\psi(p)^2p^{2k-2-2s}\big]^{-1}.$$

The factor $1+q\varepsilon p^{1-s}$ is different from 1 only if $p\nmid trN$, in which case it is $1+\psi(p)p^{k-1-s}$. Take a prime number not dividing pr, and repeat the same type of calculation with $\sum_{(n,\,pr)=1}\lambda(tn^2)n^{-s}$ and that prime number. Then we eventually obtain the desired equality of our theorem.

We now define, for a prime number p, an operator R_p acting on $\mathscr{S}_k(N,\,\psi)$ by

$$R_p=\begin{cases}T_p^{\psi} & \text{if } p|N,\\ \bar{\psi}(p)T_p^{\psi} & \text{if } p\nmid N.\end{cases}\tag{14.3}$$

Then R_p is a self-adjoint operator if $p\nmid N$. Since R_p is a constant times $T'(1,p^2)_{k,\psi}$, we see that the R_p for all prime numbers form a commutative ring.

Theorem 14.4. *Let $\mathfrak{P}$ be a set of prime numbers and f a nonzero element of $\mathscr{S}_k(N,\,\psi)$ such that $f|R_p=\xi_p f$ with $\xi_p\in\mathbf{C}$ for every $p\in\mathfrak{P}$. Then there exists a normalized Hecke eigenform $g(z)=\sum_{n=1}^{\infty}c(n)\mathbf{e}(nz)\in\mathscr{S}_k(N,\,\psi)$ such that*

$$\xi_p=\begin{cases}c(p)^2 & \text{if } p|N,\\ |c(p)|^2-p^{k-1}-p^{k-2} & \text{if } p\nmid N,\end{cases}\tag{14.4}$$

provided $p\in\mathfrak{P}$. Conversely, let $g(z)=\sum_{n=1}^{\infty}c(n)\mathbf{e}(nz)$ be a normalized Hecke eigenform in $\mathscr{S}_k(N,\,\psi)$. Then $g|R_p=\xi_p g$ with ξ_p as in (14.4) *for every prime number p.*

PROOF. Let $\delta=1$ if $p\nmid N$ and $\delta=0$ if $p|N$. Given f as in the first part of our theorem, let $Y=\{h\in\mathscr{S}_k(N,\,\psi)\mid h|R_p=\xi_p h \text{ for every } p\in\mathfrak{P}\}$. Then Y is stable under $T'(n)_{k,\psi}$ for every n, since $T'(n)_{k,\psi}$ commutes with $T'(1,\,p^2)_{k,\psi}$ for every p. Therefore Lemma 2.12 guarantees a normalized Hecke eigenform g in Y. From (13.6a) we obtain $T'(p)^2=T'(p^2)+\delta pT'(p,\,p)$. Since $T'(p^2)=T'(1,\,p^2)+\delta T'(p,\,p)$, we see that

$$T'(1,\,p^2)=T'(p)^2-\delta(p+1)T'(p,\,p).\tag{14.5}$$

Suppose $g|T'(n)_{k,\psi}=c(n)g$. Then from (14.5) and (13.9b) we see that $g|T_p^{\psi}=\alpha_p g$ with $\alpha_p=c(p)^2-\delta\psi(p)p^{k-2}(p+1)$, and so we obtain (14.4) in view of (13.10). The converse part can be easily proved by employing (14.5).

Lemma 14.5. *For every* $f \in \mathscr{S}_k(N, \psi)$, $\sigma \in \mathrm{Aut}(\mathbf{C})$, *and a prime number* p *we have* $f^\sigma \in \mathscr{S}_k(N, \psi^\sigma)$ *and* $(f|R_p)^\sigma = f^\sigma|R_p$. *In particular, if* $f|R_p = \xi_p f$ *with* $\xi_p \in \mathbf{C}$, *then* $\xi_p \in \overline{\mathbf{Q}}$ *and* $f^\sigma|R_p = \xi_p^\sigma f^\sigma$.

PROOF. This is similar to Lemma 13.13, and can be proved in the same way. The only point is that we employ the formula for $b(n)$ in Lemma 14.2 instead of (13.13a).

Lemma 14.6. *Let* $f(z) = \sum_{m=1}^{\infty} \lambda(m)\mathbf{e}(mz) \in \mathscr{S}_k(N, \psi)$ *and let* φ *be a character modulo* r; *let* t *be a positive integer with no nontrivial square factor prime to* rN. *Suppose* $f|R_p = \xi_p f$ *with* $\xi_p \in \mathbf{C}$ *for* $p \nmid r$. *Let* $g(z) = \sum_{n=1}^{\infty} c(n)\mathbf{e}(nz)$ *be a normalized Hecke eigenform in* $\mathscr{S}_k(N, \psi)$ *such that* (14.4) *holds for* $p \nmid r$. *Put* $g_\varphi(z) = \sum_{n=1}^{\infty} \varphi(n)c(n)\mathbf{e}(nz)$. *Then*

$$(14.6) \qquad L(s-k+1, \varphi\psi) \sum_{n=1}^{\infty} \varphi(n)\lambda(tn^2)n^{-s} = \lambda(t)D(s-k;\, g,\, g_\varphi) \prod_{p|t} \left[1 + (\varphi\psi)(p)p^{k-1-s}\right]^{-1},$$

where $D(s;\, *,\, *)$ *is as in* (8.23).

PROOF. By (8.32) we have

$$D(s-k;\, g,\, g_\varphi) = \sum_{n=1}^{\infty} \varphi(n)c(n)^2 n^{-s} = L(s-k+1, \varphi\psi) \sum_{n=1}^{\infty} \varphi(n)c(n^2)n^{-s}.$$

Taking $t = 1$ and g as f in Theorem 14.3 and substituting $\varphi(n)n^{-s}$ for n^{-s}, we find that

$$(*) \qquad L(2s-2k+2, \varphi^2\psi^2)D(s-k;\, g,\, g_\varphi) = L(s-k+1, \varphi\psi)^2 \prod_{p\nmid r} \left[1 - \varphi(p)\zeta_p p^{-s} + (\varphi\psi)(p)^2 p^{2k-2-2s}\right]^{-1}.$$

Substituting $\varphi(n)n^{-s}$ for n^{-s} in the original formula of Theorem 14.3, we obtain

$$L_t(2s-2k+2, \varphi^2\psi^2) \sum_{n=1}^{\infty} \varphi(n)\lambda(tn^2)n^{-s} = \lambda(t)L_t(s-k+1, \varphi\psi) \prod_{p\nmid r} \left[1 - \varphi(p)\zeta_p p^{-s} + (\varphi\psi)(p)^2 p^{2k-2-2s}\right]^{-1}.$$

Dividing this by $(*)$, we obtain (14.6).

14.7. Take f in Theorem 14.4 to be the eigenform of Theorem 8.23 and Lemma 8.24, so that $a(n) = \lambda(n)$. With α_p and β_p as in that lemma, we have

(14.7)
$$L_N(2s-2k+2,\,\psi^2)\sum_{n=1}^{\infty}\lambda(n^2)n^{-s}$$
$$= L_N(s-k+1,\,\psi)\prod_p\left[(1-\alpha_p^2p^{-s})(1-\beta_p^2p^{-s})\right]^{-1}.$$

Comparing this with the equality of Theorem 14.3, we obtain $\zeta_p = \alpha_p^2+\beta_p^2$. Since $\alpha_p+\beta_p = c_p$ and $\alpha_p\beta_p = \psi(p)p^{k-1}$, we obtain $\zeta_p = c_p^2 - 2\psi(p)p^{k-1}$, which produces $\xi_p = |c_p|^2 - p^{k-1} - p^{k-2}$. Thus, once we assume f to be a Hecke eigenform, Theorem 14.4 does not contain much new. The point of the theorem is that we have an Euler product expression for $\sum_{n=1}^{\infty}\lambda(tn^2)n^{-s}$ for a wider class of functions than Hecke eigenforms. There is another aspect in the theory, which will be discussed in the next section.

We defined in Section 5 modular forms of half-integral weight with respect to congruence subgroups of $Sp(n,\,\mathbf{Q})$. We can actually define Hecke operators on them and even associate an Euler product to an eigenfunction; see [S95]. There is a parallel theory for forms of integral weight on $Sp(n,\,\mathbf{Q})$, which, in the case $n = 1$, concerns $\sum_{n=1}^{\infty}\lambda(n^2)n^{-s}$ for a modular form $\sum_{m=1}^{\infty}\lambda(m)\mathbf{e}(mz)$ of integral weight we considered in Lemma 8.24 and are considering now.

15. The eigenspaces of R_p

15.1. In this section k denotes a positive weight that is either integral or half-integral. We mainly deal with the space of modular forms $\mathscr{S}_k(\nu,\,N/\nu;\,\psi)$, where $\nu = 1$ if $k \in \mathbf{Z}$ and $\nu = 2$ if $k \notin \mathbf{Z}$. We denote by $F(\psi)$ the field generated by the values of ψ. We begin with some easy facts.

Lemma 15.2. (i) *Let* $f(z) = \sum_{n=1}^{\infty}a_n\mathbf{e}(nz/\nu) \in \mathscr{S}_k(\nu,\,N/\nu;\,\psi)$. *Put* $f_\chi(z) = \sum_{n=1}^{\infty}\chi(n)a_n\mathbf{e}(nz/\nu)$ *with a character* χ. *(Lemma 7.13 shows that* $f_\chi \in \mathscr{S}_k(\nu,\,N'/\nu;\,\chi^2\psi)$ *with an integer* N' *determined as in that lemma.) Then* $f_\chi|T_p^{\chi^2\psi} = \chi(p^2)(f|T_p^{\psi})_\chi$ *if* $k\in\mathbf{Z}$ *and* $f_\chi|\mathfrak{T}_p^{\chi^2\psi} = \chi(p^2)(f|\mathfrak{T}_p^{\psi})_\chi$ *if* $k\notin\mathbf{Z}$, *for every prime number* p.

(ii) *The space* $\mathscr{S}_k(\nu,\,N/\nu;\,\psi)$ *is spanned by its* $F(\psi)$*-rational elements.*

PROOF. The first equality of (i) can easily be seen from the formula of Lemma 14.2, and the second one from (13.13a). To prove (ii), put $\mathscr{S}_\psi^N = \mathscr{S}_k(\nu,\,N/\nu;\,\psi)$ for simplicity. By Theorem 7.5(iib) we can find a finite subset A of $\mathscr{S}_\psi^N \cap \mathscr{M}_k(\mathbf{Q}_{\mathrm{ab}})$ that spans $\mathscr{S}_\psi^N$ over $\mathbf{Q}_{\mathrm{ab}}$. By Theorem 7.5(v) there is a finite extension K of $F(\psi)$ contained in $\mathbf{Q}_{\mathrm{ab}}$ such that every member of A is K-rational. Then $\mathscr{S}_\psi^N \cap \mathscr{M}_k(K)$ is stable under $\mathrm{Gal}\big(K/F(\psi)\big)$ by Theorem 7.5(iia). Therefore (ii) follows from Lemma 2.13.

15.3. We define an operator R_p acting on $\mathscr{S}_k(\nu,\,N/\nu;\,\psi)$ for every prime number $p\nmid N$ by

$$R_p = \begin{cases} \bar{\psi}(p)T_p^{\psi} & \text{if } k \in \mathbf{Z}, \\ \bar{\psi}(p)\mathfrak{T}_p^{\psi} & \text{if } k \notin \mathbf{Z}. \end{cases} \tag{15.1}$$

If $k \in \mathbf{Z}$, this is the same as what was defined in (14.3). We see from (13.17) and (14.2) that R_p is a self-adjoint operator.

We note an easy consequence of Lemma 15.2(i):

(15.2) $f_\chi|R_p = (f|R_p)_\chi$ *if* f_χ *is as in Lemma 15.2(i),* $p \nmid N$, *and* $\chi(p) \neq 0$.

15.4. Let f be a nonzero element of $\mathscr{S}_k(\nu,\, N/\nu;\, \psi)$ such that $f|R_p = \xi_p f$ for almost all prime numbers p. We call $\{\xi_p\}$ a system of *R*-**eigenvalues** of weight k. Take another such system $\{\xi'_p\}$ defined with respect to the same weight, but with a character that may or may not be ψ. We say that $\{\xi_p\}$ is **equivalent** to $\{\xi'_p\}$ if $\xi_p = \xi'_p$ for almost all p. Then we denote by Ξ_k the set of all equivalence classes of systems of R-eigenvalues of weight k, and denote an element of Ξ_k by a single letter ξ, which is a function defined on a set of almost all prime numbers. We then put

$$\mathfrak{S}(\xi) = \{f \in \mathscr{S}_k^* \mid f|R_p = \xi_p f \text{ for almost all } p\}, \tag{15.3}$$

$$\mathscr{S}_k^* = \bigcup_{N,\psi} \mathscr{S}_k(\nu,\, N/\nu;\, \psi), \tag{15.3a}$$

$$\mathfrak{S}_\psi(\xi) = \bigcup_{0<N\in\nu\mathbf{Z}} \mathscr{S}_k(\nu,\, N/\nu;\, \psi) \cap \mathfrak{S}(\xi), \tag{15.3b}$$

where the union in (15.3a) is taken over all possible N and ψ. We say that ξ occurs in $\mathscr{S}_k(\nu,\, N/\nu;\, \psi)$ if $\mathfrak{S}(\xi) \cap \mathscr{S}_k(\nu,\, N/\nu;\, \psi) \neq \{0\}$.

For example, take f as above and put $f(z) = \sum_{n=1}^{\infty} a_n \mathbf{e}(nz/\nu)$ and $f_\chi(z) = \sum_{n=1}^{\infty} \chi(n) a_n \mathbf{e}(nz/\nu)$ with a character χ. By Lemma 7.13, $f_\chi \in \mathscr{S}_k(\nu,\, N'/\nu;\, \psi')$ with a multiple N' of N and $\psi' = \chi^2\psi$. Then (15.2) shows that $f_\chi|R_p = \xi_p f_\chi$ for almost all p, that is,

(15.4) $f \in \mathfrak{S}(\xi) \Longrightarrow f_\chi \in \mathfrak{S}(\xi)$ *for every Dirichlet character* χ.

Let $\xi \in \Xi_k$ and $\sigma \in \mathrm{Aut}(\mathbf{C})$. Then ξ^σ as an element of Ξ_k can be defined by $(\xi^\sigma)_p = \xi_p^\sigma$ for almost all p. Indeed, take a nonzero element f of $\mathfrak{S}(\xi) \cap \mathscr{S}_k(\nu,\, N/\nu;\, \psi)$ as above. Then by Lemmas 13.13 and 14.5, $f^\sigma \in \mathscr{S}_k(\nu,\, N/\nu;\, \psi^\sigma)$ and $f^\sigma|R_p = \xi_p^\sigma f^\sigma$ for almost all p, which gives the desired fact. Thus

(15.4a) $\mathfrak{S}(\xi)^\sigma = \mathfrak{S}(\xi^\sigma)$ *for every* $\sigma \in \mathrm{Aut}(\mathbf{C})$.

Let us next discuss the connection of R-eigenvalues with Hecke eigenvalues of integral weight. We first consider the case $k \notin \mathbf{Z}$. Let $0 \neq f \in \mathfrak{S}(\xi) \cap \mathscr{S}_k(2,\, N/2;\, \psi)$. Then $f|\mathfrak{T}_p^\psi = \omega_p f$ with $\omega_p = \psi(p)\xi_p$ for p in a set $\mathfrak{P}$ that contains almost all prime numbers. As shown in the proof of Theorem 13.11, we can find ω_p even for $p \notin \mathfrak{P}$ and an element $g(z) = \sum_{n=1}^{\infty} c_n \mathbf{e}(nz)$ of

$\mathscr{M}_{2k-1}(N, \psi^2)$ such that

(15.5) $$\sum_{n=1}^{\infty} c_n n^{-s} = \prod_p \left[1 - \omega_p p^{-s} + \psi(p^2)p^{2k-2-2s}\right]^{-1}.$$

Thus ξ corresponds to the set of Hecke eigenvalues $\{\omega_p\}$ occurring in $\mathscr{M}_{2k-1}$. We have $g \in \mathscr{S}_{2k-1}$ under the condition stated in Theorem 12.8(iii).

Next suppose $k \in \mathbf{Z}$. Given a nonzero element f of $\mathfrak{S}(\xi)$ with $\xi \in \Xi_k$, take (N, ψ) so that $f \in \mathscr{S}_k(N, \psi)$. Then by Theorem 14.4 there exists a normalized Hecke eigenform $g(z) = \sum_{n=1}^{\infty} c_n \mathbf{e}(nz)$ such that (14.4) holds for almost all p. Therefore we can define Ξ_k to be the equivalence classes of $\{c_p\}$, by saying that $\{c_p\}$ is **equivalent** to $\{c'_p\}$ if $|c_p|^2 = |c'_p|^2$ for almost all p. Thus, for both integral and half-integral k, ξ corresponds to an eigenform of integral weight.

15.5. Let F be a finite algebraic extension of $\mathbf{Q}$ contained in $\mathbf{C}$. We call F totally real if every isomorphic image of F into $\mathbf{C}$ is contained in $\mathbf{R}$, and call it totally imaginary if no isomorphic image of F into $\mathbf{C}$ is contained in $\mathbf{R}$. We call F a **CM-field** if it is a totally imaginary quadratic extension of a totally real field. Let ρ denote the complex conjugation. Then the following statements, in which F is a finite algebraic extension of $\mathbf{Q}$ contained in $\mathbf{C}$, can easily be verified.

(15.6a) *F is either totally real or a CM-field if and only if $F^\rho = F$ and $\rho\sigma = \sigma\rho$ on F for every ring-injection σ of F into $\mathbf{C}$.*

(15.6b) *The composite of finitely many fields of type* (15.6a) *is also a field of the same type.*

(15.6c) *Every subfield of a field of type* (15.6a) *is also a field of the same type.*

For $\xi \in \Xi_k$ we denote by $F(\xi)$ the smallest extension of $\mathbf{Q}$ that contains ξ_p for almost all p, and by $F(\xi, \psi)$ the composite of $F(\xi)$ and $F(\psi)$.

Lemma 15.6. (i) *Given an element $g(z) = \sum_{n=1}^{\infty} c_n \mathbf{e}(nz)$ of $\mathscr{S}_k(N, \psi)$ that is a normalized Hecke eigenform, let K be the field generated over $\mathbf{Q}$ by the c_p for all $p \nmid N$. Then K is either totally real or a CM-field.*

(ii) *$F(\xi)$ is a finite totally real algebraic extension of $\mathbf{Q}$.*

(iii) *$F(\xi, \psi)$ is totally real or a CM-field.*

(iv) *$\mathfrak{S}(\xi) \cap \mathscr{S}_k(\nu, N/\nu; \psi)$ is spanned by its $F(\xi, \psi)$-rational elements.*

PROOF. By Lemma 13.13 and Theorem 7.5(v), K of (i) is a finite algebraic extension of $\mathbf{Q}$. Let L be the field generated over $\mathbf{Q}$ by $\psi(p)^{1/2}$ for all $p \nmid N$. Then L is a finite abelian extension of $\mathbf{Q}$. Put $\alpha_p = \bar{\psi}(p)^{1/2} c_p$ with any choice of $\bar{\psi}(p)^{1/2}$. From (13.10) we see that $\alpha_p \in \mathbf{R}$. Let $\sigma \in \mathrm{Aut}(\mathbf{C})$. Then g^σ is a normalized Hecke eigenform contained in $\mathscr{S}_k(N, \psi^\sigma)$, and

$\alpha_p^\sigma = [\bar{\psi}(p)^\sigma]^{1/2}c_p^\sigma \in \mathbf{R}$ for the same reason. Since KL is generated over L by the α_p for all $p \nmid N$, we see that $(KL)^\rho = KL$ and $\rho\sigma = \sigma\rho$, which proves (i).

To prove (ii), we first observe that $\xi_p \in \mathbf{R}$, since R_p is self-adjoint. Thus the point is $\bigl[F(\xi) : \mathbf{Q}\bigr] < \infty$. If $k \in \mathbf{Z}$, this follows from (14.4) and (i). If $k \notin \mathbf{Z}$, we take g as in (15.5). Since $\omega_p = \psi(p)\xi_p$, the desired finiteness follows from (i). Then (iii) follows immediately. To prove (iv), put $\mathscr{S}_\psi^N = \mathscr{S}_k(\nu, N/\nu; \psi)$. Since $\mathfrak{S}(\xi) \cap \mathscr{S}_\psi^N \cap \mathscr{M}_k(\mathbf{Q}_{\mathrm{ab}})$ is stable under R_p, we can find a finite set A of $\mathbf{Q}_{\mathrm{ab}}$-rational forms that spans $\mathfrak{S}(\xi) \cap \mathscr{S}_\psi^N$. By Theorem 7.5(v) there exists a finite algebraic extension K of $F(\xi, \psi)$ such that every member of A is K-rational; we may assume that K is normal over $F(\xi, \psi)$. Then $\mathfrak{S}(\xi) \cap \mathscr{S}_\psi^N \cap \mathscr{M}_k(K)$ is stable under $\mathrm{Gal}\bigl(K/F(\xi, \psi)\bigr)$. Therefore (iv) follows from Lemma 2.13.

16. Main theorems on arithmeticity

The aim of this section is to prove some theorems on the special values of various Dirichlet series such as $\mathscr{D}_N(s; f, g)$ of (8.28). The main idea is to compare such values with certain inner products of modular forms. We note that by Lemmas 6.4 and 7.11 every element of $\mathscr{S}_k$ (resp. $\mathscr{N}_k$) is rapidly decreasing (resp. slowly increasing) at every cusp, and so $\langle f, h\rangle$ is meaningful for every $f \in \mathscr{N}_k$ and $h \in \mathscr{S}_k$, for the reason explained in §6.5. Throughout this section, ρ denotes the complex conjugation, and $G(\chi)$ the Gauss sum of a character χ defined under the convention of §2.7. We first prove two theorems on the projection maps from $\mathscr{N}_k$ to $\mathscr{S}_k$ or its subspaces.

Theorem 16.1. *For each weight $k \in 2^{-1}\mathbf{Z}, > 0$, there exists a $\mathbf{C}$-linear map $\mathfrak{p}_k$ of $\mathscr{N}_k$ into $\mathscr{S}_k$ with the following properties:*

(16.1a) $\quad \langle f, h\rangle = \langle \mathfrak{p}_k(f), h\rangle$ *for every $f \in \mathscr{N}_k$ and $h \in \mathscr{S}_k$;*

(16.1b) $\quad \mathfrak{p}_k(f)^\sigma = \mathfrak{p}_k(f^\sigma)$ *for every $f \in \mathscr{N}_k$ and every $\sigma \in \mathrm{Aut}(\mathbf{C})$.*

PROOF. With a fixed congruence subgroup Γ and a fixed $f \in \mathscr{N}_k(\Gamma)$, the map $h \mapsto \langle f, h\rangle$ for $h \in \mathscr{S}_k(\Gamma)$ is a $\mathbf{C}$-linear map of $\mathscr{S}_k(\Gamma)$ into $\mathbf{C}$, and so there exists a unique $g \in \mathscr{S}_k(\Gamma)$ such that $\langle f, h\rangle = \langle g, h\rangle$ for every $h \in \mathscr{S}_k(\Gamma)$. Putting $g = \mathfrak{p}_k(f)$, we obtain the desired map $\mathfrak{p}_k$ of (16.1a) since the replacement of Γ by a smaller group does not change g. Indeed, take a congruence subgroup Γ' that is a normal subgroup of Γ. Then we find an element g' of $\mathscr{S}_k(\Gamma')$ such that $\langle f, h'\rangle = \langle g', h'\rangle$ for every $h' \in \mathscr{S}_k(\Gamma')$. Take any $\gamma \in \Gamma$, and write $\varphi\|\alpha$ for $\varphi\|_k\alpha$ for any function φ on $\mathfrak{H}$ and $\alpha \in G_k$. Then $g'\|\gamma \in \mathscr{S}_k(\Gamma')$, and for every $h' \in \mathscr{S}_k(\Gamma')$ we have $\langle g'\|\gamma, h'\rangle = \langle g', h'\|\gamma^{-1}\rangle = \langle f, h'\|\gamma^{-1}\rangle = \langle f\|\gamma, h'\rangle = \langle f, h'\rangle = \langle g', h'\rangle$, and so $g'\|\gamma = g'$. Thus $g' \in \mathscr{S}_k(\Gamma)$, and $g' = g$ as expected.

To prove (16.1b), given $f \in \mathscr{N}_k$, we consider the sum expression for f in Lemma 7.8(ii). Then we can put $f = g_0 + D_{k-2}g_* + cE_2$ where g_0 is the element of $\mathscr{M}_k$ in that expression, g_* is an element of $C_{k-2}(\Gamma)$ determined suitably by that expression, $c \in \mathbf{C}$, and E_2 is the function of (7.10); cE_2 is necessary only when $k = 2$. For $h \in \mathscr{S}_k$ we have $\langle D_{k-2}g_*, h\rangle = 0$ by Corollary 6.9(i), and $\langle E_2, h\rangle = 0$ by (9.50). Thus $\langle f, h\rangle = \langle g_0, h\rangle$. Next, by (9.45a) we can put $g_0 = t + r$ with $t \in \mathscr{S}_k$ and $r \in \mathscr{E}_k$. Then $\langle g_0, h\rangle = \langle t, h\rangle$ for $h \in \mathscr{S}_k$. Since $\langle f, h\rangle = \langle t, h\rangle$, we have $t = p_k(f)$. Replace f by f^σ with $\sigma \in \mathrm{Aut}(\mathbf{C})$. From Lemma 7.8(i) we see that g_0 is replaced by g_0^σ, and $g_0^\sigma = t^\sigma + r^\sigma$. We have $t^\sigma \in \mathscr{S}_k$ by Theorem 7.5(i), and $r^\sigma \in \mathscr{E}_k$ by Theorem 9.26. Thus $p_k(f^\sigma) = t^\sigma = p_k(f)^\sigma$. This proves (16.1b) and completes the proof.

Lemma 16.2. *Let $\xi \in \Xi_k$; suppose ξ occurs in $\mathscr{S}_k(\nu, N/\nu; \psi)$. Then there exist a $\mathbf{C}$-linear map $r_{k,\psi}^N$ of $\mathscr{N}_k$ into $\mathscr{S}_k(\nu, N/\nu; \psi)$ and also a $\mathbf{C}$-linear map $r_{k,\psi}^{N,\xi}$ of $\mathscr{N}_k$ into $\mathfrak{S}(\xi) \cap \mathscr{S}_k(\nu, N/\nu; \psi)$ with the following properties:*

(16.2a) $\langle f, h\rangle = \langle r_{k,\psi}^N(f), h\rangle$ *for every* $f \in \mathscr{N}_k$ *and* $h \in \mathscr{S}_k(\nu, N/\nu; \psi)$;

(16.2b) $r_{k,\psi}^N(f)^\sigma = r_{k,\psi^\sigma}^N(f^\sigma)$ *for every* $f \in \mathscr{N}_k$ *and every* $\sigma \in \mathrm{Aut}(\mathbf{C})$;

(16.2c) $\langle f, h\rangle = \langle r_{k,\psi}^{N,\xi}(f), h\rangle$ *for every* $f \in \mathscr{N}_k$ *and* $h \in \mathfrak{S}(\xi) \cap \mathscr{S}_k(\nu, N/\nu; \psi)$;

(16.2d) $r_{k,\psi}^{N,\xi}(f)^\sigma = r_{k,\psi^\sigma}^{N,\xi^\sigma}(f^\sigma)$ *for every* $f \in \mathscr{N}_k$ *and every* $\sigma \in \mathrm{Aut}(\mathbf{C})$.

PROOF. For a fixed $f \in \mathscr{N}_k$, the map $h \mapsto \langle f, h\rangle$ is a $\mathbf{C}$-linear map of $\mathscr{S}_k(\nu, N/\nu; \psi)$ into $\mathbf{C}$, and so the existence of an element $r_{k,\psi}^N(f)$ of $\mathscr{S}_k(\nu, N/\nu; \psi)$ satisfying (16.2a) is obvious. Thus (16.2b) is our problem. We have clearly $r_{k,\psi}^N(f) = r_{k,\psi}^N(g)$ if $g = p_k(f)$ with p_k of Theorem 16.1, and so it is sufficient to prove (16.2b) when $f \in \mathscr{S}_k$. Given $f \in \mathscr{S}_k$, take a multiple K of N so that $f \in \mathscr{S}_k\bigl(\Gamma(K)\bigr)$. Define the map $q : \mathscr{S}_k\bigl(\Gamma(K)\bigr) \to \mathscr{S}_k\bigl(\Gamma(N)\bigr)$ as in Lemma 7.15. Then $q(f)^\sigma = q(f^\sigma)$ for every $\sigma \in \mathrm{Aut}(\mathbf{C})$ and $\langle f, h\rangle = \langle q(f), h\rangle$ for every $h \in \mathscr{S}_k\bigl(\Gamma(N)\bigr)$. This means that to prove (16.2b), we may assume that $f \in \mathscr{S}_k\bigl(\Gamma(N)\bigr)$.

As observed in the proof of Theorem 7.5(iv), $\mathscr{S}_k\bigl(\Gamma(N)\bigr)$ is the direct sum of $\mathscr{S}_k(N, N; \chi)$ with some characters χ modulo N, and $\mathscr{S}_k(N, N; \chi)$ is orthogonal to $\mathscr{S}_k(N, N; \chi')$ if $\chi \neq \chi'$. Therefore, to prove $r_{k,\psi}^N(f)^\sigma = r_{k,\psi^\sigma}^N(f^\sigma)$, we may assume that $f \in \mathscr{S}_k(N, N; \chi)$ for some χ. Then $f^\sigma \in \mathscr{S}_k(N, N; \chi^\sigma)$ by Theorem 7.5(iia). Put $f(z) = \sum_{n=1}^\infty c(n)\mathbf{e}(nz/N)$ and $Pf = (\nu/N)\sum_{u=1}^{N/\nu} f(z+\nu u)$. By Lemma 7.14, $Pf = \sum_{n=1}^\infty c(Nn/\nu)\mathbf{e}(nz/\nu) \in \mathscr{S}_k(\nu, N; \chi)$ and $\langle Pf, h\rangle = \langle f, Ph\rangle$ for every $h \in \mathscr{S}_k\bigl(\Gamma(N)\bigr)$. Clearly $Ph = h$ if $h \in \mathscr{S}_k(\nu, N/\nu; \psi)$. Therefore $r_{k,\psi}^N(f) = 0$ if $\chi \neq \psi$, and $r_{k,\psi}^N(f) = Pf$ if $\chi = \psi$. In either case we have $r_{k,\psi}^N(f)^\sigma = r_{k,\psi^\sigma}^N(f^\sigma)$.

Next, to find $r_{k,\psi}^{N,\xi}$ satisfying (16.2c, d), let $\xi_1, \ldots, \xi_e$ be the elements of Ξ_k that occur in $\mathscr{S}_k(\nu, N/\nu; \psi)$ and let $V_i = \mathfrak{S}(\xi_i) \cap \mathscr{S}_k(\nu, N/\nu; \psi)$. Since

R_p is self-adjoint for every p, we easily see that $\langle V_i, V_j\rangle = 0$ for $i \neq j$, and $\mathscr{S}_k(\nu, N/\nu; \psi) = \bigoplus_{i=1}^{e} V_i$. For $\xi = \xi_i$ we let $r_{k,\psi}^{N,\xi}$ denote the composite of $r_{k,\psi}^{N}$ and the projection map of $\mathscr{S}_k(\nu, N/\nu; \psi)$ to V_i. Then (16.2c) is clearly satisfied, and from (16.2b), Theorem 7.5(iia), and (15.4a) we obtain (16.2d). This completes the proof.

Lemma 16.3. *Given f and g as in Lemma 8.22, suppose that they are normalized Hecke eigenforms in $\mathscr{S}_k(N, \psi)$ and $\mathscr{S}_\ell(M, \varphi)$, respectively. Put*

$$\mathfrak{D}_K(s;\, f,\, g) = L_K(2s+2-k-\ell,\, \psi\varphi) \sum_{(n,K)=1} a(n)b(n)n^{-s} \tag{16.3}$$

with a common multiple K of N and M. Then $\mathfrak{D}_K(s;\, f,\, g) \neq 0$ for Re$(s) \geq$ $(k+\ell)/2$.

PROOF. For $t \in \mathbf{R}$ put

$$A(s) = \mathfrak{D}_K(s+it;\, f,\, g)\mathfrak{D}_K(s-it;\, f_\rho,\, g_\rho)\mathfrak{D}_K(s;\, f,\, f_\rho)\mathfrak{D}_K(s;\, g,\, g_\rho),$$

where f_ρ is defined by (5.14). Define $\alpha_p, \beta_p, \gamma_p, \delta_p$ as in Lemma 8.22. Since $\log\bigl\{\prod_p(1-x_p p^{-s})^{-1}\bigr\} = \sum_{m=1}^{\infty}\sum_p m^{-1}p^{-ms}x_p^m$, we find that

$$\log A(s) = \sum_{m=1}^{\infty}\sum_{p\nmid K} m^{-1}p^{-ms}\bigl|\alpha_p^m + \beta_p^m + \xi_p^m + \eta_p^m\bigr|^2, \tag{16.4}$$

where $\xi_p = \bar{\gamma}_p p^{it}$ and $\eta_p = \bar{\delta}_p p^{it}$, provided A is holomorphic and the last double infinite series is convergent at s. Put $\kappa = (k+\ell)/2$ and define $\mathscr{D}_N$ by (8.28). Then

$$\mathfrak{D}_K(s;\, f,\, g) = \mathscr{D}_K(s-\kappa;\, f,\, g).$$

By Theorem 8.21, $\mathscr{D}_K(s;\, f,\, g)$ is holomorphic on the whole $\mathbf{C}$, except for possible simple poles at $s = 0$ and $s = -1$, which may occur only if $k = \ell$ and $\psi\varphi$ is trivial. If $k = \ell$, $\tilde{\Gamma}$ of (8.29) has $\Gamma(s+1)$ as a factor, and so $s = -1$ cannot be a pole of $\mathscr{D}_K$. Thus $s = 0$ is the only possible pole of $\mathscr{D}_K$. Therefore $\mathfrak{D}_K$ is holomorphic except for a possible simple pole at $s = \kappa$, which may occur only if $k = \ell$ and $\psi\varphi$ is trivial. Consequently,

(16.5) *A is holomorphic on* $\mathbf{C}$ *except for possible poles on* Re$(s) = \kappa$.

Let Re$(s) = \sigma$ be the line of convergence of the right-hand side of (16.4). Since $\tilde{\Gamma}$ of (8.29) has many poles where $\mathfrak{D}_K(s;\, f,\, g)$ must have zeros, we see that $\sigma \neq -\infty$. Suppose $\sigma > \kappa$; then A is holomorphic at $s = \sigma$, and for real $s > \sigma$ we have $\log A(s) \geq 0$, and so $A(s) \geq 1$. Thus $A(\sigma) \geq 1$, which means that $\log A$ is holomorphic at $s = \sigma$, but that contradicts the well-known fact that a Dirichlet series with nonnegative coefficients is not holomorphic at the real point on the line of convergence. Therefore, $\sigma \leq \kappa$. This implies that $A(s) = \exp\bigl[\log A(s)\bigr] \neq 0$ for Re$(s) > \kappa$, and so $\mathfrak{D}_K(s;\, f,\, g) \neq 0$ for Re$(s) > \kappa$.

Suppose $\mathfrak{D}_K(\kappa + it;\, f,\, g) = 0$ with $t \in \mathbf{R}$; then $\mathfrak{D}_K(\kappa - it;\, f_\rho,\, g_\rho) = 0$. Since any pole of $\mathfrak{D}_K(s;\, *,\, *)$ is at most simple, we see that A is holomorphic at $s = \kappa$. In view of (16.5), this means that A is holomorphic at every real point, in particular at $s = \sigma$. We can now repeat the above argument. To be explicit, for real $s > \sigma$ we have $\log A(s) \geq 0$, and so $A(s) \geq 1$. Thus $A(\sigma) \geq 1$, which means that $\log A$ is holomorphic at $s = \sigma$, and we obtain a contradiction. Thus $\mathfrak{D}_K(s;\, f,\, g) \neq 0$ for $\mathrm{Re}(s) = \kappa$. This completes the proof.

Theorem 16.4. *Suppose $2 \leq k \in \mathbf{Z}$; let $\xi \in \Xi_k$ and let K be a finite extension of $F(\xi,\, \psi)$ that is totally real or a CM-field (cf. §15.5 and Lemma 15.6). Also let f, h, and p be K-rational elements of $\mathfrak{S}(\xi) \cap \mathscr{S}_k(N,\, \psi)$. Then*

$$\{\langle f,\, h\rangle/\langle p,\, p\rangle\}^\sigma = \langle f^\sigma,\, h^\sigma\rangle/\langle p^\sigma,\, p^\sigma\rangle \tag{16.6}$$

for every $\sigma \in \mathrm{Aut}(\mathbf{C})$ provided $p \neq 0$. The equality holds even for an arbitrary $f \in \mathfrak{S}(\xi) \cap \mathscr{S}_k(N,\, \psi)$ if we replace f^σ on the right-hand side by $f^{\rho\sigma\rho}$.

PROOF. By Theorem 14.4 there exist a multiple M of N and a normalized Hecke eigenform $g(z) = \sum_{n=1}^\infty c(n)\mathbf{e}(nz) \in \mathscr{S}_k(N,\, \psi)$ such that for every $p \nmid M$ we have $\xi_p = |c(p)|^2 - p^{k-1} - p^{k-2}$ and $f|R_p = \xi_p f$ for every $f \in \mathscr{S}_k(N,\, \psi) \cap \mathfrak{S}(\xi)$. Take $\mu = 0$ or 1 so that $\mu - k \in 2\mathbf{Z}$. By Lemma 2.14 there exists a character ζ such that $\zeta(-1) = 1$, ζ^2 is nontrivial, and the conductor of ζ is prime to M. Let φ_0 be the primitive character associated with ζ/ψ, r the conductor of φ_0, and φ the (possibly imprimitive) character modulo Mr associated with φ_0. Then $\varphi(-1) = (-1)^\mu$. We fix the symbols N, ψ, ξ, g, M, and ζ. Then φ and r are determined.

We now take a positive integer t satisfying the following condition:

(16.7) *t has no nontrivial square factor prime to rM.*

Put then $\theta_t(z) = 2^{-1}\sum_{n\in\mathbf{Z}} \varphi(n)n^\mu \mathbf{e}(tn^2 z/2)$. By Lemmas 5.5 and 8.17, $\theta_t \in \mathscr{M}_\ell(2,\, 2tr^2M^2;\, \varphi_t)$, where $\ell = \mu + 1/2$ and $\varphi_t(n) = \varphi(n)\left(\dfrac{t}{n}\right)$. Given $0 \neq f(z) = \sum_{n=1}^\infty \lambda(n)\mathbf{e}(nz) \in \mathscr{S}_k(N,\, \psi) \cap \mathfrak{S}(\xi)$, put $f_0(z) = f(z/2)$. Then $f_0 \in \mathscr{S}_k(2,\, N;\, \psi)$ and

$$D\bigl(2^{-1}(s - k - 1/2);\, f_0,\, \theta_t\bigr) = (t/2)^{(-s-\mu)/2}\sum_{m=1}^\infty \varphi(m)\lambda(tm^2)m^{-s}.$$

By Lemma 14.6 with rM as r there, we obtain

$$\begin{aligned} L(s - k + 1,\, \psi\varphi)&\sum_{m=1}^\infty \varphi(m)\lambda(tm^2)m^{-s} \\ &= \lambda(t)D(s - k;\, g,\, g_\varphi)\prod_{p|t}\bigl[1 + (\psi\varphi)(p)p^{k-1-s}\bigr]^{-1}, \end{aligned} \tag{16.7a}$$

where $g_\varphi(z) = \sum_{n=1}^\infty \varphi(n)c(n)\mathbf{e}(nz)$. By Lemma 7.13, $g_\varphi \in \mathscr{S}_k(r^2M^2,\, \psi\varphi^2)$. Using the symbol $\mathfrak{D}_K$ of Lemma 16.3, we have, with $K = M$,

$$\mathfrak{D}_M(s;\,g,\,g_\varphi) = L(2s+2-2k,\,\psi^2\varphi^2)D(s-k;\,g,\,g_\varphi),$$

which is nonzero for $\mathrm{Re}(s) \geq k$. We evaluate our functions at $s = k$. Our choice of φ shows that $\psi^2\varphi^2$ is nontrivial, and so $L(1,\,\psi\varphi) \neq 0$ and $L(2,\,\psi^2\varphi^2) \neq 0$. Thus by Lemma 16.3, $D(0;\,g,\,g_\varphi) \neq 0$, and we can conclude that $D(-1/4;\,f_0,\,\theta_t)$ is $\lambda(t)$ times a nonzero number, whose explicit form is

$$(t/2)^{-(k+\mu)/2}\mathfrak{D}_M(k;\,g,\,g_\varphi)L(1,\,\psi\varphi)^{-1}L(2,\,\psi^2\varphi^2)^{-1}\prod_{p|t}\left[1+(\psi\varphi)(p)p^{-1}\right]^{-1}.$$

We now evaluate (8.27) at $s = -1/4$ with $(f_0,\,\theta_t)$ in place of $(f,\,g)$ there. Putting $\kappa = (k+\mu)/2$, $N_0 = 2tr^2M^2$, and $\omega = \psi\varphi_t$, we obtain

$$N_0(4\pi)^{-\kappa}\Gamma(\kappa)D(-1/4;\,f_0,\,\theta_t) = \mu(\Phi)\langle\theta_t^\rho E,\,f_0\rangle, \tag{16.8}$$

where $E(z) = E_{k-\ell}\big(z,\,(\mu-k)/2+1;\,\Gamma,\,\bar{\omega}\big)$ with $\Gamma = \Gamma(N_0,\,N_0)$. Put $p = (k-\mu)/2-1$. Then $0 \leq p \in \mathbf{Z}$ and $E(z) = E_{3/2+2p}(-p;\,\Gamma,\,\bar{\omega})$. By (8.20) we have

$$(-4\pi)^{-p}\varepsilon(0)E(z) = D^p_{3/2}E_{3/2}(0;\,\Gamma,\,\bar{\omega}),$$

where $\varepsilon(0) = \prod_{a=0}^{p-1}(3/2+a)$. Since $\bar{\omega}^2$ is nontrivial, $E_{3/2}(0;\,\Gamma,\,\bar{\omega}) \in \mathscr{M}_{3/2}(\mathbf{Q}_{\mathrm{ab}})$ by Theorem 8.15(iii), and so $\pi^{-p}E(z) \in \mathscr{N}_{k-\ell}(\mathbf{Q}_{\mathrm{ab}})$ by Lemma 7.8(i). Put $q(z) = \pi^{-p}\theta_t^\rho(2z)E(2z)$. Then $q \in \mathscr{N}_k(\mathbf{Q}_{\mathrm{ab}})$, and $\langle\theta_t^\rho E,\,f_0\rangle = 2^k\pi^p\langle q,\,f\rangle$ by (6.12a). By Lemma 16.2, $\langle q,\,f\rangle = \langle r^{N,\xi}_{k,\psi}(q),\,f\rangle$. We have shown that $\langle\theta_t^\rho E,\,f_0\rangle$ is $\lambda(t)$ times a nonzero number, and so the same holds for $\langle r^{N,\xi}_{k,\psi}(q),\,f\rangle$.

Once $N,\,\psi,\,\xi,\,M,\,g,\,\zeta,\,\varphi$, and t are fixed, Γ does not depend on f. Therefore $r^{N,\xi}_{k,\psi}(q)$ is also independent of f. Emphasizing the dependence on t and f, put $r^{N,\xi}_{k,\psi}(q) = h_t$ and $\lambda(n) = \lambda(f,\,n)$. Then $h_t \in \mathscr{S}_k(\mathbf{Q}_{\mathrm{ab}})$ by (16.2d), and

$$\langle h_t,\,f\rangle = \lambda(f,\,t)w(t,\,\psi,\,\varphi,\,g) \tag{16.9}$$

for every $f \in \mathscr{S}_k(N,\,\psi)\cap\mathfrak{S}(\xi)$ with a nonzero number $w(t,\,\psi,\,\varphi,\,g)$ given by

$$w(t,\,\psi,\,\varphi,\,g) = \frac{2^{-k}\pi^{-p}\mu(\Phi)^{-1}N_0(4\pi)^{-\kappa}\Gamma(\kappa)(t/2)^{-\kappa}\mathfrak{D}_M(k;\,g,\,g_\varphi)}{L(1,\,\psi\varphi)L(2,\,\psi^2\varphi^2)\prod_{p|t}\left[1+(\psi\varphi)(p)p^{-1}\right]}. \tag{16.9a}$$

This is independent of f. If $f \neq 0$, then $\lambda(f,\,n) \neq 0$ for some n. We can put $n = tm^2$ with m prime to rM and an integer t satisfying (16.7). Then (16.7a) shows that $\lambda(f,\,t) \neq 0$. Thus for every nonvanishing $f \in \mathscr{S}_k(N,\,\psi)\cap\mathfrak{S}(\xi)$ we can find t such that $\langle h_t,\,f\rangle \neq 0$, and so

(16.10) *The* h_t *for all* t *satisfying* (16.7) *span* $\mathscr{S}_k(N,\,\psi)\cap\mathfrak{S}(\xi)$.

Now take another $f' \in \mathscr{S}_k(N,\,\psi)\cap\mathfrak{S}(\xi)$ and another t' satisfying (16.7). Then from (16.9) and (16.9a) we obtain

$$\frac{\langle h_{t'},\,f'\rangle}{\langle h_t,\,f\rangle} = \frac{\lambda(f',\,t')}{\lambda(f,\,t)}\cdot\frac{w(t',\,\psi,\,\varphi,\,g)}{w(t,\,\psi,\,\varphi,\,g)}, \tag{16.11}$$

$$\text{(16.11a)}\qquad \frac{w(t',\,\psi,\,\varphi,\,g)}{w(t,\,\psi,\,\varphi,\,g)}=(t'/t)^{\kappa}\cdot\frac{\prod_{p|t}\left[1+(\psi\varphi)(p)p^{-1}\right]}{\prod_{p|t'}\left[1+(\psi\varphi)(p)p^{-1}\right]}.$$

Let $\sigma\in\mathrm{Aut}(\mathbf{C})$. By Theorem 7.5(iia) and (15.4a) we have $\mathscr{S}_k(N,\,\psi)^\sigma=\mathscr{S}_k(N,\,\psi^\sigma)$ and $\mathfrak{S}(\xi)^\sigma=\mathfrak{S}(\xi^\sigma)$; also, $\lambda(f,\,n)^\sigma=\lambda(f^\sigma,\,n)$. Taking $\psi^\sigma,\,\varphi^\sigma$, and f^σ in place of $\psi,\,\varphi$, and f, from (16.2d), Theorem 8.15(iii), and Lemma 7.8(i) we see that $(h_t)^\sigma$ takes the place of h_t for f^σ. Thus

$$\langle (h_t)^\sigma,\,f^\sigma\rangle=\lambda(f^\sigma,\,t)w(t,\,\psi^\sigma,\,\varphi^\sigma,\,g^\sigma).$$

From (16.11a) we see that

$$\left\{w(t',\,\psi,\,\varphi,\,g)/w(t,\,\psi,\,\varphi,\,g)\right\}^\sigma=w(t',\,\psi^\sigma,\,\varphi^\sigma,\,g^\sigma)/w(t,\,\psi^\sigma,\,\varphi^\sigma,\,g^\sigma).$$

Therefore

$$\left\{\langle h_{t'},\,f'\rangle/\langle h_t,f\rangle\right\}^\sigma=\langle (h_{t'})^\sigma,\,(f')^\sigma\rangle/\langle (h_t)^\sigma,f^\sigma\rangle.$$

Since h_t is $\mathbf{Q}_{\mathrm{ab}}$-rational, $(h_t)^{\sigma\rho}=(h_t)^{\rho\sigma}$. Let $j\in\mathscr{S}_k(N,\,\psi)\cap\mathfrak{S}(\xi)$. By (16.10), we can find a finite set T of elements satisfying (16.7) such that $j=\sum_{\tau\in T}a_\tau h_\tau$ with $a_\tau\in\mathbf{C}$. Then $\langle j,\,f'\rangle=\sum_{\tau\in T}a_\tau^\rho\langle h_\tau,\,f'\rangle$, and

$$\left\{\frac{\langle j,\,f'\rangle}{\langle h_t,\,f\rangle}\right\}^\sigma=\sum_{\tau\in T}a_\tau^{\rho\sigma}\left\{\frac{\langle h_\tau,\,f'\rangle}{\langle h_t,\,f\rangle}\right\}^\sigma=\sum_{\tau\in T}a_\tau^{\rho\sigma}\frac{\langle (h_\tau)^\sigma,\,(f')^\sigma\rangle}{\langle (h_t)^\sigma,\,f^\sigma\rangle}=\frac{\langle j^{\rho\sigma\rho},\,(f')^\sigma\rangle}{\langle (h_t)^\sigma,\,f^\sigma\rangle}.$$

In particular, for $f=h_t$ we have

$$(*)\qquad \left\{\langle j,\,f'\rangle/\langle h_t,\,h_t\rangle\right\}^\sigma=\langle j^{\rho\sigma\rho},\,(f')^\sigma\rangle/\langle (h_t)^\sigma,\,(h_t)^\sigma\rangle.$$

Take j and f' to be the same K-rational nonzero element p of $\mathscr{S}_k(N,\,\psi)\cap\mathfrak{S}(\xi)$ with K as in our theorem. Then $p^{\rho\sigma\rho}=p^\sigma$, and so

$$\left\{\langle p,\,p\rangle/\langle h_t,\,h_t\rangle\right\}^\sigma=\langle p^\sigma,\,p^\sigma\rangle/\langle (h_t)^\sigma,\,(h_t)^\sigma\rangle.$$

Dividing $(*)$ by this, we obtain

$$\left\{\langle j,\,f'\rangle/\langle p,\,p\rangle\right\}^\sigma=\langle j^{\rho\sigma\rho},\,(f')^\sigma\rangle/\langle p^\sigma,\,p^\sigma\rangle,$$

which proves (16.6), since $j^{\rho\sigma\rho}=j^\sigma$ if j is K-rational. This completes the proof of Theorem 16.4.

16.5. Let $0<k\in 2^{-1}\mathbf{Z}$ and $\Gamma=\Gamma(N,\,N)$ with a positive integer N and let ψ be a character modulo N. We put then

$$\text{(16.12)}\qquad C_k(z,\,s;\,\Gamma,\,\psi)=\begin{cases}L(2s+k,\,\psi)E_k(z,\,s;\,\Gamma,\,\psi) & \text{if } k\in\mathbf{Z},\\ L(4s+2k-1,\,\psi^2)E_k(z,\,s;\,\Gamma,\,\psi) & \text{if } k\notin\mathbf{Z}\end{cases}$$

with $E_k(\cdots)$ of (8.12); we assume $N\in 2\mathbf{Z}$ if $k\notin\mathbf{Z}$.

We are going to state some results about $C_k(z,\,\lambda;\,\Gamma,\,\psi)$ for certain $\lambda\in 2^{-1}\mathbf{Z}$ belonging to the set Λ_k defined by

(16.13a) $\Lambda_k = \{\lambda \in \mathbf{Z} \,|\, 1-k \le \lambda \le 0\}$ if $k \in \mathbf{Z}$,

(16.13b) $\Lambda_k = \{\lambda \in \mathbf{Z} \,|\, 1-k < 2\lambda \le 0\}$
$\bigcup \{\lambda \in 2^{-1}\mathbf{Z} \,|\, \lambda - k \in \mathbf{Z},\ 1-k \le \lambda < (1-k)/2\}$ if $k \notin \mathbf{Z}$.

For $\lambda \in \Lambda_k$ we put

$$C_k^*(z,\,\lambda;\,\Gamma,\,\psi) = \alpha_{k,\psi,\lambda} C_k(z,\,\lambda;\,\Gamma,\,\psi) \quad \text{with}$$

$$\alpha_{k,\psi,\lambda} = \begin{cases} G(\psi)^{-1} i^k \pi^{-k-\lambda} & \text{if } k \in \mathbf{Z}, \\ G(\psi^2)^{-1} \pi^{-3\lambda-2[k]} & \text{if } k \notin \mathbf{Z} \text{ and } \lambda \in \mathbf{Z}, \\ G(\psi)^{-1} 2^{1/2} i^{[k]} \pi^{-k-\lambda} & \text{if } k \notin \mathbf{Z} \text{ and } \lambda \notin \mathbf{Z}. \end{cases} \tag{16.14}$$

Theorem 16.6. *Suppose that ψ^2 is nontrivial if $k \notin \mathbf{Z}$ and $k + 2\lambda = 3/2$ or $1/2$. Then for every $\lambda \in \Lambda_k$ the function $C_k^*(z,\,\lambda;\,\Gamma,\,\psi)$ belongs to $\mathscr{N}_k(\mathbf{Q}_{\mathrm{ab}})$ and $C_k^*(z,\,\lambda;\,\Gamma,\,\psi)^\sigma = C_k^*(z,\,\lambda;\,\Gamma,\,\psi^\sigma)$ for every $\sigma \in \mathrm{Gal}(\mathbf{Q}_{\mathrm{ab}}/\mathbf{Q})$.*

PROOF. In this proof Γ is always the same; therefore, suppressing Γ, we write $(z,\,s;\,\psi)$ for $(z,\,s;\,\Gamma,\,\psi)$.

Case I: $k \in \mathbf{Z}$. Let $\lambda \in \Lambda_k$. First suppose $-k/2 < \lambda \le 0$; put $p = -\lambda$ and $\kappa = k - 2p$. Then $p \ge 0$ and $\kappa > 0$. By (8.20) we have

$$L(\kappa,\,\psi) D_\kappa^p E_\kappa(z,\,0;\,\psi) = (-4\pi)^{-p} \varepsilon_\kappa(0) C_k(z,\,\lambda;\,\psi).$$

Clearly $\varepsilon_\kappa(0) \in \mathbf{Q}^\times$. Thus the desired result follows from Lemma 2.10, Theorem 8.15(i), and Lemma 7.8(i).

Next suppose $1-k \le \lambda \le -k/2$. By (8.18), $2C_k(z,\,s;\,\psi) = E_k^N(z,\,s;\,\psi)$, and so the desired result for $\lambda = 1-k$ follows from Theorem 8.15(ii). If $\lambda > 1-k$, put $p = \lambda + k - 1$ and $\kappa = k - 2p$. Then $p > 0$ and $\kappa > 0$. By (8.20) we have

$$D_\kappa^p C_\kappa(z,\,1-\kappa;\,\psi) = (-4\pi)^{-p} \varepsilon_\kappa(1-\kappa) C_k(z,\,\lambda;\,\psi).$$

and $\varepsilon_\kappa(1-\kappa) \in \mathbf{Q}^\times$, and so the desired result follows from the case $\lambda = 1-k$ and Lemma 7.8(i).

Case II: $k \notin \mathbf{Z}$. Suppose $\lambda \in \mathbf{Z}$ and $1-k < 2\lambda \le 0$; put $p = -\lambda$ and $\kappa = k - 2p$. Then $0 \le p \in \mathbf{Z}$ and $3/2 \le \kappa \in 2^{-1}\mathbf{Z}$. By (8.20) we have

$$L(2[k] - 4p,\,\psi^2) D_\kappa^p E_\kappa(z,\,0;\,\psi) = (-4\pi)^{-p} \varepsilon_\kappa(0) C_k(z,\,\lambda;\,\psi).$$

We see that $0 < 2[k] - 4p \in 2\mathbf{Z}$ and $\varepsilon_\kappa(0) \in \mathbf{Q}^\times$. By our assumption, ψ^2 is nontrivial if $\kappa = 3/2$. Thus the desired result follows from Lemma 2.10, Theorem 8.15(iii), and Lemma 7.8(i).

Next suppose $\lambda - k \in \mathbf{Z}$ and $1-k \le \lambda < (1-k)/2$; put $p = \lambda + k - 1$ and $\kappa = k - 2p$. Then $0 \le p \in \mathbf{Z}$ and $1 < \kappa \in 2^{-1}\mathbf{Z}$. By (8.20) we have

$$D_\kappa^p C_k(z,\,1-\kappa;\,\psi) = (-4\pi)^{-p} \varepsilon_\kappa(1-\kappa) C_k(z,\,\lambda;\,\psi).$$

Therefore this case can be settled by the same argument as before in view of Theorem 8.15(iv). This completes the proof.

Theorem 16.7. *Let $\xi \in \Xi_k$ with $0 < k \in \mathbf{Z}$ as in §15.4. Given $f \in \mathfrak{S}(\xi) \cap \mathscr{S}_k(N, \psi)$ and $g \in \mathscr{M}_\ell(M, \varphi)$ with any positive weight $\ell < k$, a divisor M of N^∞, and a character φ modulo M, define $\mathscr{D}_N(s;\, f,\, g)$ by (8.28). For $\kappa \in 2^{-1}\mathbf{Z}$ put $\lambda = \kappa - 1 - (k-\ell)/2$. Assuming that $\lambda \in \Lambda_{k-\ell}$ with Λ of (16.13a, b), put*

$$A(\kappa;\, f,\, g) = \beta\mathscr{D}_N(\kappa;\, f,\, g) \quad \text{with} \quad \beta = \pi^{-k-\lambda}\Gamma(\lambda + k - 1)\bar{\alpha}_{k-\ell,\bar{\omega},\lambda},$$

where $\omega = \psi\varphi$ and α_{***} is as in (16.14). Let $0 \neq p \in \mathfrak{S}(\xi) \cap \mathscr{S}_k(N, \psi)$ as in Theorem 16.4. Then

$$\big[A(\kappa;\, f,\, g)/\langle p,\, p\rangle\big]^\sigma = A(\kappa;\, f^\sigma,\, g^{\rho\sigma\rho})/\langle p^\sigma,\, p^\sigma\rangle$$

for every $\sigma \in \mathrm{Aut}(\mathbf{C})$.

PROOF. From (8.27) we obtain

$$N_0(4\pi)^{-\mu}\Gamma(\mu)\mathscr{D}_N(\kappa;\, f,\, g) = \mu(\Phi)\langle g_\rho C,\, f\rangle,$$

where $\mu = \lambda + k - 1$, N_0 is a positive multiple of N that divides MN, $\Phi = \Gamma\backslash\mathfrak{H}$, and $C(z) = C_{k-\ell}(z,\, \lambda;\, \Gamma,\, \bar{\omega})$ with $\Gamma = \Gamma(N_0,\, N_0)$. Since $\mu(\Phi) \in \pi\mathbf{Q}^\times$ by (6.6), putting $C^*(z) = C^*_{k-\ell}(z,\, \lambda;\, \Gamma,\, \bar{\omega})$, we have

$$R\pi^{-\mu-1}\Gamma(\mu)\bar{\alpha}_{k-\ell,\bar{\omega},\lambda}\mathscr{D}_N(\kappa;\, f,\, g) = \langle g_\rho C^*,\, f\rangle$$

with a constant $R \in \mathbf{Q}^\times$ independent of f and g. Therefore, from (16.2c) we see that $\beta\mathscr{D}_N(\kappa;\, f,\, g) = \langle r(g_\rho C^*),\, f\rangle$ with $r = r^{N,\xi}_{k-\ell,\psi}$ and β defined as above. Thus we obtain the desired result from Theorem 16.4 and (16.2d).

16.8. Given the space $\mathscr{S}_k(N, \psi)$ with $k \in \mathbf{Z}$, let $\mathscr{S}'_k(N, \psi)$ denote its subspace spanned by the functions $h(tz)$ with $h \in \mathscr{S}_k(M, \psi)$ for all integers t and M such that tM divides N and the conductor of ψ divides M. Let $\mathscr{S}^0_k(N, \psi)$ be the orthogonal complement of $\mathscr{S}'_k(N, \psi)$ in $\mathscr{S}_k(N, \psi)$. We call a normalized Hecke eigenform f in $\mathscr{S}_k(N, \psi)$ **primitive** if it belongs to $\mathscr{S}^0_k(N, \psi)$, and call N the **conductor** of f. If f is primitive, then clearly f^σ is primitive for every $\sigma \in \mathrm{Aut}(\mathbf{C})$. The basic facts on primitive forms (often called *newforms*) can be found in Atkin-Lehner [AL70] (for trivial ψ), Casselman [C73], and Miyake [Mi71].

Let g be a nonzero element of $\mathscr{S}_k(N, \psi)$ such that $g|T'(p)_{k,\psi} = c_p g$ for almost all p. Then we can find a primitive element f contained in $\mathscr{S}_k(N, \psi)$ such that $f|T'(p)_{k,\psi} = c_p f$ for almost all p. We then say that f is **associated** to g.

For an arbitrary $f(z) = \sum_{n=1}^\infty c(n)\mathbf{e}(nz) \in \mathscr{S}_k\big(\Gamma_1(N)\big)$ and a primitive character χ put

$$D(s;\, f,\, \chi) = \sum_{n=1}^\infty \chi(n)c(n)n^{-s}, \tag{16.15}$$

(16.16) $A(m;\, f,\, \chi) = (\pi i)^{-m} G(\chi)^{-1} D(m;\, f,\, \chi) \qquad (0 < m < k,\ m \in \mathbf{Z}).$

Since $\sum_{n=1}^{\infty} \chi(n)c(n)\mathbf{e}(nz) \in \mathscr{S}_k$, from Theorem 8.2 we see that $D(s;\, f,\, \chi)$ is an entire function, and so its value at any $m \in \mathbf{Z}$ is meaningful.

Theorem 16.9. *Given a primitive $f \in \mathscr{S}_k(N,\, \psi)$ and a primitive character χ, denote by K_f (resp. K_χ) the field generated over $\mathbf{Q}$ by the Fourier coefficients of f (resp. the values of χ). Then for every $\sigma \in \mathrm{Aut}(\mathbf{C})$ we can define two nonzero complex numbers $u_+(f^\sigma)$ and $u_-(f^\sigma)$ with the following properties:*

(i) *$u_\pm(f^\sigma)^\rho = \pm u_\pm(f^{\sigma\rho})$, where ρ is the complex conjugation.*

(ii) *$A(m;\, f,\, \chi) \in u_\pm(f)K_f K_\chi$ if $\chi(-1) = \pm(-1)^m$ for every $m \in \mathbf{Z}$, $0 < m < k$.*

(iii) *$\left[A(m;\, f,\, \chi)/u_\pm(f)\right]^\sigma = A(m;\, f^\sigma,\, \chi^\sigma)/u_\pm(f^\sigma)$ for every $\sigma \in \mathrm{Aut}(\mathbf{C})$ if $m \in \mathbf{Z}$, $0 < m < k$, and $\chi(-1) = \pm(-1)^m$.*

(iv) Put $E(f) = i^{1-k}\pi G(\psi)\langle f,\, f\rangle$. Then $E(f) \in u_+(f)u_-(f)K_f$ and

$$\left[E(f)/\{u_+(f)u_-(f)\}\right]^\sigma = E(f^\sigma)/\{u_+(f^\sigma)u_-(f^\sigma)\}$$

for every $\sigma \in \mathrm{Aut}(\mathbf{C})$.

This was given in [S77] as an application of [S76]. In fact, the results of [S76] can be derived from Theorem 16.4 by taking f there to be any normalized Hecke eigenform and h to be an Eisenstein series, without assuming f to be primitive.

If $k = 2$, the constants $u_\pm(f^\sigma)$ are periods of the differential form $f^\sigma dz$ on $\Gamma_1(N)\backslash\mathfrak{H}$. For details, we refer the reader to [S77, Theorem 3].

Lemma 16.10. (i) *Let f be a normalized Hecke eigenform in $\mathscr{S}_k(N,\, \psi)$. Then $D(s;\, f, \chi) \neq 0$ for $\mathrm{Re}(s) \geq (k+1)/2$ for every character χ.*

(ii) *Let f be a nonzero element of $\mathscr{S}_2(\Gamma_1(N))$ and let $0 < M \in \mathbf{Z}$. Then there exists a primitive character φ, whose conductor is prime to M, such that $D(1;\, f,\, \varphi) \neq 0$ and $\varphi(-1)$ has a given signature.*

Assertion (i) can be proved by the same technique as in the proof of Lemma 16.3. For details, we refer the reader to [S76, Proposition 2] and [S78, Proposition 4.16]. As for (ii), the proof is given in [S77, Theorem 2] when f is primitive. The condition that f is primitive is unnecessary, as explained in the last four lines of page 213 and the first four lines of page 214 in [S77]. This type of nonvanishing for the zeta functions associated to the forms on $GL_2(F)$ with an arbitrary number field F was given by Rohrlich in [Ro89].

16.11. Let $0 \neq f \in \mathfrak{S}(\xi) \cap \mathscr{S}_k(2,\, N/2;\, \psi)$ with half-integral k. We assume the following condition:

(16.17) *$k \geq 3/2$: if $k = 3/2$, then $\langle f,\, \theta_1(z,\, \mu)\rangle = 0$ for every $\mu \in \mathscr{L}(\mathbf{Q})$, where θ_1 is as in* (12.9b).

Then by Theorems 12.8 and 13.11 there exists a Hecke eigenform $g(z) = \sum_{n=1}^{\infty} c(n)\mathbf{e}(nz) \in \mathscr{S}_{2k-1}(N, \psi^2)$ such that (15.5) holds with $\omega_p = \psi(p)\xi_p$ for almost all p, as already explained in §15.4. Choosing g suitably, we may assume that g is primitive. Clearly g is uniquely determined by f.

Theorem 16.12. *Let f be as above and g the primitive form in $\mathscr{S}_{2k-1}$ determined by f as above. For $q \in \mathscr{N}_k$ put*

(16.18) $$I(q, f) = 2^{1/2} i^{[k]-1} \pi G(\psi)\langle q, f\rangle.$$

Let $u_{\pm}(f)$ be as in Theorem 16.9. then

(16.18a) $$\big[I(q, f)/u_{-}(g)\big]^{\sigma} = I(q^{\rho\sigma\rho}, f^{\sigma})/u_{-}(g^{\sigma})$$

for every $\sigma \in \mathrm{Aut}(\mathbf{C})$.

PROOF. Our argument is similar to the proof of Theorem 16.4. We take a multiple M of N such that $f|R_p = \xi_p f$ for every $f \in \mathfrak{S}(\xi) \cap \mathscr{S}_k(2, N/2; \psi)$ and every $p \nmid M$. We also take a primitive character φ of conductor r and a positive integer t such that

(16.19) $\varphi(-1) = 1$ *and every prime factor of M divides r;*

(16.20) t *has no nontrivial square factor prime to r.*

Such a φ of course exists. Put $\theta_t(z) = 2^{-1}\sum_{n\in\mathbf{Z}} \varphi(n)\mathbf{e}(tn^2z/2)$. By Lemmas 5.5 and 8.17 we see that $\theta_t \in \mathscr{M}_{\ell}(2, 2tr^2; \varphi_t)$, where $\ell = 1/2$ and $\varphi_t(n) = \varphi(n)\left(\dfrac{t}{n}\right)$. Let $0 \neq f(z) = \sum_{n=1}^{\infty} \lambda(n)\mathbf{e}(nz) \in \mathfrak{S}(\xi) \cap \mathscr{S}_k(2, N/2; \psi)$. By Theorem 13.11 we have

(∗) $$L(s-[k]+1, \chi_t\varphi)\sum_{m=1}^{\infty} \varphi(m)\lambda(tm^2)m^{-s} = \lambda(t)D(s; g, \varphi),$$

where χ_t is defined by $\chi_t(n) = \psi(n)\left(\dfrac{t}{n}\right)$. Also, we have

(∗∗) $$D\big(2^{-1}(s-[k]-1); f, \theta_t\big) = (t/2)^{-s/2}\sum_{m=1}^{\infty} \varphi(m)\lambda(tm^2)m^{-s}.$$

Combining (∗) with (∗∗), we obtain

(#) $$\lambda(t)(t/2)^{-s/2}D(s; g, \varphi) = L\big(s-[k]+1, \chi_t\varphi\big)D\big(2^{-1}(s-[k]-1); f, \theta_t\big).$$

The left-hand side of (∗∗) at $s = 2k-2$ is $D(s_0; f, \theta_t)$ with $s_0 = [k]/2 - 1$. We now evaluate (8.27) with θ_t as g there at $s = s_0$. Then

(16.21) $$N_0(4\pi)^{-\kappa}\Gamma(\kappa)D(s_0; f, \theta_t) = \mu(\Phi)\langle \theta_t^{\rho}E, f\rangle,$$

where $N_0 = tr^2N$, $\kappa = k-1$, $\Phi = \Gamma\backslash\mathfrak{H}$, and $E(z) = E_{[k]}\big(z, 0; \Gamma, \bar{\omega}\big)$ with $\Gamma = \Gamma(N_0, N_0)$ and $\omega = \psi\varphi_t$.

We first assume that $k > 3/2$; the case $k = 3/2$ will be treated later. Then $2k-2 \geq k$, and so $D(2k-2, g, \varphi) \neq 0$ by Lemma 16.10(i), which combined with (#) shows that $D(s_0;\, f,\, \theta_t) \neq 0$ provided $\lambda(t) \neq 0$, since $L(s - [k] + 1,\, \chi_t\varphi) \neq 0$ for $s = 2k - 2$. Now $E \in \mathscr{N}_{[k]}(\mathbf{Q}_{\mathrm{ab}})$ by Theorem 8.15(i). We then proceed as in the proof of Theorem 16.4. To be explicit, put $h_t = r_{k,\psi}^{N,\xi}(\theta_t^\rho E)$ with $r_{k,\psi}^{N,\xi}$ of Lemma 16.2 and $\lambda(n) = \lambda(f,\, n)$. Then $h_t \in \mathscr{S}_k(\mathbf{Q}_{\mathrm{ab}})$ by (16.2d), and

$$\langle h_t,\, f\rangle = \lambda(f,\, t)w_t(\psi,\, \varphi,\, g) \tag{16.22}$$

for every $f \in \mathscr{S}_k(2,\, N/2,\, \psi) \cap \mathfrak{S}(\xi)$ with a nonzero number $w_t(\psi,\, \varphi,\, g)$ independent of f given by

$$w_t(\psi,\, \varphi,\, g) = R(2t)^{1/2}\pi^{-[k]} \cdot \frac{D(2k-2;\, g,\, \varphi)}{L([k],\, \chi_t\varphi)},$$

where R is an element of $\mathbf{Q}^\times$ independent of f. If $f \neq 0$, then $\lambda(f,\, n) \neq 0$ for some n. We can put $n = tm^2$ with m prime to r and an integer t satisfying (16.20). Then $(*)$ shows that $\lambda(f,\, t) \neq 0$. Thus for every nonvanishing $f \in \mathscr{S}_k(2,\, N/2,\, \psi) \cap \mathfrak{S}(\xi)$ we can find t such that $\langle h_t,\, f\rangle \neq 0$, and so we obtain

(16.23) *The h_t for all t satisfying* (16.20) *span* $\mathscr{S}_k(2,\, N/2,\, \psi) \cap \mathfrak{S}(\xi)$.

Let $\sigma \in \mathrm{Aut}(\mathbf{C})$; take f^σ, g^σ, and φ^σ in place of f, g, and φ, but with the same t. By Theorem 8.15(i), E^σ takes the place of E, and by (16.2d), h_t^σ takes the place of h_t. Define $P_N(m,\, \chi)$ by (2.19). Then $L([k],\, \chi_t\varphi) = G(\chi_t\varphi)(\pi i)^{[k]}P_r([k],\, \chi_t\varphi)$. Thus employing (16.16), we have

$$w_t(\psi,\, \varphi,\, g) = R(2t)^{1/2}i^{[k]-1}\pi^{-1} \cdot \frac{G(\varphi)A(2k-2;\, g,\, \varphi)}{G(\chi_t\varphi)P_r([k],\, \chi_t\varphi)}.$$

By (2.4a) and Lemma 2.8 we have

$$\bigl[t^{1/2}G(\psi)G(\varphi)/G(\chi_t\varphi)\bigr]^\sigma = t^{1/2}G(\psi^\sigma)G(\varphi^\sigma)/G(\chi_t^\sigma\varphi^\sigma).$$

Put $B(\psi) = 2^{1/2}i^{[k]-1}\pi G(\psi)$. Since $(-1)^{2k-2} = -1$, from Theorem 16.9(ii) and Lemma 2.10 we obtain

$$\bigl[B(\psi)w_t(\psi,\, \varphi,\, g)/u_-(g)\bigr]^\sigma = B(\psi^\sigma)w_t(\psi^\sigma,\, \varphi^\sigma,\, g^\sigma)/u_-(g^\sigma).$$

Combining this with (16.22), we obtain

$$\bigl[B(\psi)\langle h_t,\, f\rangle/u_-(g)\bigr]^\sigma = B(\psi^\sigma)\langle h_t^\sigma,\, f^\sigma\rangle/u_-(g^\sigma).$$

Given $h \in \mathfrak{S}(\xi) \cap \mathscr{S}_k(2,\, N/2;\, \psi)$, we can put, by (16.23), $h = \sum_{t\in T} a_t h_t$ with a finite set T of elements t satisfying (16.20) and $a_t \in \mathbf{C}$. Then $\langle h,\, f\rangle = \sum_{t\in T} a_t^\rho\langle h_t,\, f\rangle$ and $\langle h^{\rho\sigma\rho},\, f^\sigma\rangle = \sum_{t\in T} a_t^{\rho\sigma}\langle h_t^\sigma,\, f^\sigma\rangle$. Therefore

$$\bigl[B(\psi)\langle h,\, f\rangle/u_-(g)\bigr]^\sigma = B(\psi^\sigma)\langle h^{\rho\sigma\rho},\, f^\sigma\rangle/u_-(g^\sigma).$$

This proves (16.18a) for $h \in \mathfrak{S}(\xi) \cap \mathscr{S}_k(2,\, N/2;\, \psi)$. Given an arbitrary $q \in \mathscr{N}_k$, we have $\langle q,\, f\rangle = \langle h,\, f\rangle$ with $h = r_{k,\psi}^{N,\xi}(q)$, where $r_{k,\psi}^{N,\xi}$ is as in Lemma 16.2. In view of (16.2d) we obtain (16.18a).

Next let us assume that $k = 3/2$. In this case $2k-2 = 1$ and $D(1;\, g,\, \varphi)$ may be 0. If $D(1;\, g,\, \varphi) \neq 0$, then our argument in the case $k > 3/2$ can be repeated. Suppose $D(1;\, g,\, \varphi) = 0$ with φ satisfying (16.19). Put $g_\varphi(z) = \sum_{n=1}^{\infty} \varphi(n)c(n)\mathbf{e}(nz)$. Then, by Lemma 7.13, $g_\varphi \in \mathscr{S}_k(r^2N, \psi^2\varphi^2)$. By Lemma 16.10(ii) we can find a character φ_1, whose conductor is prime to rM, such that $D(1;\, g_\varphi,\, \varphi_1) \neq 0$ and $\varphi_1(-1) = 1$. Since $D(1;\, g_\varphi,\, \varphi_1) = D(1;\, g,\, \varphi\varphi_1)$, taking $\varphi\varphi_1$ in place of φ, we can employ our reasoning in the case $k > 3/2$ for $k = 3/2$. This completes the proof.

Remark. Theorem 16.12 is essentially the same as [S81, Theorem 1]. However, the constant appearing in the definition of $I(q,\, f)$ in (16.18) is different from that for the corresponding quantity in [S81]. The difference is caused by the difference of the definition of modular forms of half-integral weight, as explained in §5.2. There is another point that should be remembered: in the definition of $D(s;\, f,\, g)$ in (8.23) the exponent is $-s-(k+\ell)/2$ instead of the simpler $-s$ chosen in [S81].

Theorem 16.13. *Let k and ℓ be half-integral weights such that $k > \ell$ and $k \geq 5/2$; let $f \in \mathfrak{S}(\xi) \cap \mathscr{S}_k(2,\, N/2;\, \psi)$ with $\xi \in \Xi_k$ and $h \in \mathscr{M}_\ell(2,\, N/2;\, \varphi)$. Given an integer m such that $m-k+\ell \in 2\mathbf{Z}$ and $-k-\ell \leq m \leq k-\ell-2$ put*

$$B(m;\, f,\, h) = 2^{1/2} i^{[\ell]-1} G(\varphi)^{-1} \pi^{-m-[k]-1} \mathscr{D}_N(m/2;\, f,\, h).$$

Let g be the primitive element of $\mathscr{S}_{2k-1}(N,\, \psi^2)$ determined by f as in §16.11, and $u_-(g)$ the constant defined in Theorem 16.9. Then

$$\big[B(m;\, f,\, h)/u_-(g)\big]^\sigma = B(m;\, f^\sigma,\, h^\sigma)/u_-(g^\sigma)$$

for every $\sigma \in \mathrm{Aut}(\mathbf{C})$.

PROOF. Let $\lambda = 1 + (m-k+\ell)/2$ and $\kappa = (m+k+\ell)/2$. From (8.27) we obtain

$$N(4\pi)^{-\kappa}\Gamma(\kappa)\mathscr{D}_N(m/2;\, f,\, h) = \mu(\Phi)\langle h^\rho C,\, f\rangle,$$

where $C(z) = C_{k-\ell}(z,\, \lambda;\, \Gamma,\, \bar{\omega})$ with $\omega = \psi\varphi$. Put $\beta = N(4\pi)^{-\kappa}\Gamma(\kappa)\mu(\Phi)^{-1}\bar{\alpha}$ with $\alpha = \alpha_{k-\ell,\bar{\omega},\lambda}$ given by (16.14). We see that $\lambda \in \mathbf{Z}$ and $1-k = \ell \leq \lambda \leq 0$, and so λ belongs to the set $\Lambda_{k-\ell}$ of (16.13a), and Theorem 16.6 is applicable. We have $\beta\mathscr{D}_N(m/2;\, f,\, h) = \langle h^\rho C^*,\, f\rangle$ with $C^*(z) = C^*_{k-\ell}(z,\, \lambda;\, \Gamma,\, \bar{\omega})$. Since $\kappa + k \in \mathbf{Z}$, we have $\Gamma(\kappa) \in \pi^{1/2}\mathbf{Q}^\times$. Also, $\overline{G(\bar{\omega})}^{-1} = \omega(-1)G(\omega)^{-1}$. With $I(q,\, f)$ as in (16.18) we have $\gamma\mathscr{D}_N(m/2;\, f,\, h) = I(h^\rho C^*,\, f)$ with $\gamma = 2^{1/2} i^{[k]-1}\pi G(\psi)\beta$. Calculating γ explicitly and then applying Lemma 2.8 to $G(\psi)G(\varphi)/G(\psi\varphi)$, we obtain the desired formula from Theorem 16.12.

16.14. We add here two remarks.

1. In addition to Theorems 16.7 and 16.13 there is one more case of the values of $\mathscr{D}_N(s;\, f,\, h)$, that is, the case with f of half-integral weight k and h

of integral weight $\le k$. In this case we can state the results similar to Theorem 16.13 and prove them in the same manner as an application of Theorems 16.6 and 16.12. We leave the details to the reader, as no new ideas are required. Indeed, they were given in [S81, Theorem 3], though the reader is warned of the difference in formulation noted in the remark before Theorem 16.13.

In addition to this, we can also investigate $\mathscr{D}(m;\, f,\, \chi)$ for $\mathscr{D}$ of (8.30). This is essentially a special case of Theorem 16.7 with a certain theta function in place of g, but there are some exceptional cases which require some nontrivial calculation. We refer the reader to [S91, pp. 604–605] and the paper of J. Sturm cited there.

2. Let K be an imaginary quadratic field. Given $\lambda \in \mathscr{L}(K)$ and $0 < \kappa \in \mathbf{Z}$, put

$$f(z) = \sum_{\xi \in K} \lambda(\xi)\xi^{\kappa}\mathbf{e}(\xi\bar{\xi}z) \qquad (z \in \mathfrak{H}).$$

Then we can show that $f \in \mathscr{S}_{\kappa+1}$. In this case we can connect $< f,\, f >$ and the special values $D(m,\, f,\, \chi)$ with $h(\tau)$ with $\tau \in K \cap \mathfrak{H}$ and $h \in \mathscr{M}_{\nu}(\mathbf{Q}_{\mathrm{ab}})$ for a suitable ν. For this we refer the reader to [S76, §5] and [S07, §13].

17. Hilbert modular forms

17.1. The theory of modular forms can be developed with respect to congruence subgroups of $SL_2(F)$ with any totally real algebraic number field F of finite degree. Such forms are traditionally called **Hilbert modular forms.** Practically all the results we presented in this book for modular forms on $\mathfrak{H}$, including those of half-integral weight, can be extended to the case of Hilbert modular forms. Let us now briefly explain the basics of this topic, emphasizing its difference from the case over $\mathbf{Q}$.

With a fixed F, we denote by $\mathfrak{g}$ the maximal order of F and $\mathfrak{d}$ the different of F relative to $\mathbf{Q}$. We denote by $\mathbf{a}$ the set of all archimedean primes of F. For each $v \in \mathbf{a}$ we denote by F_v the v-completion of F which is naturally identified with $\mathbf{R}$. We put $F_{\mathbf{a}} = \prod_{v \in \mathbf{a}} F_v$. Similarly we put $SL_2(F)_{\mathbf{a}} = \prod_{v \in \mathbf{a}} SL_2(F_v)$, and we let $SL_2(F)_{\mathbf{a}}$ act on $\mathfrak{H}^{\mathbf{a}}$ in an obvious fashion. For each element $\alpha \in SL_2(F)$ we can assign an element of $SL_2(F)_{\mathbf{a}}$ whose components are all equal to α, which acts on $\mathfrak{H}^{\mathbf{a}}$. In this way we can let $SL_2(F)$ act on $\mathfrak{H}^{\mathbf{a}}$.

For every integral ideal $\mathfrak{n}$ in F we put

$$\Gamma(\mathfrak{n}) = \bigl\{\alpha \in SL_2(\mathfrak{g}) \,\big|\, \alpha - 1 \prec \mathfrak{n}\bigr\}. \tag{17.1}$$

By a **congruence subgroup** of $SL_2(F)$ we mean a subgroup of $SL_2(F)$ that has $\Gamma(\mathfrak{n})$ as a subgroup of finite index for some $\mathfrak{n}$. To define modular forms, we need the notion of a weight. By an **integral weight** we mean an element of $\mathbf{Z}^{\mathbf{a}}$, and by a **half-integral weight** we mean an element $(k_v)_{v \in \mathbf{a}}$

of $\mathbf{Q}^{\mathbf{a}}$ such that $k_v - 1/2 \in \mathbf{Z}$ for every $v \in \mathbf{a}$. Thus, if $[F : \mathbf{Q}] = 3$ for example, $(-7/2, 1/2, 9/2)$ is a half-integral weight, but $(3, 7/2, 4)$ is not. For a half-integral k we put $[k] = (k_v - 1/2)_{v\in\mathbf{a}}$.

For $c = (c_v)_{v\in\mathbf{a}} \in \mathbf{C}^{\mathbf{a}}$ we put $\mathbf{e}_{\mathbf{a}}(c) = \mathbf{e}\big(\sum_{v\in\mathbf{a}} c_v\big)$ and also

$$\theta(z) = \sum_{a\in\mathfrak{g}} \mathbf{e}_{\mathbf{a}}(a^2 z/2) \qquad (z \in \mathfrak{H}^{\mathbf{a}}), \tag{17.2}$$

$$\Gamma_0 = \big\{\gamma \in SL_2(F) \,\big|\, a_\gamma \in \mathfrak{g},\, b_\gamma \in 2\mathfrak{d}^{-1},\, c_\gamma \in 2\mathfrak{d},\, d_\gamma \in \mathfrak{g}\big\}. \tag{17.3}$$

Then for every $\gamma \in \Gamma_0$ there exists a function $h_\gamma(z)$ on $\mathfrak{H}^{\mathbf{a}}$ such that

$$\theta(\gamma z) = h_\gamma(z)\theta(z), \tag{17.4a}$$

$$h_\gamma(z)^4 = \prod_{v\in\mathbf{a}} j(\gamma_v,\, z_v)^2, \tag{17.4b}$$

where $j(\beta,\, w) = c_\beta w + d_\beta$ for $\beta \in SL_2(\mathbf{R})$ and $w \in \mathfrak{H}$; see [S85a]. Then we define a factor of automorphy $j_\gamma^k(z)$ for a weight k, $\gamma \in \Gamma_0$, and $z \in \mathfrak{H}^{\mathbf{a}}$ by

$$j_\gamma^k(z) = \begin{cases} \prod_{v\in\mathbf{a}} j(\gamma_v,\, z_v)^{k_v} & \text{if } k \in \mathbf{Z}^{\mathbf{a}}, \\ h_\gamma(z) j_\gamma^{[k]}(z) & \text{if } k \notin \mathbf{Z}^{\mathbf{a}}. \end{cases} \tag{17.5}$$

Given a congruence subgroup Γ of $SL_2(F)$ contained in Γ_0 and a weight k, we denote by $\mathscr{M}_k(\Gamma)$ the set of all holomorphic functions f on $\mathfrak{H}^{\mathbf{a}}$ such that $f(\gamma z) = j_\gamma^k(z) f(z)$ for every $\gamma \in \Gamma$, and we call such an f a **Hilbert modular form** of weight k with respect to Γ. Here we assume that $F \neq \mathbf{Q}$. In fact, if $F \neq \mathbf{Q}$, then we can show that every such f has an expansion

$$f(z) = \sum_{\xi\in\mathfrak{a}} c(\xi)\mathbf{e}(\xi z) \tag{17.6}$$

with a fractional ideal $\mathfrak{a}$ in F and $c(\xi) \in \mathbf{C}$ such that $c(\xi) = 0$ if $\xi_v < 0$ for some $v \in \mathbf{a}$. In other words, we need condition (3.4d) only when $F = \mathbf{Q}$.

Basically we can extend all definitions and results in the case $F = \mathbf{Q}$ to the case $F \neq \mathbf{Q}$. The generalization of $D(s,\, f)$ of (8.3) is

$$[\mathfrak{g}^\times : \mathfrak{t}]^{-1} \sum_{\xi\in F^\times/\mathfrak{t}} c(\xi) \prod_{v\in\mathbf{a}} \xi_v^{-s-k_v/2}, \tag{17.7}$$

where $\mathfrak{t}$ is a subgroup of $\mathfrak{g}^\times$ of finite index that makes the last sum meaningful; the existence of such a $\mathfrak{t}$ can be shown.

We encounter some phenomena which do not exist in the case $F = \mathbf{Q}$. For example, given $\sigma \in \mathrm{Aut}(\mathbf{C})$, we can define a function f^σ on $\mathfrak{H}^{\mathbf{a}}$ by

$$f^\sigma(z) = \sum_{\xi\in\mathfrak{a}} c(\xi)^\sigma \mathbf{e}(\xi z). \tag{17.8}$$

This is indeed a Hilbert modular form, but its weight is a transform of k by σ in a natural way, and so it is not necessarily k.

Without going further, we merely mention some references. In the paper [Kl28] Kloostermann gave some basic results on Hilbert modular forms of weight k in the case where the k_v for all $v \in \mathbf{a}$ are equal, and treated holomorphic Eisenstein series. The case of more general integral weights was discussed in [S78], which includes the generalization of Theorem 16.9. This was originally published in the *Duke Mathematical Journal,* but the typesetter and copyeditor made an incredible number of mistakes. Though corrections were later published in the same journal, I advise the reader to read the revised version included in my *Collected Papers,* vol. III. The factor of automorphy h_γ as in (17.4a, b) was established in [S85a], and Eisenstein series were discussed in [S85a] and [S85b]. The generalizations of Theorems 11.3 and 12.8 were given in [S87]. The contents of §§14, 15, and 16 of the present book can be viewed as special cases of [S91] which discusses the same problems in the Hilbert modular case.

APPENDIX

A1. Proof of various facts

A1.1. Equality (3.7a) can be proved as follows. Since $f(x+iy)$ is a C^∞ function of x invariant under $x \mapsto x+m$ for every $m \in M$, it has a Fourier expansion

$$\text{(A1.1)} \qquad f(x+iy) = \sum_{h\in L} c_h(y)\mathbf{e}\big(\mathrm{tr}(hx)\big)$$

with C^∞ functions c_h of y, by virtue of a general principle on the Fourier expansion of a function with C^∞ parameters; also termwise partial differentiation of (A1.1) can be justified. For the proof of these facts the reader is referred to any textbook on Fourier analysis in $\mathbf{R}^n$; they are also proved in [S07, Theorem A2.2]. Put $b_h(y) = c_h(y)\mathbf{e}\big(-i\cdot\mathrm{tr}(hy)\big)$. Then $f(z) = \sum_{h\in L} b_h(y)\mathbf{e}\big(\mathrm{tr}(hz)\big)$. Take the variable $z_{\mu\nu} = x_{\mu\nu}+iy_{\mu\nu}$ and apply $\partial/\partial\bar{z}_{\mu\nu} = 2^{-1}(\partial/\partial x_{\mu\nu} + i\partial/\partial y_{\mu\nu})$ to the last equality. Then

$$0 = \sum_{h\subset L}(i/2)(\partial b_h/\partial y_{\mu\nu})\mathbf{e}\big(\mathrm{tr}(hz)\big),$$

and so $\partial b_h/\partial y_{\mu\nu} = 0$ for every (μ, ν), which means that b_h is a constant. Thus we obtain (3.7a).

A1.2. We next prove Lemma 3.3. First we observe that $f(x+iy) = \sum_{h\in L} c(h)\mathbf{e}\big(i\cdot\mathrm{tr}(hy)\big)\mathbf{e}\big(\mathrm{tr}(hx)\big)$, and so

$$\text{(A1.2)} \qquad \mathbf{e}\big(i\cdot\mathrm{tr}(hy)\big)c(h) = A\int_{S_{\mathbf{a}}/M} f(x+iy)\mathbf{e}\big(-\mathrm{tr}(hx)\big)dx,$$

where $S_{\mathbf{a}} = S_n(\mathbf{R})$ and $A = \mathrm{vol}(S_{\mathbf{a}}/M)^{-1}$. Taking $y = (2\pi)^{-1}1_n$ in (A1.2), we obtain $|c(h)| \le B\exp\big(\mathrm{tr}(h)\big)$ with a constant B independent of h. Now from (3.6b) we obtain $c(h) = c({}^t aha)$ for every $a \in U$, and so

$$(*) \qquad |c(h)| \le B\exp\big(\mathrm{tr}({}^t aha)\big) \text{ for every } a \in U.$$

Now suppose $n > 1$; let h be an element of L that is not nonnegative. Our task is to show that $c(h) = 0$. We can find $x = (x_i)_{i=1}^n \in \mathbf{Q}^n$ such that ${}^t xhx < 0$. Replacing x by $x+u$ with a "small" vector u, we may assume

G. Shimura, *Modular Forms: Basics and Beyond*, Springer Monographs in Mathematics, DOI 10.1007/978-1-4614-2125-2, © Springer Science+Business Media, LLC 2012 that $x_1x_2 \neq 0$. Multiplying x by a positive integer, we may also assume that $x \in \mathbf{Z}^n$. Let $y = [-x_2 \ x_1 \ 0 \ \cdots \ 0]$ and $b = xy$; here y is a row vector and so $b \in \mathbf{Z}_n^n$. Then $b^2 = 0$ since $yx = 0$, and

$$(**) \qquad \mathrm{tr}({}^t bhb) = \mathrm{tr}({}^t y \cdot {}^t xhxy) = y \cdot {}^t y \cdot {}^t xhx = (x_1^2 + x_2^2) \cdot {}^t xhx < 0.$$

Put $a = (1+b)^m$ with $0 < m \in \mathbf{Z}$. Since $(1+b)(1-b) = 1$, we have $1 + b \in GL_n(\mathbf{Z})$, and so $a \in U$ if $m \in N\mathbf{Z}$ with a suitably large integer N. For such an m we have

$$\mathrm{tr}({}^t aha) = \mathrm{tr}\big({}^t(1+mb)h(1+mb)\big) = p + mq + m^2 r$$

with p, q, $r \in \mathbf{R}$, which are independent of m; in particular, $r = \mathrm{tr}({}^t bhb) < 0$, as shown in $(**)$. Now by $(*)$ we have $|c(h)| \leq B\exp(p + mq + m^2 r)$ for $0 < m \in N\mathbf{Z}$. Making m large, we find that $c(h) = 0$ as expected.

Lemma A1.3. (i) *$\Gamma(1)$ is generated by the elements of the forms of (4.31), and consequently $\Gamma(1)$ is generated by ι and $P \cap \Gamma(1)$.*

(ii) *Let $\Gamma' = \{\gamma \in \Gamma(1) \mid b_\gamma \equiv c_\gamma \equiv 0 \pmod{2\mathbf{Z}_n^n}\}$. Then Γ' is generated by the elements of the forms*

$$\text{(A1.3)} \qquad \begin{bmatrix} a & 0 \\ 0 & d \end{bmatrix}, \quad \begin{bmatrix} 1 & b \\ 0 & 1 \end{bmatrix}, \quad \begin{bmatrix} 1 & 0 \\ c & 1 \end{bmatrix}, \quad b \equiv c \equiv 0 \pmod{2\mathbf{Z}_n^n}.$$

(iii) *Let Γ^* be the subgroup of $\Gamma(1)$ generated by the elements of the forms*

$$\text{(A1.4)} \qquad \begin{bmatrix} a & 0 \\ 0 & d \end{bmatrix}, \quad \begin{bmatrix} 0 & -1 \\ 1 & 0 \end{bmatrix}, \quad \begin{bmatrix} 1 & b \\ 0 & 1 \end{bmatrix}, \quad b \equiv 0 \pmod{2\mathbf{Z}_n^n}.$$

Then $\Gamma' \subset \Gamma^*$.

PROOF. We first prove (ii). For $x \in \mathbf{Z}^n$, let $[x]$ denote the greatest common divisor of its components. We put $[0] = 0$. Also, for $a \in \mathbf{Z}_n^n$ let a^j denote its jth column and a_i^j its (i, j)-entry. Now let $\gamma = \begin{bmatrix} a & b \\ c & d \end{bmatrix} \in \Gamma'$. Our idea is to reduce $[a^1]$ and $[c^1]$ by multiplying by elements of the forms listed in (A1.3). Clearly $[c^1]$ is even, and so $[a^1]$ is odd, since $\gamma \in SL_{2n}(\mathbf{Z})$. First suppose $[a^1] < [c^1]$. Considering $\begin{bmatrix} u & 0 \\ 0 & v \end{bmatrix}\begin{bmatrix} a & b \\ c & d \end{bmatrix}$ instead of $\begin{bmatrix} a & b \\ c & d \end{bmatrix}$ with a suitable $u = {}^t v^{-1} \in GL_n(\mathbf{Z})$, we may assume that $a_1^1 > 0$, $a_2^1 = \cdots = a_n^1 = 0$. Then a_1^1 is odd and $< [c^1]$. For each k we can find an integer s_k^1 such that $|c_k^1 + 2s_k^1 a_1^1| \leq a_1^1$. Take any $s \in S_n(\mathbf{Z})$ whose $(1, k)$-entry is such s_k^1, and put $\begin{bmatrix} 1 & 0 \\ 2s & 1 \end{bmatrix}\begin{bmatrix} a & b \\ c & d \end{bmatrix} = \begin{bmatrix} a & b \\ p & q \end{bmatrix}$. Then we find $[p^1] \leq [a^1]$, and so $[p^1] < [a^1]$, since $[p^1]$ is even. Next assume that $0 < [c^1] < [a^1]$. Then, first considering vc with a suitable $v \in GL_n(\mathbf{Z})$, and then $\begin{bmatrix} 1 & 2s \\ 0 & 1 \end{bmatrix}\gamma$ with a suitable s, we can reduce this to the case $[a^1] < [c^1]$. Repeating these procedures, we obtain an element, written again as $\gamma = \begin{bmatrix} a & b \\ c & d \end{bmatrix}$, with $c^1 = 0$. Then $[a^1] = 1$. For the same reason as above, we may assume that $a_1^1 = 1$, $a_2^1 = \cdots = a_n^1 = 0$. Since ${}^t ad - {}^t cb = 1$, we see that $d_1^1 = 1$. Then left multiplication by $\mathrm{diag}[u, v]$ with a suitable $v \in GL_n(\mathbf{Z})$ produces $d^1 = {}^t(1 \ \ 0 \ \ \cdots \ \ 0)$ without changing a^1 and c^1. Furthermore, left multiplication by $\begin{bmatrix} 1 & 2s \\ 0 & 1 \end{bmatrix}$ with a suitable s produces $b^1 = 0$. We obtain in this way an element of the form

$$\begin{bmatrix} 1 & r & 0 & s \\ 0 & a' & 0 & b' \\ 0 & t & 1 & u \\ 0 & c' & 0 & d' \end{bmatrix}$$

with a', b', c', d' of size $n-1$. From the relations ${}^t bd = {}^t db$ and ${}^t ac = {}^t ca$ we obtain $s = t = 0$, and from ${}^t da - {}^t bc = 1$ we obtain $r = u = 0$. Thus our matrix in question is of the form

$$\text{(A1.5)} \qquad \begin{bmatrix} 1 & 0 & 0 & 0 \\ 0 & a' & 0 & b' \\ 0 & 0 & 1 & 0 \\ 0 & c' & 0 & d' \end{bmatrix}$$

with $\begin{bmatrix} a' & b' \\ c' & d' \end{bmatrix} \in Sp(n-1, \mathbf{Z})$, $b' \equiv c' \equiv 0 \pmod{2\mathbf{Z}_{n-1}^{n-1}}$. The proof of (ii) is therefore completed by induction on n, since if $\begin{bmatrix} a' & b' \\ c' & d' \end{bmatrix}$ is of a type belonging to (A1.3), then so is the matrix of (A1.5).

To prove (i) and (iii), we note

$$\text{(A1.6)} \qquad \begin{bmatrix} 0 & -1 \\ 1 & 0 \end{bmatrix} = \begin{bmatrix} 1 & -1 \\ 0 & 1 \end{bmatrix} \begin{bmatrix} 1 & 0 \\ 1 & 1 \end{bmatrix} \begin{bmatrix} 1 & -1 \\ 0 & 1 \end{bmatrix},$$

$$\text{(A1.7)} \qquad \begin{bmatrix} 1 & 0 \\ -b & 1 \end{bmatrix} = \begin{bmatrix} 0 & -1 \\ 1 & 0 \end{bmatrix} \begin{bmatrix} 1 & b \\ 0 & 1 \end{bmatrix} \begin{bmatrix} 0 & -1 \\ 1 & 0 \end{bmatrix}^{-1}.$$

In view of (A1.7) we obtain (iii) immediately from (ii). As for (i) we employ the same type of argument as in the proof of (ii). Since we have ι in (4.31), we can use it in addition to the matrices of (A1.3) without congruence conditions, again in view of (A1.7). Now left multiplication by ι changes (a, c) into $(-c, a)$. We first assume that $0 < [a^1] \le [c^1]$ and repeat the above argument with the following modification: take $s_k^1 \in \mathbf{Z}$ so that $0 \le c_k^1 + s_k^1 a_1^1 < a_1^1$, and use s instead of $2s$. Then we can reduce the problem to the case $[c^1] \le [a^1]$, and further to the case $c_1 = 0$, and eventually to (A1.5). If $\begin{bmatrix} a' & b' \\ c' & d' \end{bmatrix} = \iota_{n-1}$, then (A1.5) does not belong to the three types of (4.31), but applying (A1.6) to ι_{n-1} and employing (A1.7), we can justify our induction.

A1.4. As noted in §4.5, formulas (4.12) and (4.13) are equivalent to (4.17) and (4.18). Therefore our result of §4.9 shows that (4.15) holds for γ of

(4.31). Thus, by Lemma A1.3(i) we have

$$\varphi({}^t\mu_\gamma(z)^{-1}u,\ \gamma z;\ r,\ s) = \zeta \cdot j_\gamma(z)^{1/2}\varphi(u,\ z;\ r'',\ s'') \tag{A1.8}$$

for every $\gamma \in \Gamma(1)$ with some r'', s'', and $\zeta \in \mathbf{T}$, since it is easy to see that (A1.8) is "associative" with respect to successive applications of elements of $\Gamma(1)$. Thus our task is to determine r'' and s''. For this purpose we first note that if $w \in \mathfrak{H}_n$, $f(u) = \varphi(u,\ w;\ r,\ s)$, and $\ell = wp' + q'$ with $p',\ q' \in \mathbf{Z}^n$, then

$$f(u+\ell) = f(u)\mathbf{e}\big(2^{-1} \cdot {}^tp'q' - {}^tsp' + {}^trq' + {}^t\overline{\ell}(w-\overline{w})^{-1}(u+2^{-1}\ell)\big). \tag{A1.9}$$

This follows from (4.4). Observe that r and s are determined modulo $\mathbf{Z}^n$ by this formula. For $\gamma \in \Gamma(1)$, $z \in \mathfrak{H}_n$, and $m = zp + q$ with $p,\ q \in \mathbf{Z}^n$ put $\ell = {}^t\mu_\gamma(z)^{-1}m$, $w = \gamma(z)$, and

$$g(u) = f\big({}^t\mu_\gamma(z)^{-1}u\big) = \varphi\big({}^t\mu_\gamma(z)^{-1}u,\ w;\ r,\ s\big).$$

If $\gamma = \begin{bmatrix} a & b \\ c & d \end{bmatrix}$, then $\gamma^{-1} = \begin{bmatrix} {}^td & -{}^tb \\ -{}^tc & {}^ta \end{bmatrix}$ by (1.2b), $\mu_\gamma(z)^{-1} = \mu(\gamma^{-1},\ w)$ by (1.14), and $z = {}^tz = {}^t(-{}^tcw + {}^ta)^{-1} \cdot {}^t({}^tdw - {}^tb)$, and so

$$\ell = {}^t(-{}^tcw + {}^ta)(zp + q) = {}^t({}^tdw - {}^tb)p + {}^t(-{}^tcw + {}^ta)q. \tag{A1.9a}$$

Thus $\ell = wp' + q'$ with $p' = dp - cq$ and $q' = aq - bp$. From (4.11) we obtain $(w - \overline{w})^{-1} = \mu_\gamma(z)(z - \overline{z})^{-1} \cdot {}^t\overline{\mu_\gamma(z)}$, and so from (A1.9) we obtain $g(u + m) = g(u)\mathbf{e}(X)$ with

$$X = 2^{-1} \cdot {}^tp'q' - {}^tsp' + {}^trq' + {}^t\overline{m}(z - \overline{z})^{-1}(u + 2^{-1}m).$$

Since ${}^tda - {}^tbc = 1$ and ${}^tp\sigma p - {}^t\{\sigma\}p \in 2\mathbf{Z}$ for every $p \in \mathbf{Z}^n$ and $\sigma \in S_n(\mathbf{Z})$, a straightforward calculation shows that

$$2^{-1} \cdot {}^tp'q' - {}^tsp' + {}^trq' \equiv 2^{-1} \cdot {}^tpq - {}^ts_1p + {}^tr_1q \pmod{\mathbf{Z}}$$
$$\text{with} \qquad \begin{bmatrix} r_1 \\ s_1 \end{bmatrix} = {}^t\gamma \begin{bmatrix} r \\ s \end{bmatrix} + \frac{1}{2}\begin{bmatrix} \{{}^tac\} \\ \{{}^tbd\} \end{bmatrix}.$$

Thus

$$g(u+m) = g(u)\mathbf{e}\big(2^{-1} \cdot {}^tpq - {}^ts_1p + {}^tr_1q + {}^t\overline{m}(z-\overline{z})^{-1}(u+2^{-1}m)\big). \tag{A1.10}$$

If r'' and s'' are as in (A1.8), then they are determined modulo $\mathbf{Z}^n$ by formula (A1.9) with f replaced by g, which is formula (A1.10). Therefore $r'' \equiv r_1$ and $s'' \equiv s_1 \pmod{\mathbf{Z}^n}$. This combined with (4.5) proves (i) of Theorem 4.4.

Next assume that both $\{{}^tac\}$ and $\{{}^tbd\}$ belong to $2\mathbf{Z}^n$. Then we can put

$$\varphi\big({}^t\mu_\gamma(z)^{-1}u,\ \gamma z;\ 0,\ 0\big) = \lambda_\gamma j_\gamma(z)^{1/2}\varphi(u,\ z;\ 0,\ 0) \tag{A1.11}$$

with a constant $\lambda_\gamma \in \mathbf{T}$. For $k = zr + s$ with $r,\ s \in \mathbf{R}^n$, from (4.3) we obtain

$$\theta(u + k,\ z;\ 0,\ 0) = \mathbf{e}\big(-2^{-1} \cdot {}^trzr - {}^tr(u + s)\big)\theta(u,\ z;\ r,\ s),$$

which combined with (4.2) shows that

$$\mathbf{e}\big(-{}^t\bar{k}(z-\bar{z})^{-1}(u+k/2)\big)\varphi(u+k,\, z;\, 0,\, 0) = \mathbf{e}(2^{-1}Y)\varphi(u,\, z;\, r,\, s)$$
$$\text{with}\quad Y = -2\cdot{}^t\bar{k}(z-\bar{z})^{-1}(u+k/2) + {}^t(u+k)(z-\bar{z})^{-1}(u+k) - {}^trzr - 2\cdot{}^tr(u+s) - {}^tu(z-\bar{z})^{-1}u.$$

We can easily verify that $Y = -{}^trs$, and so we obtain

$$\text{(A1.12)}\qquad \varphi(u,\, z;\, r,\, s) = \mathbf{e}\big(2^{-1}A(u,\, z;\, r,\, s)\big)\varphi(u+k,\, z;\, 0,\, 0)$$
$$\text{with}\quad A(u,\, z;\, r,\, s) = {}^trs - {}^t(\bar{z}r+s)(z-\bar{z})^{-1}(2u+zr+s).$$

Put $w = \gamma(z)$, $v = \mu_\gamma(z)^{-1}u$, and $k' = \mu_\gamma(z)^{-1}k$. Taking $k = zr+s$ in place of $m = zp+q$ in (A1.9a), we find that $k' = wr'+s'$ with $r' = dr - cs$ and $s' = as - br$. Therefore, combining (A1.11) and (A1.12), we obtain

$$\begin{aligned}\varphi\big(\gamma(u,\, z);\, r',\, s'\big) &= \mathbf{e}\big(2^{-1}A(\gamma(u,\, z);\, r',\, s')\big)\varphi(v+k',\, w;\, 0,\, 0)\\ &= \mathbf{e}\big(2^{-1}A(\gamma(u,\, z);\, r',\, s')\big)\lambda_\gamma j_\gamma(z)^{1/2}\varphi(u+k,\, z;\, 0,\, 0)\\ &= \mathbf{e}\big(2^{-1}B\big)\lambda_\gamma j_\gamma(z)^{1/2}\varphi(u,\, z;\, r,\, s)\\ &\qquad \text{with}\quad B = A(\gamma(u,\, z);\, r',\, s') - A(u,\, z;\, r,\, s).\end{aligned}$$

We have $B = {}^tr's' - {}^trs - {}^t\bar{k}(z-\bar{z})^{-1}(2u+k) + {}^t\bar{k}'(w-\bar{w})^{-1}(2v+k')$. Using again (4.11), we see that the last two terms cancel each other. Since $\begin{bmatrix} r' \\ s' \end{bmatrix} = {}^t\gamma^{-1}\begin{bmatrix} r \\ s \end{bmatrix}$, exchanging $(r,\, s)$ for $(r',\, s')$ and writing $h_\gamma(z)$ for $\lambda_\gamma j_\gamma(z)^{1/2}$, we obtain (ii) of Theorem 4.4, or rather (4.18).

Finally, to prove (iii) of Theorem 4.4, we observe that $\lambda_\gamma^4 = 1$ for the first two types of elements of (A1.3). As for the third type, making substitutions $z \mapsto z - 2c$ and $z \mapsto -z^{-1}$ in (4.34a), we find that

$$\theta\big(z(2cz+1)^{-1}\big) = \pm\det(2cz+1)^{1/2}\theta(z)\quad \textit{if}\quad c \in S_n(\mathbf{Z})$$

with θ of (4.19). Thus $\lambda_\gamma = \pm 1$ for the third type. (Theorem 4.7(2), which will be proven in §A1.6, gives a stronger result.) The proof of Theorem 4.4 is now complete.

A1.5. Let us now prove (1) of Theorem 4.7. To simplify the notation, suppress the subscript γ, and put $L = \mathbf{Z}^n$, $A = L/{}^tdL$, $B = L/dL$, and $s[x] = {}^txsx$ for $s \in \mathbf{Q}_n^n$ and $x \in \mathbf{Q}^n$. Since $c \cdot {}^td = d \cdot {}^tc$, we see that $c \cdot {}^tdL \subset dL$. Therefore $x \mapsto cx$ sends A into B. Since $\gamma \in SL_{2n}(\mathbf{Z})$, we have $cL + dL = L$, and so the map is surjective. Comparing the orders of the groups, we find that the map gives an isomorphism of A onto B. If $y = cx$, we have, by (1.2a),

$$\begin{aligned}bd^{-1}[y] &= {}^tcbd^{-1}c[x] = ({}^tad - 1)d^{-1}c[x]\\ &= {}^tac[x] - d^{-1}c[x] \equiv -d^{-1}c[x] \pmod 2,\end{aligned}$$

since the diagonal elements of ${}^t ac$ are even. This shows that the two sums of (4.25) are the same.

From (4.22a) we see that $h_{\alpha\gamma} = h_\gamma$ for $\alpha = \mathrm{diag}[a,\, d] \in \Gamma(1)$ and every $\gamma \in \Gamma^\theta$. Therefore, to prove (2) and the first equality of (1), we may assume that $\det(d) > 0$. Under this assumption, put $w = {}^t d^{-1} z(cz+d)^{-1}$ and $p = bd^{-1}$. Then, by (1.2a),

$$\begin{aligned}(\gamma z - p)(cz + d) &= az + b - bd^{-1}(cz + d) = az - b \cdot {}^t c \cdot {}^t d^{-1} z \\ &= (a \cdot {}^t d - b \cdot {}^t c) \cdot {}^t d^{-1} z = {}^t d^{-1} z,\end{aligned}$$

and so $\gamma(z) = w + p$. With θ of (4.19) we thus have

(A1.13) $$h_\gamma(z)\theta(z) = \theta(\gamma z) = \theta(w + p) = \sum_{x\in L} \mathbf{e}\big((1/2)(w + p)[x]\big).$$

Putting $x = v + dg$ with $v \in B$ and $g \in L$, we find

(A1.14) $$\begin{aligned}\theta(w + p) &= \sum_v \sum_g \mathbf{e}\big((p/2)[v + dg] + (w/2)[v + dg]\big) \\ &= \sum_v \mathbf{e}\big((p/2)[v]\big)\theta(0,\, z(cz + d)^{-1}d;\, d^{-1}v,\, 0),\end{aligned}$$

since $p[v + dg] \equiv p[v] \pmod 2$. Now (4.34) shows that

$$\det(-iz)^{1/2}\theta(0,\, z;\, r, s) = \mathbf{e}({}^t rs)\theta(0,\, -z^{-1};\, -s,\, r).$$

Put $z = i\tau 1_n$ with $0 < \tau \in \mathbf{R}$ and observe that

$$\lim_{\tau\to 0} \tau^{n/2}\theta(0,\, i\tau 1_n;\, r,\, s) = \mathbf{e}({}^t rs)\delta(s),$$

where $\delta(s) = 1$ or 0 according as $s \in L$ or $s \notin L$. Taking the limit of $\tau^{n/2}$ times (A1.13) combined with (A1.14) as τ tends to 0, we obtain (4.25).

A1.6. To prove (2) of Theorem 4.7, given $\gamma \in \Gamma^\theta$, assume that $\det(d) - 1 \in 2\mathbf{Z}$. In view of (4.21) and (4.22a), replacing γ by $\gamma \cdot \mathrm{diag}[e,\, e]$ with $e = \mathrm{diag}[-1,\, 1_{n-1}]$ if necessary, we may assume that $\det(d) > 0$. Put $f = \det(d)$, $g = fd^{-1}$, and $s = -fd^{-1}c$. Then $g \prec \mathbf{Z}$, $s \in S_n(\mathbf{Z})$, f is odd, $-fs = gd \cdot {}^t c \cdot {}^t g$, and $\{d \cdot {}^t c\} \in 2\mathbf{Z}^n$ by (4.6), and so by (4.7) the diagonal elements of s are even. Denote by λ the first sum of (4.25) and put $\sigma = h_\gamma(z)^2/j_\gamma(z)$. Then $\sigma = \lambda^2/\det(d)$. Now

$$\lambda = \sum_{x\in A} \mathbf{e}\big(s[x]/(2f)\big) = [{}^t dL : fL]^{-1} \sum_{x\in L/fL} \mathbf{e}\big(s[x]/(2f)\big).$$

By Lemma 2.3 we can find $u \in \mathbf{Z}_n^n$ such that $\det(u)$ is a positive integer prime to f and ${}^t usu - \mathrm{diag}[r_1,\, \ldots,\, r_n] \prec f\mathbf{Z}$ with $r_i \in \mathbf{Z}$. We can take r_i to be the $(i,\, i)$-entry of ${}^t usu$. Then $r_i \in 2\mathbf{Z}$ by (4.7), and

$$\lambda = f^{1-n} \prod_{\nu=1}^{n} \sum_{x=1}^{f} \mathbf{e}\big(r_\nu x^2/(2f)\big).$$

Put $r_\nu/f = 2b_\nu/a_\nu$ with relatively prime integers a_ν and b_ν; take $a_\nu > 0$. Then

$$\sum_{x=1}^{f} \mathbf{e}\big(r_\nu x^2/(2f)\big) = (f/a_\nu)\sum_{x=1}^{a_\nu} \mathbf{e}(x^2 b_\nu/a_\nu) = f\left(\frac{b_\nu}{a_\nu}\right)\varepsilon(a_\nu)a_\nu^{-1/2}$$

by Theorem 2.6, where $\varepsilon(a)$ is ε_a of (0.6). Thus we obtain

$$\lambda = f\prod_{\nu=1}^{n}\left(\frac{b_\nu}{a_\nu}\right)\varepsilon(a_\nu)a_\nu^{-1/2}.$$

Since $cL + dL = L$, we have $sL + fL = gL$. For a prime number p put $L_p = \mathbf{Z}_p^n$. Then $gL_p = L_p$ if $p\nmid f$. If $p|f$, then ${}^t uL_p = uL_p = L_p$, and hence ${}^t ugL_p = {}^t usuL_p + fL_p$. From this we easily see that the elementary divisors of g are $\{(f, r_\nu)\}_{\nu=1}^{n}$. Since $a_\nu = |f/(f, r_\nu)|$ and $d = fg^{-1}$, we thus know that the a_ν are exactly the elementary divisors of d and

$$\lambda = \det(d)^{1/2}\prod_{\nu=1}^{n}\left(\frac{b_\nu}{a_\nu}\right)\varepsilon(a_\nu),$$

so that $\sigma = \lambda^2/\det(d) = \left(\dfrac{-1}{\det(d)}\right)$, which proves (2) of Theorem 4.7.

A2. Whittaker functions

A2.1. We need some **Whittaker** (or **confluent hypergeometric**) **functions:**

(A2.1) $$\tau(y;\,\alpha,\,\beta) = \int_0^\infty e^{-yt}(1+t)^{\alpha-1}t^{\beta-1}dt,$$

(A2.2) $$V(y;\,\alpha,\,\beta) = e^{-y/2}\,\Gamma(\beta)^{-1}\,y^\beta\,\tau(y;\,\alpha,\,\beta).$$

Here $0 < y \in \mathbf{R}$ and $(\alpha, \beta) \in \mathbf{C}^2$. The integral of (A2.1) is convergent for $\mathrm{Re}(\beta) > 0$, and so defines a holomorphic function of (α, β) under that condition; also it can be shown that $V(y;\,\alpha,\,\beta)$ can be continued to a holomorphic function of (α, β) on the whole $\mathbf{C}^2$. We have also

(A2.3) $$\big[\tau(y;\,\alpha,\,\beta)/\Gamma(\beta)\big]_{\beta=0} = 1.$$

For these and the following two lemmas, the reader is referred to [S07, Section A3].

Lemma A2.2. (i) *For every compact subset K of $\mathbf{C}^2$ there exist two positive constants A and B depending only on K such that*

$$|e^{y/2}V(y;\,\alpha,\,\beta)| \le A\big(1 + y^{-B}\big) \quad \textit{if} \quad (\alpha,\,\beta) \in K.$$

(ii) $V(y;\,1-\beta,\,1-\alpha) = V(y;\,\alpha,\,\beta)$.

Lemma A2.3. *If* $\mathrm{Re}(\alpha+\beta) > 1$ *and* $z = x + iy \in \mathfrak{H}$, *then*

$$\sum_{m\in\mathbf{Z}}(z+m)^{-\alpha}(\overline{z}+m)^{-\beta} = i^{\beta-\alpha}(2\pi)^{\alpha+\beta}\sum_{n\in\mathbf{Z}}\mathbf{e}(nx + i|n|y)g_n(y;\,\alpha,\,\beta),$$

where g_n *is given by*

$$\text{(A2.4)}\qquad \Gamma(\alpha)\Gamma(\beta)g_n(y;\,\alpha,\,\beta) = \begin{cases} n^{\alpha+\beta-1}\,\tau(4\pi ny;\,\alpha,\,\beta) & \text{if } n>0, \\ |n|^{\alpha+\beta-1}\,\tau(4\pi|n|y;\,\beta,\,\alpha) & \text{if } n<0, \\ \Gamma(\alpha+\beta-1)(4\pi y)^{1-\alpha-\beta} & \text{if } n=0. \end{cases}$$

Here, for $v \in \mathbf{C}^\times$ and $\alpha \in \mathbf{C}$ we define v^α by

$$v^\alpha = \exp(\alpha\,\log(v)), \qquad -\pi < \mathrm{Im}[\log(v)] \le \pi.$$

Then $v^{\alpha+\beta} = v^\alpha v^\beta$, $v^{m\alpha} = (v^\alpha)^m$ for $m \in \mathbf{Z}$, and $(uv)^\alpha = u^\alpha v^\alpha$ provided $\arg(u)$, $\arg(v)$, and $\arg(u)+\arg(v)$ are all contained in the interval $(-\pi,\,\pi]$ for suitable choices of $\arg(u)$ and $\arg(v)$.

From Lemma A2.2(ii) we obtain

$$\text{(A2.5)}\qquad \tau(y;\,1-\beta,\,1-\alpha)/\Gamma(1-\alpha) = y^{\alpha+\beta-1}\tau(y;\,\alpha,\,\beta)/\Gamma(\beta).$$

This combined with (A2.3) gives

$$\text{(A2.6)}\qquad \tau(y;\,1,\,\beta)/\Gamma(\beta) = y^{-\beta}.$$

Lemma A2.4. (i) *Given* $(\alpha,\,\beta) \in \mathbf{C}^2$ *and* $A \in \mathbf{R}^\times$, *put* $\sigma = \alpha-\beta$ *and* $\lambda = \beta(1-\alpha)$. *Define a function* $f_A(y)$ *for* $0 < y \in \mathbf{R}$ *by*

$$\text{(A2.7)}\qquad f_A(y) = \begin{cases} V(2Ay;\,\alpha,\,\beta) & \text{if } A>0, \\ |2Ay|^{-\sigma}V(|2A|y;\,\beta,\,\alpha) & \text{if } A<0. \end{cases}$$

Then f_A *satisfies the differential equation*

$$\text{(A2.8)}\qquad y^2f''(y) + \sigma yf'(y) + (\lambda + A\sigma y - A^2y^2)f(y) = 0.$$

Moreover, if f *is a solution of (A2.8) and* $f(y) = O(y^B)$ *with* $B \in \mathbf{R}$ *as* $y \to \infty$, *then* f *is a constant multiple of* f_A.

(ii) *With the same notation as in* (i), *define a function* φ_A *on* $\mathfrak{H}$ *by*

$$\text{(A2.9)}\qquad \varphi_A(x+iy;\,\sigma,\,\lambda) = e^{iAx}f_A(y).$$

Let ε *and* δ_σ *be as in (6.13a, b). Then*

$$\text{(A2.10)}\qquad \varepsilon\varphi_A(z;\,\sigma,\,\lambda) = \begin{cases} \lambda(4Ai)^{-1}\varphi_A(z;\,\sigma-2,\,\lambda+2-\sigma) & \text{if } A>0, \\ (4Ai)^{-1}\varphi_A(z;\,\sigma-2,\,\lambda+2-\sigma) & \text{if } A<0, \end{cases}$$

$$\text{(A2.11)}\qquad \delta_\sigma\varphi_A(z;\,\sigma,\,\lambda) = \begin{cases} iA\varphi_A(z;\,\sigma+2,\,\lambda+\sigma) & \text{if } A>0, \\ (\lambda+\sigma)iA\varphi_A(z;\,\sigma+2,\,\lambda+\sigma) & \text{if } A<0. \end{cases}$$

PROOF. Since $(1+t)^\alpha = (1+t)^{\alpha-1} + (1+t)^{\alpha-1}t$, from (A2.1) we obtain

$$(*1)\qquad \tau(y;\,\alpha+1,\,\beta) = \tau(y;\,\alpha,\,\beta) + \tau(y;\,\alpha,\,\beta+1).$$

Also, we easily see that

$$(*2)\qquad (\partial/\partial y)\tau(y;\,\alpha,\,\beta) = -\tau(y;\,\alpha,\,\beta+1),$$

(∗3) $$(\partial/\partial y)^2\tau(y;\,\alpha,\,\beta)=\tau(y;\,\alpha,\,\beta+2).$$

Since $\int_0^\infty(\partial/\partial t)\big[e^{-yt}(1+t)^\alpha t^\beta\big]dt=0$ for $\mathrm{Re}(\beta)>0$, we have

(∗4) $$\beta\tau(y;\,\alpha+1,\,\beta)=y\tau(y;\,\alpha+1,\,\beta+1)-\alpha\tau(y;\,\alpha,\,\beta+1)$$

for such β, and even for any β by meromorphic continuation.

Now we have

$$\begin{aligned}y(\partial/\partial y)^2\tau(y;\,\alpha,\,\beta)&=y\tau(y;\,\alpha,\,\beta+2)\quad\text{by }(*3)\\&=y\tau(y;\,\alpha+1,\,\beta+1)-y\tau(y;\,\alpha,\,\beta+1)\quad\text{by }(*1)\\&=\beta\tau(y;\,\alpha+1,\,\beta)+(\alpha-y)\tau(y;\,\alpha,\,\beta+1)\quad\text{by }(*4)\\&=\beta\tau(y;\,\alpha,\,\beta)+(\beta+\alpha-y)\tau(y;\,\alpha,\beta+1)\quad\text{by }(*1).\end{aligned}$$

Thus we obtain

$$\big\{y(\partial/\partial y)^2+(\alpha+\beta-y)\partial/\partial y-\beta\big\}\tau(y;\,\alpha,\,\beta)=0.$$

From this we easily see that f_A is a solution of (A2.8).

Let f be a solution of (A2.8) such that $f(y)=O(y^B)$ as $y\to\infty$. Putting $f'=df/dy$, we have

$$(y^\sigma f')'=y^\sigma(f''+\sigma y^{-1}f')=y^\sigma(A^2-A\sigma y^{-1}-\lambda y^{-2})f=O(y^C)$$

with $C\in\mathbf{R}$ as $y\to\infty$. It follows that $y^\sigma f'$, as well as f', is $O(y^D)$ with $D\in\mathbf{R}$. Put $h=f_Af'-f'_Af$. Then $h'=f_Af''-f''_Af=-\sigma y^{-1}h$, and so $h=ay^{-\sigma}$ with a constant a. From Lemma A2.2(i) and (∗2) we see that both f_A and f'_A are $O(e^{-|A|y/2})$ as $y\to\infty$. Therefore we have $a=0$, which means that f/f_A is a constant. This completes the proof of (i). Formulas (A2.10) and (A2.11) can be verified by employing (∗1), (∗2), and (∗4).

Lemma A2.5. *For every compact subset K of $\mathbf{C}^2$ we have*

$$\lim_{y\to\infty}e^{y/2}V(y;\,\alpha,\,\beta)=1$$

uniformly for $(\alpha,\,\beta)\in K$.

PROOF. From (∗4) we obtain

$$V(y,\,\alpha+1,\,\beta)=V(y,\,\alpha+1,\,\beta+1)-\alpha y^{-1}V(y,\,\alpha,\,\beta+1).$$

Since this is consistent with the desired formula, it is sufficient to prove it for α in a compact subset K_1 of $\mathbf{C}$ and β in a compact subset K_2 of $\big\{\beta\in\mathbf{C}\,\big|\,\mathrm{Re}(\beta)>0\big\}$. If $\mathrm{Re}(\beta)>0$, we have

$$e^{y/2}V(y;\,\alpha,\,\beta)=\Gamma(\beta)^{-1}\int_0^\infty e^{-x}(1+y^{-1}x)^{\alpha-1}x^{\beta-1}dx,$$

and so

$$e^{y/2}V(y;\,\alpha,\,\beta)-1=\Gamma(\beta)^{-1}\int_0^\infty e^{-x}\big[(1+y^{-1}x)^{\alpha-1}-1\big]x^{\beta-1}dx.$$

We can find two positive numbers A and B such that

$$\left|(1+y^{-1}x)^{\alpha-1}\right|+1 \le Ax^B \quad \text{for } x \ge 1,\ y \ge 1, \text{ and } \alpha \in K_1.$$

Indeed, take $m > 1$ so that $\mathrm{Re}(\alpha) < m$ for every $\alpha \in K_1$. Then for $x \geq 1$, $y \geq 1$, and $\alpha \in K_1$ we have $|(1+y^{-1}x)^{\alpha-1}| \leq (2x)^m$. Now, given $\varepsilon > 0$, we can find $C \geq 1$ such that

$$A|\Gamma(\beta)^{-1}| \int_C^\infty e^{-x}|x^{B+\beta-1}|dx \le \varepsilon \quad \text{for } \beta \in K_2.$$

For every $\eta > 0$, we can find (a small) $h > 0$ such that

$$|t^{\alpha-1} - 1| < \eta \quad \text{for } \alpha \in K_1 \text{ and } |t-1| \le h.$$

Take $\eta = \varepsilon/M$ with $M = \mathrm{Max}_{\beta\in K_2}\left|\Gamma\big(\mathrm{Re}(\beta)\big)/\Gamma(\beta)\right|$, and let $D = C/h$. Then we can find $D \geq 1$ such that

$$\left|(1+y^{-1}x)^{\alpha-1} - 1\right| \le \varepsilon/M \quad \text{for } x \le C,\ y \ge D, \text{ and } \alpha \in K_1.$$

Then, for $y \geq D$, $\alpha \in K_1$, and $\beta \in K_2$ we have

$$\left|\Gamma(\beta)^{-1}\int_0^\infty e^{-x}\big[(1+y^{-1}x)^{\alpha-1}-1\big]x^{\beta-1}dx\right|$$

$$\le \varepsilon M^{-1}\left|\Gamma(\beta)^{-1}\int_0^C e^{-x}x^{\beta-1}dx\right| + A\left|\Gamma(\beta)^{-1}\int_C^\infty e^{-x}|x^{B+\beta-1}|dx\right| \le 2\varepsilon.$$

This proves our lemma.

A3. Eisenstein series of half-integral weight

A3.0. In this section we denote by μ the **Moebius function.** This is defined for $m \in \mathbf{Z}$, > 0, and $\mu(m) \neq 0$ if and only if m is square-free, in which case $\mu(m) = (-1)^r$, where r is the number of prime factors of m; in particular, $\mu(1) = 1$. We have

(A3.0a) $$\sum_{d|m} \mu(d) = \begin{cases} 1 & \text{if } m = 1, \\ 0 & \text{if } m > 1, \end{cases}$$

(A3.0b) $$\sum_{n=1}^\infty \mu(n)\theta(n)n^{-s} = \prod_p \left[1 - \theta(p)p^{-s}\right]$$

for every $\mathbf{C}$-valued multiplicative function θ defined for $0 < m \in \mathbf{Z}$, where p runs over all prime numbers.

Lemma A3.1. *Let χ_0 be a primitive character modulo r, and χ a character modulo rs with $0 < s \in \mathbf{Z}$ such that $\chi(n) = \chi_0(n)$ for n prime to s. Then for any integer q we have*

$$\sum_{n=1}^{rs} \chi(n)\mathbf{e}(nq/rs) = G(\chi_0) \sum_{0<c|(s,q)} c\mu(s/c)\chi_0(s/c)\bar{\chi}_0(q/c),$$

where $G(\chi_0)$ *denotes the Gauss sum of (2.3),* and (s, q) the greatest common divisor of s and q. We put $G(\chi_0) = 1$ if χ_0 is trivial, as we did in *(2.15)*.

PROOF. For $0 < c \in \mathbf{Z}$ we have

$$(*) \qquad \sum_{n=1}^{rc} \chi_0(n)\mathbf{e}(nq/rc) = \begin{cases} 0 & \text{if } c \nmid q, \\ c\bar{\chi}_0(q/c)G(\chi_0) & \text{if } c|q. \end{cases}$$

Indeed, the left-hand side equals

$$\sum_{m=1}^{r} \chi_0(m) \sum_{a=1}^{c} \mathbf{e}\big((m+ar)q/rc\big) = \sum_{m=1}^{r} \chi_0(m)\mathbf{e}(mq/rc) \sum_{a=1}^{c} \mathbf{e}(aq/c),$$

which is 0 if $c \nmid q$, and is $c\bar{\chi}_0(q/c)G(\chi_0)$ by (2.3a) if $c|q$. By (A3.0a) we have

$$\sum_{n=1}^{rs} \chi(n)\mathbf{e}(nq/rs) = \sum_{n=1}^{rs} \Bigg(\sum_{d|(n,s)} \mu(d) \Bigg) \chi_0(n)\mathbf{e}(nq/rs).$$

Putting $n = md$ and $s = cd$, we see that the last sum over n equals

$$\sum_{d|s} \mu(d)\chi_0(d) \sum_{m=1}^{rc} \chi_0(m)\mathbf{e}(mq/rc),$$

which combined with $(*)$ gives the desired result.

A3.2. Our next aim is to investigate the analytic nature of $E_k(z, s; \Gamma, \psi)$ of (8.12) with $k \notin \mathbf{Z}$. Let $\gamma = \begin{bmatrix} * & * \\ cN & d \end{bmatrix} \in \Gamma = \Gamma(M, N)$ with $d > 0$. By (4.40), $j_\gamma^k(z)^{-1} = (cNz + d)^{-k}\varepsilon_d \left(\dfrac{2cN}{d}\right)$, and so by Lemma 8.7 and (8.12a) we have

$$\text{(A3.1)} \quad E_k(z, s; \Gamma, \psi) = y^s \sum_{d=1}^{\infty} \psi(d) \sum_{c \in \mathbf{Z}} \left(\frac{2cN}{d}\right) \varepsilon_d (cNz + d)^{-k} |cNz + d|^{-2s}.$$

Notice that the terms for even d are 0, which we keep in mind in our calculation. Fixing N and ψ, we put

$$\text{(A3.2)} \qquad E'(z, s) = (-iNz)^{-k} E_k\big(-(Nz)^{-1}, s; \Gamma, \psi\big),$$

since this is easier than the original $E_k(z, s)$. From (A3.1) we easily obtain

$$\text{(A3.3)} \qquad E'(z, s) = i^k y^s N^{-k-s} E^*(z, s) \quad \text{with}$$

$$E^*(z, s) = \sum_{d=1}^{\infty} \sum_{b \in \mathbf{Z}} \psi(d) \left(\frac{-2Nb}{d}\right) \varepsilon_d (dz + b)^{-k} |dz + b|^{-2s}.$$

Putting $b = dm + \ell$ with $m \in \mathbf{Z}$ and $1 \le \ell \le d$, we obtain

$$E^*(z, s) = \sum_{d=1}^{\infty} \psi(d)\varepsilon_d d^{-k-2s} \sum_{\ell=1}^{d} \left(\frac{-2N\ell}{d}\right) \sum_{m \in \mathbf{Z}} \left(z + \tfrac{\ell}{d} + m\right)^{-k} \left|z + \tfrac{\ell}{d} + m\right|^{-2s}.$$

Applying Lemma A2.3 to the last sum with $\alpha = k + s$ and $\beta = s$, we find that

$$\text{(A3.4)} \qquad E^*(z,\,s) = i^{-k}(2\pi)^{2s+k}\sum_{n\in\mathbf{Z}}\alpha_n(s)\mathbf{e}(nx+i|n|y)g_n(y;\,k+s,\,s)$$

$$\text{with}\quad \alpha_n(s) = \sum_{d=1}^{\infty}\left(\frac{-2N}{d}\right)\varepsilon_d d^{-k-2s}\psi(d)\sum_{\ell=1}^{d}\left(\frac{\ell}{d}\right)\mathbf{e}(\ell n/d).$$

If $n=0$, the last sum over ℓ is nonzero only if $d=u^2$ with an odd $u>0$, in which case $\varepsilon_d = 1$. Denoting Euler's function by φ, we see that the last sum $\sum_{\ell=1}^{d}$ equals $\varphi(u^2) = u^2\prod_{p|u}(1-p^{-1}) = u^2\sum_{v|u}\mu(v)v^{-1}$ by (A3.0b). Therefore, putting $u = vw$, we have

$$\alpha_0(s) = \sum_{v,w}\psi(v^2w^2)(vw)^{2-2k-4s}\mu(v)v^{-1}$$
$$= \sum_{w}\psi(w^2)w^{2-2k-4s}\sum_{v}\mu(v)\psi(v^2)v^{1-2k-4s}.$$

Thus in view of (A3.0b) we obtain

$$\text{(A3.5)} \qquad \alpha_0(s) = L_N(4s+2k-2,\,\psi^2)/L_N(4s+2k-1,\,\psi^2).$$

The formula for α_n with $n\neq 0$ can be given as follows.

Lemma A3.3. *Let t be a positive or negative square-free integer. Put $\kappa = 2k$ and $\lambda = 1/2-k$, and define primitive characters ω_1 and ω_2 by*

$$\omega_1(a) = \left(\frac{2tN}{a}\right)\psi(a) \quad \textit{for } (a,\,tN)=1,$$
$$\omega_2(a) = \psi(a)^2 \qquad \textit{for } (a,\,N)=1.$$

Then, for $n = tm^2$ with $0<m\in\mathbf{Z}$, we have

$$L_N(4s-2\lambda,\,\omega_2)\alpha_n(s) = L_N(2s-\lambda,\,\omega_1)\beta_n(s) \quad \textit{with}$$
$$\beta_n(s) = \sum\mu(a)\omega_1(a)\omega_2(b)a^{\lambda-2s}b^{2-\kappa-4s},$$

where the last sum is extended over all ordered pairs of positive integers a, b prime to N such that ab divides m.

PROOF. Let r and u be odd positive integers. Assuming r to be square-free, put $G_{n,r,u} = \sum_{m=1}^{ru^2}\left(\frac{m}{ru^2}\right)\mathbf{e}(mn/ru^2)$. By Lemma A3.1 and (2.4a) we have

$$G_{n,r,u} = \varepsilon_r r^{1/2}\sum_{0<c|(u^2,n)} c\mu(u^2/c)\left(\frac{u^2/c}{r}\right)\left(\frac{n/c}{r}\right).$$

Put $u^2 = ac$. Since $\mu(a)=0$ unless a is square-free, we can put $u = ab$ with $0<b\in\mathbf{Z}$. Thus

$$G_{n,r,u} = \varepsilon_r r^{1/2}\sum_{a,b}ab^2\mu(a)\left(\frac{n/b^2}{r}\right),$$

where $(a,\,b)$ is taken under the conditions $ab = u$ and $ab^2|n$, and so

$$\alpha_n(s) = \sum_{r,u} \left(\frac{-2N}{ru^2}\right)\psi(ru^2)\mu(r)^2(ru^2)^{-k-2s}\varepsilon_r G_{n,r,u}$$

$$= \sum_{r,a,b} \left(\frac{-2N}{r}\right)\varepsilon_r^2\mu(r)^2\psi(ra^2b^2)(ra^2b^2)^{-k-2s}r^{1/2}ab^2\mu(a)\left(\frac{n/b^2}{r}\right),$$

where (r, a, b) is taken under the conditions that r and a are square-free, $(rab, N) = 1$, and $ab^2|n$. Put $n = tm^2$ as in our lemma. Then $ab^2|n$ only if $b|m$. Put $m = bh$; then $a|th$. Put $\omega'(r) = \left(\frac{2tN}{r}\right)\psi(r)$. Then we see that

$$\varepsilon_r^2\left(\frac{-2N}{r}\right)\left(\frac{n/b^2}{r}\right)\psi(r) = \begin{cases} 0 & \text{if } (th, r) \neq 1, \\ \omega'(r) & \text{if } (th, r) = 1, \end{cases}$$

and so

$$\alpha_n(s) = \sum_{b|m} \psi(b)^2 b^{2-\kappa-4s} \sum_{a|th} \mu(a)\psi(a)^2 a^{2\lambda-4s} \sum_{(r,th)=1} \mu(r)^2\omega'(r)r^{\lambda-2s}.$$

By (A3.0b) we have

$$\sum_{a|th} \mu(a)\psi(a)^2 a^{2\lambda-4s} = \prod_{p|th} \left(1 - \psi(p)^2 p^{2\lambda-4s}\right),$$

$$\sum_{(r,th)=1} \mu(r)^2\omega'(r)r^{\lambda-2s} = \prod_{p\nmid thN} \left(1 + \omega_1(p)p^{\lambda-2s}\right)$$

$$= \frac{L_N(2s-\lambda,\, \omega_1)}{L_N(4s-2\lambda,\, \omega_2)} \prod_{p|th,\, p\nmid N} \frac{1 - \omega_1(p)p^{\lambda-2s}}{1 - \omega_2(p)p^{2\lambda-4s}}.$$

Notice that $\omega_1(p) = 0$ if $p|t$ and $p \nmid N$. Therefore we obtain

$$L_N(4s-2\lambda,\, \omega_2)\alpha_n(s) = L_N(2s-\lambda,\, \omega_1)\sum_{b|m} \psi(b)^2 b^{2-\kappa-4s} \prod_{p|h,\, p\nmid N} \left[1-\omega_1(p)p^{\lambda-2s}\right].$$

The formula of our lemma follows immediately from this.

Lemma A3.4. *Let* $R(s, \chi) = (r/\pi)^{(s+\nu)/2}\Gamma\big((s+\nu)/2\big)L(s, \chi)$ *with a nontrivial primitive character* χ *modulo* r *and* $\nu = 0$ *or* 1 *determined by* $\chi(-1) = (-1)^\nu$. *Then for any compact subset* J *of* $\mathbf{R}$ *there exists a constant* C_J *independent of* r *and* χ *such that* $|R(s, \chi)| \le C_J\, r^{2+|\sigma|/2}$ *for* $\mathrm{Re}(s) = \sigma \in J$.

PROOF. If $\mathrm{Re}(s) = \sigma > 1$, we have, by (8.10a),

$$|R(1-s,\, \bar\chi)| = |R(s,\, \chi)| \le (r/\pi)^{(\sigma+\nu)/2}\Gamma\big((\sigma+\nu)/2\big)\zeta(\sigma).$$

Therefore it is sufficient to prove the desired estimate for $-1 < \sigma < 2$. For this we use (8.10d) (with χ in place of ψ), which can be written

$$R(s, \chi) = P(s, \chi) + W(\chi)P(1-s, \bar\chi),$$

$$P(s, \chi) = \int_1^\infty g(y, \chi)y^{(\nu+s-2)/2}dy,$$

where $g(y, \chi) = \sum_{n=1}^\infty \chi(n)n^\nu \exp(-\pi n^2 y/r)$. Observe that

$$|g(y, \chi)| \le \sum_{n=1}^{\infty} ne^{-\pi n y/r} = e^{-\pi y/r}(1 - e^{-\pi y/r})^{-2}.$$

Substituting tr/π for y, we obtain

$$|P(s, \chi)| \le (r/\pi)^{(\sigma+\nu)/2} \int_{\pi/r}^{\infty} e^{-t}(1 - e^{-t})^{-2} t^{(\sigma+\nu-2)/2} dt.$$

To prove our estimate, we first assume $r > \pi$. Decompose the last integral into two parts over the intervals $(1, \infty)$ and $(\pi/r, 1)$. The first part is a continuous function in σ independent of r and χ. As for the second part, we have $e^{-t}(1 - e^{-t})^{-2} \le At^{-2}$ for $0 \le t \le 1$ with a constant A. Thus

$$\int_{\pi/r}^{1} e^{-t}(1 - e^{-t})^{-2} t^{(\sigma+\nu-2)/2} dt \le A \int_{\pi/r}^{1} t^{(\sigma+\nu)/2-3} dt \le B + Cr^{2-(\sigma+\nu)/2}$$

if $-1 \le \sigma \le 2$, with constants B and C independent of r and χ. Therefore $|P(s, \chi)| \le Dr^2$ for $-1 \le \sigma \le 2$ with a constant D independent of r and χ. Next, if $r < \pi$, then the estimate of the integral over $(0, \infty)$ gives the desired result. Once $P(s, \chi)$ is majorized, then replacing (s, χ) by $(1 - s, \chi)$, we obtain the estimate of $P(1 - s, \chi)$. Adding these, we can complete the proof.

Theorem A3.5. *Let $\kappa = 2k$ and $\lambda = 1/2 - k$. For $z \in \mathfrak{H}$ and $s \in \mathbf{C}$ put*

$$F'(z, s) = L_N(4s - 2\lambda, \omega_2)E'(z, s) \cdot \begin{cases} \Gamma(s)\Gamma\big(s + (1 - \lambda - \lambda_0)/2\big) & (\kappa \le 1), \\ \Gamma(s + k)\Gamma\big(s + (\lambda_0 - \lambda)/2\big) & (\kappa \ge -1), \end{cases}$$

where $\lambda_0 = 0$ or 1 according as λ is even or odd. Then $(2s - \lambda - 1)F'(z, s)$ can be continued as a holomorphic function to the whole s-plane. Moreover, for any compact subset K of $\mathbf{C}$ there exist two positive constants u and v depending on K such that

(A3.6) $$\big|(2s - \lambda - 1)F'(z, s)\big| \le u(y^v + y^{-v}) \qquad (y = \mathrm{Im}(z))$$

for every $s \in K$ and every $z \in \mathfrak{H}$. The factor $2s - \lambda - 1$ is unnecessary either if $(|\kappa| + 1)/2$ is even or ψ^2 is nontrivial.

Remark. If $\kappa = \pm 1$, the two expressions for the product of two gamma factors are identical.

PROOF. We first consider the case $\kappa \ge 1$. By (A2.4), (A3.3), (A3.4), (A3.5), and Lemma A3.3, we have

(A3.7) $$(2\pi)^{-k-2s} N^{k+s} y^{-s} F'(z, s) = \sum_{n \in \mathbf{Z}} \mathbf{e}(nx + |n|iy)|n|^{2s+k-1} A_n(y, s),$$

where we understand that $|0|^{2s+k-1} = 1$, and

(A3.7a) $$A_n(y, s) = L_N(2s - \lambda, \omega_1)\beta_n(s)\Gamma\big(s + (\lambda_0 - \lambda)/2\big) \cdot \Gamma(s)^{-1} \cdot \begin{cases} \tau(4\pi n y;\ s + k,\ s) & (n > 0), \\ \tau(4\pi |n| y;\ s,\ s + k) & (n < 0), \end{cases}$$

(A3.7b) $$A_0(y, s) = L_N(4s+\kappa-2, \omega_2)\Gamma\bigl(s+(\lambda_0-\lambda)/2\bigr) \cdot \Gamma(s)^{-1}\Gamma(2s+k-1)(4\pi y)^{1-k-2s}.$$

Clearly, $A_n(y, s)$ for every $n \in \mathbf{Z}$ is meromorphic on the whole s-plane. Now we have

(*) $$\frac{\Gamma\bigl(s+(\lambda_0-\lambda)/2\bigr)}{\Gamma(s)} = \prod_{a=1}^{c}(s+a-1),$$

where $c = (\lambda_0 - \lambda)/2$. We have also

$$\begin{aligned}\Gamma(2s+k-1)&L_N(4s+\kappa-2, \omega_2)\\ &= \Gamma(2s+k-1)L(4s+\kappa-2, \omega_2)\prod_{p|N}\bigl[1-\omega_2(p)p^{2-\kappa-4s}\bigr].\end{aligned}$$

Therefore the only possible pole of $A_0(y, s)$ may occur at $s = \lambda/2 + 1/4$ or $(\lambda+1)/2$ when ω_2 is trivial. The pole at $s = \lambda/2 + 1/4$ is cancelled by $1 - 2^{2-\kappa-4s}$; the pole at $s = (\lambda+1)/2$ is cancelled by the factor $s + c - 1$ if λ is odd, and by the factor $2s - \lambda - 1$ if λ is even. Thus $(2s - \lambda - 1)A_0(y, s)$ is entire, and by Lemma A3.4, is bounded by $g(y^h + y^{-h})$ with constants g and h depending only on K. The factor $2s - \lambda - 1$ is unnecessary if λ is odd or ω_2 is nontrivial. If λ is even and ω_2 is trivial, then the residue of $A_0(y, s)$ at $s = (\lambda+1)/2$ is an element of $y^{-1/2}\mathbf{Q}^\times$.

To study $A_n(y, s)$ for $n \neq 0$, first note that $\beta_n(s)$ is entire, and $|\beta_n(s)| \leq \gamma|n|^{|\delta|\mathrm{Re}(s)+\varepsilon}$ with constants γ, δ, ε independent of n. Let $n = tm^2$ as in Lemma A3.3, and let $\omega_1(-1) = (-1)^\eta$ with $\eta = 0$ or 1. Since $\psi(-1) = (-1)^{[k]} = (-1)^\lambda$, we see that $\lambda - \eta$ is even if and only if $n > 0$. Suppose $n > 0$. Then $\eta = \lambda_0$ and

(*+) $$A_n(y, s) = \Gamma\bigl(s+(\eta-\lambda)/2\bigr)\beta_n(s)L(2s-\lambda, \omega_1)\prod_{p|N}\bigl[1-\omega_1(p)p^{\lambda-2s}\bigr] \cdot \Gamma(s)^{-1}\tau(4\pi ny;\, s+k,\, s).$$

By Lemma A2.2(i), the last product $\Gamma(s)^{-1}\tau(4\pi ny; s+k, s)$, when $s \in K$, is bounded by

$$C(4\pi ny)^{-\mathrm{Re}(s)}\mathrm{Max}\bigl(1, (4\pi ny)^B\bigr),$$

where B and C are constants that depend on K but not on n. The first line of factors of (*+) is an entire function of s, except when ω_1 is trivial and $s = \lambda/2$ or $s = (\lambda+1)/2$. (This can happen only if λ is even.) But either pole is cancelled by the factor $1 - 2^{\lambda-2s}$ or $2s - \lambda - 1$. Therefore $(2s - \lambda - 1)A_n(y, s)$ is entire, and by Lemma A3.4, $|(2s - \lambda - 1)A_n(y, s)| \leq un^v(y^w + y^{-w})$ for $s \in K$ with constants u, v, w depending only on K.

Next suppose $n < 0$. Then $\lambda - \eta$ is odd, and so $\lambda_0 + \eta = 1$. Thus

(*−) $$A_n(y, s) = \beta_n(s)\Gamma\bigl(s+(\eta-\lambda)/2\bigr)L(2s-\lambda, \omega_1)\prod_{p|N}\bigl[1-\omega_1(p)p^{\lambda-2s}\bigr] \cdot \Gamma(s+k)^{-1}\tau(4\pi|n|y;\, s,\, s+k)$$

$$\cdot \frac{\Gamma(s+k)\Gamma\big(s+(\lambda_0-\lambda)/2\big)}{\Gamma\big(s+(\eta-\lambda)/2\big)\Gamma(s)}.$$

Notice that $0 \le k-(\eta-\lambda)/2 = (\lambda_0-\lambda)/2 \in \mathbf{Z}$. Therefore, by the same reasoning as in the case $n > 0$, we see that $(2s-\lambda-1)A_n(y, s)$ is entire. Actually the factor $2s-\lambda-1$ is unnecessary. Indeed, the pole at $s=(\lambda+1)/2$ may occur only if ω_1 is trivial, in which case λ is odd, and so $0 < \lambda_0-\lambda \in 2\mathbf{Z}$ and $\Gamma\big(s+(\lambda_0-\lambda)/2\big)/\Gamma(s) = 0$ at $s=(\lambda+1)/2$. Thus $A_n(y, s)$ is entire for $n < 0$ and $\big|(2s-\lambda-1)A_n(y, s)\big| \le u'|n|^{v'}(y^{w'}+y^{-w'})$ for $s \in K$ with constants u', v', w' depending only on K. Taking the infinite sum of (A3.7), we obtain the desired result for $k > 0$.

The case $k < 0$ can be treated in a similar fashion. However, our real aim is to prove Theorem 8.14, which is our task in the next subsection.

A3.6. Put $k' = -k$ and $\lambda' = 1/2 - k'$; let λ_0' be 0 or 1 according as λ' is even or odd. Assuming that $k > 0$, denote by $F_{k'}^*$ the function F^* defined with k' and $\chi_0\overline{\psi}$ in place of k and ψ, where χ_0 is the primitive character modulo 4. Then we can easily verify that $\lambda' = 1-\lambda$, $\lambda_0' = 1-\lambda_0$, and from (8.13a) we obtain

(A3.8) $$F_{k'}^*(z, s) = y^k\overline{F^*(z, \bar{s}-k)}.$$

This reduces the proof of Theorem 8.14 to the case $k > 0$. Thus we assume $k > 0$ in this subsection. Let the symbols be as in that theorem. Given $\alpha \in SL_2(\mathbf{Q})$, put $\gamma = \begin{bmatrix} 0 & -1 \\ N & 0 \end{bmatrix}\alpha$. From (A3.2) we see that $j_\alpha(z)^{-k}E_k(\alpha z, s) = i^k j_\gamma(z)^{-k}E'(\gamma z, s)$ with a suitable branch of j_γ^{-k}, and so

(A3.9) $$j_\alpha(z)^{-k}F^*(\alpha z, s) = i^k(2s-\lambda-1)j_\gamma(z)^{-k}F'(\gamma z, s).$$

Therefore the first part of Theorem 8.14 concerning analytic continuation of F^* follows immediately from Theorem A3.5. Thus the remaining point is the estimate of $|j_\alpha(z)^{-k}F^*(\alpha z, s)|$. If $c_\gamma = 0$, the desired fact follows from (A3.6). Suppose $c_\gamma \ne 0$. Then $\mathrm{Im}(\gamma z) = y|c_\gamma z + d_\gamma|^{-2} \le c_\gamma^{-2}y^{-1}$, and so if y is sufficiently large, then $\mathrm{Im}(\gamma z) < 1$ and (A3.6) shows that

$$\big|(2s-\lambda-1)j_\gamma(z)^{-k}F'(\gamma z, s)\big| \le 2uy^{-v}|c_\gamma z + d_\gamma|^{2v+k}$$

for $s \in K$. This proves that $F^*(z, s)$ is slowly increasing at every cusp locally uniformly in s.

It remains to prove that $F^*(z, s)$ is nonvanishing at $s = (\lambda+1)/2$ if $(|\kappa|+1)/2$ is odd and ψ^2 is trivial. In view of (A3.8) we may assume that $k > 0$. Then λ is even. As noted in the proof of Theorem A3.5, $A_0(y, s)$ has nonzero residue at $s = (\lambda+1)/2$, which gives the desired result. This completes the proof of Theorem 8.14.

A3.7. Let us now prove the last part of Theorem 8.10(ii), which states that $F(z, s)$ is slowly increasing at every cusp, locally uniformly in s. Our technique is similar to and simpler than that in the proof of Theorem A3.5. We need to examine the Fourier expansion of $j_\gamma(z)^{-k}F(\gamma z, s)$ for every $\gamma \in \Gamma(1)$. In view of (8.14a) we have only to check the expansion for a fixed (p, q). Suppose $k > 0$ for simplicity; then we use the result in [S07, p. 134]. As noted in lines 7 and 8 from the bottom of that page, the expressions there are meaningful for every $s \in \mathbf{C}$, and so we have

$$(\#)\quad F(z, s) = a(s)y^s + b(s)y^{1-k-s} + \frac{c(s)y^s}{\Gamma(s)} \sum_{0 \neq t \in \mathbf{Z}} \mathbf{e}\big((tx + i|t|y)/N\big)D_t(y, s)$$

for $z = x + iy$, and

$$\big|D_t(y, s)\big| \leq 2 \sum_{0<n|t} n^{2\sigma+k-1} \cdot \begin{cases} \big|\tau(4\pi ty/N;\, s+k,\, s)\big| & \text{if } t > 0, \\ \big|\tau(4\pi|t|y/N;\, s,\, s+k)\big| & \text{if } t < 0 \end{cases}$$

for $\sigma = \mathrm{Re}(s)$, where $a(s)$, $b(s)$, and $c(s)$ are entire functions of s, and τ is as in (A2.1). (The properties of $a(s)$ and $b(s)$ can be seen from [S07, Theorem 3.4]; $c(s) = gh^s$ with some constants g and h.) Let K be a compact subset of $\mathbf{C}$. Employing Lemma A2.2(i), we see that the last term of (#) for $s \in K$ is majorized by

$$2|c(s)|\left|1 + \frac{\Gamma(s+k)}{\Gamma(s)}\right| \sum_{t=1}^{\infty} \alpha\big(1 + (4\pi ty/N)^\beta\big)t^{2\sigma+k}e^{-2\pi ty/N},$$

where α and β are positive constants depending only on K. Then we easily see that $F(z, s)$ for $s \in K$ satisfies (6.9a). This proves the case $k > 0$. The case $k \leq 0$ can be handled in a similar way.

There is an alternative proof. Indeed, $\mathfrak{E}_k^N(z, s-k;\, p, q)$, up to some easy factors, can be obtained as the integral $\int_0^\infty \Psi(z, t)t^{s-1}dt$, where

$$\Psi(z, t) = \sum_{0 \neq (m,n) \in \mathbf{Z}^2} \lambda(m, n)(m\bar{z} + n)^k \exp\big(-\pi t|mz+n|^2/y\big)$$

with $\lambda \in \mathscr{L}(\mathbf{Q}^2)$; see [S07, pp. 64–65], [S73a, pp. 462–463]. We can make an estimate of Ψ in an elementary way, and we eventually obtain the desired property of $F(z, s)$. This was done in [S73a, pp. 463–464].

A3.8. We will now prove (iii) and (iv) of Theorem 8.15 which concern the case $3/2 \leq k \notin \mathbf{Z}$. We assume that $k > 3/2$ or ψ^2 is nontrivial. Then $E_k(z, s)$ is finite at $s = 0$. (Indeed, the factor $2s - \lambda - 1$ in Theorem 8.14 is necessary only if ψ^2 is trivial, and it is 0 at $s = 0$ only if $k = 3/2$.) By (A3.3), $E'(z, 0) = i^kN^{-k}E^*(z, 0)$, and $E^*(z, 0)$ can be obtained from (A3.4). By (A2.3) and (A2.4) the nth term of $E^*(z, 0)$ for $n > 0$ is $(-2\pi i)^k\alpha_n(0)n^{k-1}\Gamma(k)^{-1}\mathbf{e}(nz)$, but it is 0 for $n < 0$, since $\tau(y;\, \alpha, k)$ is finite. The term for $n = 0$ is also 0, since $\Gamma(2s+k-1)\Gamma(s+k)^{-1}\Gamma(s)^{-1} = 0$ for $s = 0$. Thus $E^*(z, 0)$ has a Fourier expansion of the form $\sum_{n=1}^{\infty} c_n\mathbf{e}(nz)$. Since it is slowly increasing at every cusp, it belongs to $\mathscr{M}_k$ by Lemma 6.4(i). By (8.13), $E_k(z, 0) \in \mathscr{M}_k(M, N; \bar{\psi})$. To prove a more precise result as stated in Theorem 8.15(iii), put

$$E'(z, 0) = i^k N^{-k} E^*(z, 0) = \sum_{n=1}^{\infty} A_n(\psi)\mathbf{e}(nz)$$

with $A_n(\psi)$, which, by (A2.3), (A2.4), Lemma A3.3, and (A3.4) can be given as

$$A_n(\psi) = (2\pi/N)^k n^{k-1} \Gamma(k)^{-1} \beta_n(0) L_N([k], \omega_1)/L_N(2[k], \omega_2).$$

Employing the symbol $P_N(m, \chi) = G(\chi)^{-1}(\pi i)^{-m} L_N(m. \chi)$ of (2.19), we can put

$$A_n(\psi) = n^{-1}(2\pi n/N)^k \Gamma(k)^{-1} \beta_n(0)(\pi i)^{-[k]} \cdot \frac{G(\omega_1)P_N([k], \omega_1)}{G(\omega_2)P_N(2[k], \omega_2)}.$$

Since $\Gamma(k) \in \pi^{1/2}\mathbf{Q}^{\times}$, we see that $\pi^{k-[k]}\Gamma(k)^{-1} \in \mathbf{Q}^{\times}$. Let $\chi_t(a) = \left(\frac{2tN}{a}\right)$ with t as in Lemma A3.3. Take a square-free positive integer t_0 such that $2tN/t_0$ is a square. (We are considering only positive n.) Then $G(\chi_t) = t_0^{1/2}$, and so $(2n/N)^k \in G(\chi_t)\mathbf{Q}^{\times}$. Thus

$$i^{[k]} A_n(\psi) = R_0 \beta_n(0) G(\chi_t) G(\psi\chi_t) \cdot \frac{P_N([k], \psi\chi_t)}{G(\psi^2) P_N(2[k], \psi^2)}$$

with a rational number R_0 independent of ψ. Multiplying by $G(\psi)$, we obtain

$$i^{[k]} G(\psi) A_n(\psi) = R_0 \beta_n(0) \cdot \frac{G(\chi_t)G(\psi\chi_t)}{G(\psi)} \cdot \frac{G(\psi)G(\psi)}{G(\psi^2)} \cdot \frac{P_N([k], \psi\chi_t)}{P_N(2[k], \psi^2)}.$$

Take $\sigma \in \mathrm{Gal}(\mathbf{Q}_{\mathrm{ab}}/\mathbf{Q})$ and apply σ to each factor. By Lemmas 2.8 and 2.10 we find the images of the last three factors; $\beta_n(0)^\sigma$ can easily be found from Lemma A3.3. We eventually find that

$$\left[i^{[k]} G(\psi) A_n(\psi)\right]^\sigma = i^{[k]} G(\psi^\sigma) A_n(\psi^\sigma).$$

Thus writing $E'(\psi)$ for $E'(z, 0)$, we obtain

$$\left[i^{[k]} G(\psi) E'(\psi)\right]^\sigma = i^{[k]} G(\psi^\sigma) E'(\psi^\sigma).$$

Returning to $E_k(z, 0; \Gamma, \psi)$, we see from (A3.2) that $E'(\psi) = f^X$ with $f(z) = E_k(z, 0; \Gamma, \psi)$ and the operator X defined in Lemma 7.6(ii). Define a character φ by $\varphi(d) = \left(\frac{-1}{d}\right)^{[k]}$, and take an integer s prime to MN so that $\mathbf{e}(1/MN)^\sigma = \mathbf{e}(s/MN)$. Then $(i^{[k]})^\sigma = \varphi(s) i^{[k]}$, and so by (2.17), $E'(\psi)^\sigma = \varphi(s)\psi(s)^\sigma E'(\psi^\sigma)$. Since $f \in \mathscr{M}_k(M, N; \bar{\psi})$ as noted above, Lemma 7.6(ii) shows that $\varphi(s)\psi(s)^\sigma (f^\sigma)^X = (f^X)^\sigma = E'(\psi)^\sigma$, and so $(f^\sigma)^X = E'(\psi^\sigma)$, which means that $f^\sigma = E_k(z, 0; \Gamma, \psi^\sigma)$, that is,

(A3.10) $$\left[E_k(z, 0; \Gamma, \psi)\right]^\sigma = E_k(z, 0; \Gamma, \psi^\sigma).$$

This completes the proof of Theorem 8.15(iii).

A proof of (A3.10) was given in [St80]. However, the methods of the paper are very involved, and the exposition is sketchy with many undefined symbols, and therefore it is almost impossible to follow. Here we have given a simpler proof with different ideas. The proof of [St80] uses the results on $E'(z, s)$ in [S75], which we reproduced here as Lemma A3.3, (A3.7), and (A3.7a, b). What we need in addition is Lemma 7.6(ii) and our discussion on the behavior of $A_n(\psi)$ under σ, whereas [St80] requires at least eight pages of calculations. For these reasons, we merely let the reader know the existence of the paper, with no further comments.

A3.9. To prove (iv) of Theorem 8.15, put $s_k = 1 - k$. By (A3.2) we can reduce the problem to F' whose Fourier expansion is given by (A3.7). Let us first show that $A_n(y, s_k) = 0$ if $n < 0$. Since $s_k \neq (\lambda + 1)/2$, the first line of $(*-)$ is finite at $s = s_k$, and the same is true for the second line. As for the third line, since $\lambda \leq -1$, we have $0 \geq 1 + (\lambda - \lambda_0)/2 = s_k + (\eta - \lambda)/2$, and so $\Gamma\bigl(s + (\eta - \lambda)/2\bigr)$ has a pole at $s = s_k$. This shows that $A_n(y, s_k) = 0$ for $n < 0$.

As to A_n for $n > 0$, we note that by (A2.6) the factor $\tau(4\pi ny;\, s + k,\, s)/\Gamma(s)$ at $s = s_k$ equals $(4\pi ny)^{k-1}$. Returning to $E'(z, s)$, put

$$D(z,\, s;\, \psi) = L(4s + 2k - 1,\, \psi^2)E'(z,\, s).$$

Then from (A3.7) we see that

$$2^{-k}N\pi^{-1}D(z,\, s_k;\, \psi) = \sum_{n=0}^{\infty} B_n(\psi)\mathbf{e}(nz),$$

where $B_0(\psi) = L(1 - 2[k],\, \psi^2)$ and $B_n(\psi) = L(1 - [k],\, \omega_1)\beta_n(s_k)$ if $n > 0$. Notice that $\omega_1(-1) = \psi(-1) = (-1)^{[k]}$. The function is nonzero, since $B_0(\psi) \neq 0$. From Lemma 2.10 and the formula for β_n in Lemma A3.3 we see that $B_n(\psi) \in \mathbf{Q}_{\mathrm{ab}}$ and $B_n(\psi)^\sigma = B_n(\psi^\sigma)$ for every $n \geq 0$ and every $\sigma \in \mathrm{Gal}(\mathbf{Q}_{\mathrm{ab}}/\mathbf{Q})$. Thus $\bigl[2^k\pi^{-1}D(z,\, s_k;\, \psi)\bigr]^\sigma = 2^k\pi^{-1}D(z,\, s_k;\, \psi^\sigma)$. Now let $C_k^*(\Gamma,\, \psi)$ be as in Theorem 8.5(iv) and let the symbol X be as in Lemma 7.6(ii). By (A3.2) we have $E'(z,\, s) = E_k(z,\, s;\, \Gamma,\, \psi)^X$, and so $C_k^*(\Gamma,\, \psi)^X = 2^k i^{[k]}G(\psi)^{-1}\pi^{-1}D(z,\, s_k;\, \psi)$. Given $\sigma \in \mathrm{Gal}(\mathbf{Q}_{\mathrm{ab}}/\mathbf{Q})$, take an integer t prime to N so that $\mathbf{e}(1/N)^\sigma = \mathbf{e}(t/N)$. Let a character χ be defined by $\chi(d) = \left(\frac{-1}{d}\right)^{[k]}\bar{\psi}(d)$. Then Lemma 7.6(ii) shows that $(f^\sigma)^X = \chi(t)^\sigma(f^X)^\sigma$ for $f = C_k^*(\Gamma,\, \psi)$, since $\bar{\psi}$ here is ψ there. Observe that $\chi(t)^\sigma\bigl[i^{[k]}G(\psi)^{-1}\bigr]^\sigma = i^{[k]}G(\psi^\sigma)^{-1}$ by (2.17). Therefore

$$\begin{aligned}\bigl[C_k^*(\Gamma,\, \psi)^\sigma\bigr]^X &= \chi(t)^\sigma\bigl[2^k i^{[k]}G(\psi)^{-1}\pi^{-1}D(z,\, s_k;\, \psi)\bigr]^\sigma \\ &= 2^k i^{[k]}G(\psi^\sigma)^{-1}\pi^{-1}D(z,\, s_k;\, \psi^\sigma) = C_k^*(\Gamma,\, \psi^\sigma)^X,\end{aligned}$$

and so $C_k^*(\Gamma,\, \psi)^\sigma = C_k^*(\Gamma,\, \psi^\sigma)$. This completes the proof of Theorem 8.15(iv).

A3.10. Let us now prove Theorem 8.16(i). We assume that $k = 1/2$ and ψ^2 is trivial, and so $\lambda = 0$ and $(\lambda+1)/2 = 1/2$. Since $\psi(-1) = (-1)^{[k]}$, we have $\psi(a) = \left(\frac{q}{a}\right)$ with a square-free positive integer q. The question is the value of $F^*(z, s)$ at $s = 1/2$, but we first look at F' of (A3.7). In the proof of Theorem A3.5 we have seen that $A_n(y, s)$ with $n < 0$ is finite at $s = (\lambda+1)/2$. Let $R = \prod_{p|N}(1 - p^{-1})$. From (A3.7b) we see that $A_0(y, s)$ has residue $Ry^{-1/2}/8$ at $s = 1/2$.

As for A_n with $n > 0$, we need to investigate the residue of $(*+)$ at $s = 1/2$, which is nonzero if and only if ω_1 is trivial, which is the case if and only if $2tNq$ is a square, that is, $2Nq = t\nu^2$ with an integer ν. Thus t is determined by N and ψ. For $n = tm^2$ as in Lemma A3.3 put $m_0 = m/(m, N)$. Then

$$\beta_n(1/2) = \sum_{ab|m_0} (ab)^{-1} = \sum_{c|m_0} c^{-1} \sum_{a|c} \mu(a) = 1$$

by (A3.0a). Thus the first line of (A3.7a) has residue $R\pi^{1/2}/2$ at $s = 1/2$. The second line of $(*+)$ gives $(4\pi ny)^{-1/2}$ by (A2.6). Thus, by (A3.7) the residue of $F'(z, s)$ at $s = 1/2$ is

$$(2\pi)^{3/2}N^{-1/2}y^{1/2}\sum_{m\in\mathbf{Z}} \mathbf{e}(tm^2z)Ry^{-1/2}/8 = (8N)^{-1/2}\pi^{3/2}R\sum_{m\in\mathbf{Z}} \mathbf{e}(tm^2z).$$

(Notice that $\sum_{m\in\mathbf{Z}} \mathbf{e}(tm^2z) = 1 + 2\sum_{m=1}^{\infty} \mathbf{e}(tm^2z)$.) Now

$$F^*(z, s) = (2s-1)(-iNz)^{-1/2}F'\bigl(-(Nz)^{-1}, s\bigr).$$

Put $A = (8N)^{-1/2}\pi^{3/2}R$, $v = 2t/N$, and $f(z) = A\sum_{m\in\mathbf{Z}} \mathbf{e}(vm^2z/2)$. Then the residue of $F'\bigl(-(Nz)^{-1}, s\bigr)$ at $s = 1/2$ is $f(-z^{-1})$, and so $F^*(z, 1/2) = 2(-iNz)^{-1/2}f(-z^{-1})$. Using the notation of Lemma 7.11, we have $f(z) = A\theta_v(z, \lambda)$, where λ is the characteristic function of $\mathbf{Z}$, viewed as an element of $\mathscr{L}(\mathbf{Q})$. Therefore, by that lemma, $F^*(z, 1/2) = \pi^{3/2}\theta_v(z, \nu)$ with a $\mathbf{Q}_{\mathrm{ab}}$-valued element ν of $\mathscr{L}(\mathbf{Q})$. This proves Theorem 8.16(i).

If F^* is defined without the factor $L_N(4s - 2\lambda, \psi^2)$, then the value belongs to $\pi^{-1}\mathscr{M}_{1/2}(\mathbf{Q}_{\mathrm{ab}})$.

REFERENCES

[AL70] A. O. L. Atkin and J. Lehner, Hecke operators on $\Gamma_0(m)$, *Math. Ann.* 185 (1970), 134–160.

[C73] W. Casselman, On some results of Atkin and Lehner, *Math. Ann.* 201 (1973), 301–314.

[H27] E. Hecke, Theorie der Eisensteinschen Reihen höhere Stufe und ihre Anwendung auf Funktionentheorie und Arithmetik, *Abh. Math. Sem. Hamburg* 5 (1927), 199–224 (= *Mathematische Werke,* 461–486).

[H36] E. Hecke, Über die Bestimmung Dirichletscher Reihen durch ihre Funktionalgleichung, *Math. Ann.* 112 (1936), 664–699 (= *Mathematische Werke,* 591–626).

[H37] E. Hecke, Über Modulfunktionen und die Dirichletschen Reihen mit Eulerscher Produktentwicklung I, II, *Math. Ann.* 114 (1937), 1–28, 316–351 (= *Mathematische Werke,* 644–707).

[Kl28] H. D. Kloostermann, Theorie der Eisensteinschen Reihen von mehreren Varänderlichen, *Abh. Math. Sem. Hamburg* 6 (1928), 163–188.

[KP92] A. Krazer and F. Prym, *Neue Grundlagen einer Theorie der Allgemeinen Thetafunktionen,* Teubner, Leipzig, 1892.

[Ma49] H. Maass, Über eine neue Art von nichtanalytischen automorphen Funktionen und die Bestimmung Dirichletscher Reihen durch Funktionalgleichungen, *Math. Ann.* 121 (1949), 141–183.

[Ma53] H. Maass, Die Differentialgleichungen in der Theorie der elliptischen Modulfunktionen, *Math. Ann.* 125 (1953), 235–263.

[Mi71] T. Miyake, On automorphic forms on GL_2 and Hecke operators , *Ann. of Math.* 94 (1971), 174–189.

[N77] S. Niwa, On Shimura's trace formula, *Nagoya Math. J.* 66 (1977), 183–202.

[P82, 84] T-y, Pei, Eisenstein series of weight 3/2: I, II, *Trans. Amer. Math. Soc.* 274 (1982), 573–606, 283 (1984), 589–603.

[Ra39] R. A. Rankin, Contributions to the theory of Ramanujan's function $\tau(n)$ and similar arithmetical functions I, II. *Proc. Cambridge Phil. Soc.* 35 (1939), 351–372.

[Ro89] D. Rohrlich, Nonvanishing of L-functions for $GL(2)$, *Inv. math.* 97 (1989), 381-403.

[S71] G. Shimura, *Introduction to the Arithmetic Theory of Automorphic Functions,* Publications of the Mathematical Society of Japan, No. 11, Iwanami Shoten and Princeton University Press, Princeton, NJ, 1971 (paperback edition with errata, 1994).

[S73a] G. Shimura, On modular forms of half integral weight, *Ann. of Math.* 97 (1973), 440–481 (= *Collected Papers,* II, 532–573).

[S73b] G. Shimura, Modular forms of half integral weight, in *Proceedings of the International Summer School of Modular Functions of One Variable, Antwerp, 1972,* Lecture Notes in Mathematics, vol. 320 (Springer, Heidelberg, 1973), 57–74.

[S74] G. Shimura, On the trace formula for Hecke operators, *Acta mathematica,* 132 (1974), 245–281 (= *Collected Papers,* II, 596–632).

[S75] G. Shimura, On the holomorphy of certain Dirichlet series, *Proc. London Math. Soc. 3rd ser.* 31 (1975), 79–98 (= *Collected Papers,* II, 633–652).

[S76] G. Shimura, The special values of the zeta functions associated with cusp forms, *Commun. Pure Appl. Math.* 29 (1976), 783–804 (= *Collected Papers,* II, 740–761).

[S77] G. Shimura, On the periods of modular forms, *Math. Ann.* 229 (1977), 211–221 (= *Collected Papers,* II, 813–823).

[S78] G. Shimura, The special values of the zeta functions associated with Hilbert modular forms, *Duke Math. J.* 45 (1978), 637–679, Corrections, *Duke Math. J.* 48 (1981), 697. A revised version is given in Collected Papers, III, 75–114.

[S81] G. Shimura, The critical values of certain zeta functions associated with modular forms of half-integral weight, *J. Math. Soc. Japan,* 33 (1981), 649–672 (= *Collected Papers,* III, 217–240).

[S85a] G. Shimura, On Eisenstein series of half-integral weight, *Duke Math. J.* 52 (1985), 281–314 (= *Collected Papers,* III, 610–643).

[S85b] G. Shimura, On the Eisenstein series of Hilbert modular groups, *Revista Matemática Iberoamericana* 1 (1985), 1–42 (= *Collected Papers,* III, 644–685).

[S87] G. Shimura, On Hilbert modular forms of half-integral weight, *Duke Math. J.* 55 (1987), 765–838 (= *Collected Papers,* III, 774–847).

[S90] G. Shimura, Invariant differential operators on hermitian symmetric spaces, *Ann. of Math.* 132 (1990), 237–272 (= *Collected Papers,* IV, 68–103).

[S91] G. Shimura, The critical values of certain Dirichlet series attached to Hilbert modular forms, *Duke Math. J.* 63 (1991), 557–613 (= *Collected Papers,* IV, 134–190).

[S93] G. Shimura, On the transformation formulas of theta series, *Amer. J. Math,* 115 (1993), 1011–1052. (= *Collected Papers,* IV, 191–232).

[S94] G. Shimura, Differential operators, holomorphic projection, and singular forms, *Duke Math. J.* 76 (1994), 141–173 (= *Collected Papers,* IV, 351–383).

[S95] G. Shimura, Zeta functions and Eisenstein series on metaplectic groups, *Inv. math.* 121 (1995), 21–60 (= *Collected Papers,* IV, 430–469).

[S97] G. Shimura, *Euler Products and Eisenstein Series,* CBMS Regional Conference Series in Mathematics, No. 93, American Mathematical Society, Providence, RI, 1997.

[S98] G. Shimura, *Abelian Varieties with Complex Multiplication and Modular Functions,* Princeton University Press, Princeton, NJ, 1998.

[S00] G. Shimura, *Arithmeticity in the Theory of Automorphic Forms,* Mathematical Surveys and Monographs vol. 82, American Mathematical Society, Providence, RI, 2000.

[S02] G. Shimura, Collected papers, I–IV, Springer, New York, 2002.

[S04] G. Shimura, *Arithmetic and Analytic Theories of Quadratic Forms and Clifford Groups,* Mathematical Surveys and Monographs vol. 109, American Mathematical Society, Providence, RI, 2004.

[S07] G. Shimura, *Elementary Dirichlet Series and Modular Forms,* Springer Monographs in Mathematics, Springer, New York, 2007.

[S10] G. Shimura, *Arithmetic of Quadratic Forms,* Springer Monographs in Mathematics, Springer, New York, 2010.

[St80] J. Sturm, Special values of zeta functions and Eisenstein series of half integral weight, *Amer. J. of Math.* 102 (1980), 219–240.

[St82] J. Sturm, Theta series of weight 3/2, *J. of Number Theory* 14 (1982), 353–361.

INDEX